Electronics

Kenneth Close, B.Sc., Ph.D., F.Inst.P.,
Principal Lecturer in Physics, The Polytechnic of Central London

and

John Yarwood, M.Sc., F.Inst.P.,
Professor of Physics, The Polytechnic of Central London

Published 1976

Published by University Tutorial Press Ltd.
9/10 Great Sutton Street, London EC1V 0DA

All rights reserved. No portion of the book may be reproduced by any process without written permission from the publishers

© K. J. Close and John Yarwood 1976

ISBN   0 7231 0608 8   Full bound
ISBN   0 7231 0732 7   Limp bound

Printed in Great Britain by J. W. Arrowsmith Ltd., Bristol

Preface

This book is intended for the student reading for a Degree or a Diploma in Science or Engineering of which a substantial part is devoted to Electronics. Recognising that the needs of the science student differ from those of the engineering student, especially with regard to applications of the subject, nevertheless, in the first two years of the usual undergraduate course, a great deal of the coverage needed to appreciate the fundamentals of modern electronic devices and circuits is substantially the same.

The emphasis is on semiconductor devices, circuits and systems where the signal fundamental frequency is below 100 kHz, though aspects of higher frequency work are included. High-frequency and microwave electronics are excluded because their study involves time-dependent aspects of the functioning of active and passive components which are probably best left to the more specialised later years of the course, or studied in communication electronics.

The demands made on the reader's knowledge of mathematics are not severe: in general, not beyond those required for the final year at secondary school and certainly not beyond the scope of a first year mathematics course for a science or engineering degree. Much of this text can therefore well be understood by a reader interested in electronics but who is not aiming at a specific qualification of degree standard; indeed, it is hoped that the interested pupil or teacher in a secondary school will gain much from study of this text.

Particular features of this book are the great emphasis on worked examples, the many unworked exercises given at the end of each chapter, and the frequent references to relevant experimental work and to practical electronic circuits with specific component values. It is felt that the study of modern electronics has been rendered more complex for the beginner than need be by the development of mathematical analysis which is not readily transferred by the student into the consideration of practical circuit design. To avoid this undue emphasis on analysis, the worked example relating to a practical problem is regarded as more rewarding.

The authors are indebted to students at The Polytechnic of Central London who have been subjected to many of the ideas developed in this book both in lecture programmes and in the laboratory, and to Mrs Jennifer Roberts, who typed the manuscript.

K. J. Close
J. Yarwood

Note on abbreviations and symbols used in the text

The description of electronic devices and circuits inevitably involves the repeated use of such terms as 'direct current', 'alternating current' 'electromotive force', 'potential difference' etc. These commonly used terms are often abbreviated, in these examples respectively to dc, ac, emf, pd. In writing such an abbreviation, lower case letters are used and full-stops are omitted.

Within the text where the values of passive components are given the usual decimal notation is employed and the commonly accepted abbreviations. On diagrams, where component values are given, decimal notation is avoided in accordance with recommended standards practice. In this connection, a few examples enable the procedure to be understood:

	In text	On diagram
Resistance values		
units	5.6Ω	5R6
thousands	6.3 kΩ	6k3
millions	7.4 MΩ	7M4

	In text	On diagram
Capacitance values		
	100μF	100

Where the value of the capacitance is given on a diagram in some sub-division of the farad other than the microfarad (μF), it is quoted in the same way as in the text, thus, for example, picofarad becomes pF, nanofarad becomes nF.

With regard to symbols, conventional standards practice is used, with some small alterations where clarity of exposition would seem to be desirable. As far as is practicable, lower case letters are used where varying quantities are involved whereas capital letters are used for steady quantities.

Contents

		Page
1	Physics of semiconductors	7
2	Semiconductor diodes	24
3	Field effect transistors	42
4	Junction or bipolar transistor	69
5	Operational amplifiers	102
6	Waveform generators	132
7	Logic circuits	155
8	Opto electronics	173
9	Power control utilising thyristors	193

Appendix A
    The construction of semiconductor devices and circuits   215

Appendix B
    Feedback in amplifiers   230

Appendix C
    Miller theorem   232

Index   234

1 · Physics of Semiconductors

1.1 The periodic table of the elements

There are 92 elements (excluding the transuranic ones which have played no part in the development of semiconductor devices). Of these 92, the relatively few which are of importance in semiconductor physics and technology are silicon and germanium in Group IVA of the Periodic Table, boron, aluminium, gallium and indium in Group IIIA, phosphorus, arsenic and antimony in Group VA, zinc and cadmium in Group IIB, oxygen and sulphur in Group VIA. To gain some appreciation of the relation between these elements and others, table 1.1 shows the Periodic Table of the Elements, excluding the rare earth ones, which are of little immediate importance here. The number beneath each of the chemical symbols for the element is the atomic number Z.

1.2 Group IVA elements: silicon and germanium

The elements within Group IV A of the Periodic Table (table 1.1) are carbon, silicon, germanium, tin and lead. Of these silicon and, to a lesser extent, germanium are the most important elements in the manufacture of semiconductor devices. These Group IVA elements all have a valence of 4, i.e. they are tetravalent. This is made clear by examining the arrangements in shells and sub-shells of the electrons surrounding the nuclei in these elements (table 1.2).

Recalling that the maximum possible number of electrons in a shell of principal quantum n is $2n^2$ whereas this maximum number in a sub-shell of quantum number l is $2(2l+1)$, it is seen that, in the case of silicon, the K-shell ($n=1$) and the sub-shells $2s$ and $2p$ ($l=0$ and 1 respectively) are filled, whereas for germanium, the sub-shells $3s$, $3p$ and $3d$ are filled also.

These filled (i.e. closed) electron shell structures with the corresponding nuclei form the inner cores of the atoms, closest to the nucleus and tightly bound with energies of several electron-volts. When the temperature of a material is raised to T K, the most probable energy amongst the molecules is kT where k is the Boltzmann constant, equal to $(1/11,600)$eV or 8.63×10^{-5}eV. Hence at a temperature of 300 K (27°C) the most probable thermal energy of the atoms of an element is only 0.026eV, much less than that required to disturb the inner core electron structure. Consequently, the electrons in these closed inner shells do not take part in electrical conduction at normal ambient temperatures.

In the case of both silicon and germanium there remains, outside these inner cores (closed shells around the nucleus) four electrons. These are the valence electrons which form the covalent bonds in the crystal lattice structures of these elements (fig. 1.1).

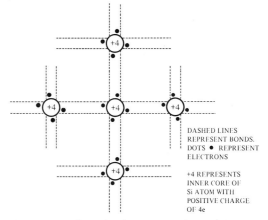

Fig. 1.1 A two-dimensional representation of ions with covalent bonds in the crystal lattice of tetravalent silicon or germanium

Suppose there are N atoms of silicon* where N is a very large number (as the Avogadro constant is

---

*Silicon is the most important element used in the construction of semiconductor devices, so silicon will be referred to in most cases henceforth. Germanium (atomic mass 72.6) could also be considered but this element has taken second place in importance.

Table 1.1 The Periodic Table of the Elements

Group IA	Group IIA	Group IIIB	Group IVB	Group VB	Group VIB	Group VIIB	Group VIII			Group IB	Group IIB	Group IIIA	Group IVA	Group VA	Group VIA	Group VIIA	Inert gases
H 1																	He 2
Li 3	Be 4											B 5	C 6	N 7	O 8	F 9	Ne 10
Na 11	Mg 12											Al 13	Si 14	P 15	S 16	Cl 17	A 18
K 19	Ca 20	Sc 21	Ti 22	V 23	Cr 24	Mn 25	Fe 26	Co 27	Ni 28	Cu 29	Zn 30	Ga 31	Ge 32	As 33	Se 34	Br 35	Kr 36
Rb 37	Sr 38	Y 39	Zr 40	Nb 41	Mo 42	Tc 43	Ru 44	Rh 45	Pd 46	Ag 47	Cd 48	In 49	Sn 50	Sb 51	Te 52	I 53	Xe 54
Cs 55	Ba 56	La 57*	Hf 72	Ta 73	W 74	Re 75	Os 76	Ir 77	Pt 78	Au 79	Hg 80	Tl 81	Pb 82	Bi 83	Po 84	At 85	Rn 86
Fr 87	Ra 88	Ac 89	Th 90	Pa 91	U 92												

*The rare earth elements with atomic number Z from 58 to 71 inclusive are omitted.

Table 1.2 Electron shells and sub-shells in the Group IVA elements

Element	Atomic No. Z	K	L		M			N				O				P		
		1s	2s	2p	3s	3p	3d	4s	4p	4d	4f	5s	5p	5d	5f	6s	6p	6d	6f
Carbon (C)	6	2	2	2														
Silicon (Si)	14	2	2	6	2	2												
Germanium (Ge)	32	2	2	6	2	6	10	2	2									
Tin (Sn)	50	2	2	6	2	6	10	2	6	10		2	2					
Lead (Pb)	82	2	2	6	2	6	10	2	6	10	14	2	6	10		2	2	

6.025×10^{26} molecules per kilomole, there is this number of atoms of silicon in 28.1 kg of the element, since the atomic mass of silicon is 28.1). In a gas or vapour state at normal pressures and temperatures, these atoms are separated from one another by distances which are large compared with their diameters; the mean free path is about 10^{-6} m to 10^{-5} m whereas the atomic diameter is about 2×10^{-10} m. Any one atom is therefore not influenced significantly as regards its valence electron structure by its comparatively distant neighbouring atoms. The atoms may be considered as isolated from one another, and the valence electrons will occupy discrete energy levels within the atom. For silicon, there will be two such distinct levels, one occupied by the s sub-shell electrons (of which there are $2N$ as each of the N atoms contains two s sub-shell valence electrons) and the other by p sub-shell electrons of which there are also $2N$ although there will be $6N$ possible states because 6 is the maximum possible number of electrons in a *p* sub-shell. Thus, only one-third of the possible states available to *p* sub-shell electrons will be occupied.

If the mean separation between neighbouring atoms is decreased (e.g. the pressure of the silicon vapour is raised and/or its temperature is lowered) the valence electron energy levels of any one will become increasingly affected by the proximity of neighbouring atoms. The atoms will have a mean separation; about this mean, a range of separations will prevail. The result will be that the single energy levels (for mean separations greatly in excess of atomic diameter) become energy bands, particularly as the mean separations are decreased to become comparable with atomic diameters. As the atomic spacing decreases to the extent where these energy bands over-lap, the $6N$ upper states (electrons in a *p* sub-shell) merge with the $2N$ lower states (*s* sub-shell electrons) giving $8N$ states of which half are occupied by the $4N$ (two *s* and two *p*) valence electrons available.

The crystalline form of silicon (and of germanium) is like that of diamond (the crystalline form of carbon). This may be regarded either as a cubic arrangement composed of eight inter-penetrating simple cubic lattices or as two face-centred cubic lattices displaced from one another by one-quarter of a diagonal of the main cube, giving the tetrahedral arrangement shown in fig. 1.2. In such a crystal of silicon, the valence electrons form the covalent bands with complete overlap between the energy band structures, with any one valence electron equally attached to neighbouring atoms. Now the $4N$ available electrons are all in the lower valence band in $4N$ states whereas the remaining $4N$ states are in a higher energy band — the so-called *conduction band*. If E_V is the energy level at the top of the valence band and E_C is the energy level at the bottom of the conduction band, there is a *forbidden gap* (the band gap) of energy width $E_G = E_C - E_V$ (fig. 1.3). For an electron in the crystal to be able to leave the valence band and enter the conduction band, it must be provided with an energy of at least E_G. At a temperature of 300 K, E_G is 1.1 eV for silicon and 0.72 eV for germanium.

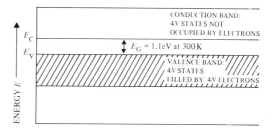

Fig. 1.3 The energy-band structure for single crystal silicon

1.3 Electrical conduction in intrinsic silicon

In pure (intrinsic) silicon, current carriers are not available until an electron is able to leave the valence band and enter the conduction band. This demands that an electron in the valence band must have imparted to it an energy which exceeds the band gap of energy width E_G, which is the energy needed to break an electron away from its covalent bonding in the crystal. At a temperature of 300 K, the most probable kinetic energy of an electron is considerably less than 1.1 eV, the value of E_G for silicon. However, there is a distribution of energies amongst the electrons at any given temperature so that, even at 300 K or below, a small fraction (increasing with temperature) will have energies exceeding 1.1 eV. When an electron leaves the valence band to enter the conduction band on receiving sufficient energy (e.g. by thermal effect or by bombardment with external particles, including photons) it will leave behind in the valence band a vacancy, i.e. a *positive hole*, usually referred to as a 'hole', designated by *h* to distinguish it from an electron *e*. In intrinsic silicon, the electrons and holes are clearly produced in equal numbers. They may be regarded as particles of equal

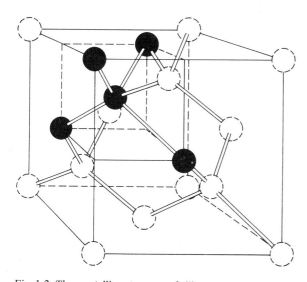

Fig. 1.2 The crystalline structure of silicon

but opposite charge where for an electron the charge is $-e$ whereas for a hole it is $+e$ (e = the electronic charge = 1.6×10^{-19} C).

There are few thermally generated electrons and holes at normal ambient temperatures, but they result in silicon at 300 K having an electrical resistivity of 2300 ohm m.

On applying a uniform electric field E to a specimen of intrinsic silicon, the drift current density J is given by

$$J = \sigma E = e(n\mu_e + p\mu_h)E \qquad (1.1)$$

where σ is the electrical conductivity (in siemen), in one cubic metre of silicon there are n conduction electrons, and p holes, whilst μ_e and μ_h are respectively the mobilities (unit: $\mathrm{ms^{-1}/Vm^{-1}} = \mathrm{m^2 s^{-1} V^{-1}}$) in intrinsic silicon of electrons and holes.

1.4 The effective masses of electrons and holes

The free electron has a mass of 9.1×10^{-31} kg and the hole in the free state is regarded as having the same mass. However, mass is basically concerned with the acceleration given to a particle (or body) in a given field of force and the electrons and holes within solid materials are not 'free' as they are within a gas at low pressure.

The full treatment of the concept of effective mass in the case of electrons and holes in crystalline media is complex. A simplified quantum-mechanical account given here confines the treatment to one dimension along an x-axis. The equation for a travelling wave is

$$F(x) = F_0 \exp[j(kx - \omega t)] \qquad (1.2)$$

in which $k = 2\pi/\lambda$ and $\omega = 2\pi\nu$, λ being the wavelength and ν the frequency.

The group velocity v_g of such a wave (taken in the wave theory of 'particle' motion to be equal to the 'particle' velocity) is known to be given by

$$v_g = \frac{d\nu}{d(\frac{1}{\lambda})} = \frac{d\omega}{dk} \qquad (1.3)$$

Moreover, $W = h\nu = h\omega/2\pi = \hbar\omega \qquad (1.4)$

where W is the energy of the 'particle' represented by a travelling wave of frequency ν.

Therefore $v_g = \frac{1}{\hbar} \frac{dW}{dk} \qquad (1.5)$

When an electron of charge $-e$ moves initially with velocity v_g in an electric field E, the energy dW acquired in a time interval dt is equal to the work done on the electron in displacing it through the distance $v_g \, dt$. This work done is the force $-eE$ acting on the electron multiplied by the displacement $-v_g \, dt$ (the negative sign is introduced here because the displacement of a negative charge is in the opposite direction to the field). Consequently,

$$dW = -eE v_g \, dt$$

Substituting for v_g from equation (1.5) gives

$$dW = \frac{-eE}{\hbar}\left(\frac{dE}{dk}\right) dt$$

Therefore, $dk = \left(\frac{-eE}{\hbar}\right) dt$

Hence the external force acting on the electron is

$$-eE = \hbar \frac{dk}{dt}$$

This force divided by the acceleration $\frac{dv_g}{dt}$ imparted must be the effective mass m^* of the electron.

Therefore, $m^* = \frac{\hbar \, dk}{dt} \Big/ \frac{dv_g}{dt} \qquad (1.6)$

Differentiation with respect to t of equation (1.5)

$$\frac{dv_g}{dt} = \frac{1}{\hbar}\frac{d^2 W}{dk \, dt} = \frac{1}{\hbar}\left(\frac{d^2 W}{dk^2}\frac{dk}{dt}\right) \qquad (1.7)$$

Substitution from equation (1.7) in equation (1.6) gives

$$m^* = \hbar^2 \Big/ \left(\frac{d^2 W}{dk^2}\right) \qquad (1.8)$$

A similar argument can be put forward for the effective mass of a hole. It is seen that these effective masses depend on the manner in which the energy of the electron (or hole) changes with the de Broglie wavelength λ of the particle.

1.5 The concentrations of current carriers in intrinsic silicon

The Fermi-Dirac distribution is

$$N(E)\,dE = \frac{4\pi}{h^3}(2m_e)^{3/2} E^{1/2}\,dE / [\exp\{(E-E_F)/kT\}+1] \quad \ldots\ldots (1.9)$$

where $N(E)dE$ is the number of electrons each of mass m_e in unit volume of a conductor which have energies in the range E to $(E+dE)$, E_F is the Fermi energy, h is the Planck constant, k is the Boltzmann constant and T K is the temperature.

In an intrinsic semiconductor such as silicon, the concentration of electrons in the conduction band is that number n per unit volume with energies exceeding E_C, the energy level at the bottom of the conduction band. For convenience, take this level to be zero, i.e. put $E_C = 0$. Then

$$n = \frac{4\pi}{h^3}(2m_e)^{3/2} \int_0^\infty E^{1/2}\,dE/[\exp\{(E-E_F)/kT\}+1] \quad (1.10)$$

At 300 K, kT is only 0.026 eV, whereas $E - E_F$ is a few eV. So $\exp\{(E-E_F)kT\}$ is much greater than 1. Putting $x = E/kT$ in equation (1.10) and ignoring the unity in the denominator, gives

$$n = \frac{4\pi}{h^3}(2m_e)^{3/2}(kT)^{3/2}\exp(E_F/kT)\int_0^\infty x^{1/2}\exp(-x)\,dx$$

$$= 2\left(\frac{2\pi m_e kT}{h^2}\right)^{3/2}\exp(E_F/kT) \quad (1.11)$$

For the concentration p of holes (each of mass m_h) in the valence band, equation (1.10) is of the form:

$$p = \frac{4\pi}{h^3}(2m_h)^{3/2}\int_{-\infty}^{-E_G}(-E_G - E)^{1/2}\,dE / [\exp\{(E_F - E)/kT\}+1]$$

because the top of the valence band is at an energy E_G (the band gap width) *below* the bottom of the conduction band, which has been taken to be at zero energy level. Again, at 300 K, $(-E_G - E)/kT \gg 1$. Putting $-(E_G - E)/kT = y$ gives

$$p = \frac{4\pi}{h^3}(2m_h)^{3/2}(kT)^{3/2}\exp\left[-(E_F + E_G)/kT\right] \times$$

$$\int_0^\infty y^{1/2}\exp(-y)\,dy$$

$$= 2\left(\frac{2\pi m_h kT}{h^2}\right)^{3/2}\exp\left[-(E_F + E_G)/kT\right] \quad (1.12)$$

In unit volume of intrinsic silicon, n and p are equal because the presence of an electron in the conduction band is a consequence of this electron's departure from the valence band to leave a hole there. From equations (1.11) and (1.2) this equality leads to

$$\left(\frac{2\pi m_e kT}{h^2}\right)^{3/2}\exp(E_F/kT) = \left(\frac{2\pi m_h kT}{h^2}\right)^{3/2}\exp\left[-(E_F + E_G)/kT\right]$$

Therefore, $\left(\dfrac{m_e}{m_h}\right)^{3/2} = \dfrac{\exp[-E_F + E_G)/kT]}{\exp(E_F/kT)}$

Therefore, $^{3}/_{2}\log(m_e/m_h) = (1/kT)(-2E_F - E_G)$

If the masses of the electron and hole are taken to be equal, $m_e = m_h$ so $2E_F = -E_G$. The Fermi energy level E_F will then be half-way through the band gap (the forbidden energy gap) i.e. half-way between the top of the valence band and the bottom of the conduction band.

At usual ambient temperatures in the case of silicon and most intrinsic semiconductors, it is correct that the Fermi level is not far from the mid-way position in the forbidden energy gap, so the effective masses of the hole and the electron may be taken to be equal. There are some semiconductors, however, for which the effective mass of the hole is considerably greater than that of the electron. An example is indium antimonide (In Sb), a 3 – 5 semiconductor, (*see* section 1.13) for which m^*_h/m^*_e is about 20. The Fermi energy level is therefore near the bottom of the conduction band for In Sb.

Putting $E_F = -E_G/2$ (since E_C is taken to be zero) and $m_e = m_h$ in equations (1.11) and (1.12) gives

$$n = p = 2\left(\frac{2\pi m_e kT}{h^2}\right)^{3/2}\exp[-E_G/2kT] \quad (1.13)$$

12 Electronics

Substituting $k = 1.38 \times 10^{-23}$ JK^{-1}, $h = 6.625 \times 10^{-34}$ Js and $m_e = m_h = 9.1 \times 10^{-31}$ kg, gives

$$n = p = n_i = 4.83 \times 10^{21} \, T^{3/2} \exp(-E_G/2kT) \quad (1.14)$$

where, for 1m³ of intrinsic semiconductor material of which the Fermi-level is half-way through the forbidden energy gap, n and p are, respectively, the numbers of electrons and holes and n_i is the intrinsic concentration.

EXAMPLE 1.5 *For intrinsic silicon at 300 K calculate the concentration of electrons in the conduction band given that the width of the forbidden energy gap is 1.1 eV. Hence find the resistivity of intrinsic silicon at 300 K knowing that the mobilities in intrinsic silicon of electrons and holes are respectively $0.13 \, m^2 s^{-1} V^{-1}$ and $0.05 \, m^2 s^{-1} V^{-1}$. the electron charge is 1.6×10^{-19} C and the effective mass of the electron is 9.1×10^{-31} kg.*

In equation (1.14), $T = 300$, $E_G = 1.1$ eV and $k = (1/11600)$ eV K^{-1}.

Therefore n
$= 4.82 \times 10^{21} \, (300)^{3/2} \exp[-1.1/(2 \times 300/11600)]$
$= 4.82 \times 10^{21} \times 5196 \exp(-21.3)$
$= 4.82 \times 10^{21} \times 5196 \times 5.623 \times 10^{-10}$
$= 1.41 \times 10^{16}$ electrons/m³.

From equation (1.1) the electrical conductivity is

$$\sigma = e(n \mu_e + p\mu_n)$$

Therefore, σ
$= 1.6 \times 10^{-19} \, (1.41 \times 10^{16} \times 0.13 + 1.41 \times 10^{16} \times 0.05)$
$= 1.6 \times 1.41 \times 0.18 \times 10^{-3} = 4.05 \times 10^{-4} \, \Omega^{-1} \, m^{-1}$

Therefore, resistivity $\rho = 1/\sigma = 2460 \, \Omega$ m.

1.6 Extrinsic *n*-type and *p*-type semiconductors

To alter substantially the electrical conductivities of the intrinsic semiconducting elements silicon and germanium they are *doped* with a very small percentage of selected elements. This doping produces extrinsic (impure) semiconductors of which there are two kinds: *n*-type and *p*-type.

To produce an *n*-type semiconductor, a *donor* technique is employed so as to provide excess electrons (negative charges) which are relatively loosely bound in the crystal lattice structure. This is brought about by the controlled addition to the tetravalent silicon (or germanium) — the host element — of an element with five valence electrons in the outermost shell structure (a pentavalent element). Suitable pentavalent elements are from Group V of the Periodic Table (table 1.1). Those which also have properties compatible with silicon (or germanium) and so are suitable as donors are all from Group VA: they are phosphorus, arsenic and antimony.

A *p*-type semiconductor is produced by doping with a suitable trivalent element giving an *acceptor* technique in which a small fraction of the silicon (or germanium) atoms is replaced by atoms of valence three, so each having one electron less in the outermost shell structure than the pentavalent silicon (or germanium). The result is an extrinsic semiconductor in which there is an excess of positive holes, resulting in *p*-type material. Suitable compatible trivalent elements for this purpose are all from Group IIIA of the Periodic Table: they are boron, aluminium, gallium and indium (table 1.1).

The amount of the dopant addition required is exceedingly small: for example 1 donor atom in 10^8 atoms of germanium increases by a factor of 12 the electrical conductivity at 30°C. As the impurity atoms added are in very low concentrations compared with those of the host element (silicon or germanium) they will be relatively widely separated in the crystal structure and hence will occupy a very narrow energy band, tantamount to a discrete level. To aim at such

Table 1.3

Properties of pure silicon and germanium

	Si	Ge
Atomic number, Z	14	32
Atomic mass, M	28.1 amu	72.6 amu
Density, ρ	2330 kg m^{-3}	5320 kg m^{-3}
Atoms/m³	5×10^{28}	4.4×10^{28}
E_G at 300 K	1.1 eV	0.72 eV
$n = p$ at 300 K	1.5×10^{16}/m³	2.5×10^{19}/m³
Resistivity at 300 K	2300Ωm	0.45Ωm
μ_e at 300 K	0.13 m²s^{-1}V^{-1}	0.38 m²s^{-1}V^{-1}
μ_h at 300 K	0.05 m²s^{-1}V^{-1}	0.18 m²s^{-1}V^{-1}

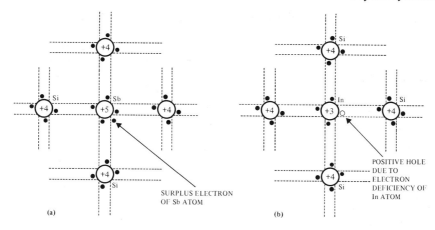

Fig. 1.4 Crystal lattice of silicon in which (a) a silicon atom is replaced by a pentavalent impurity in the form of antimony (Sb) and (b) a silicon atom is replaced by trivalent indium

low concentration controlled dopant additions, the host material needs to be of very high purity (appendix A).

Two dimensional diagrams illustrating *n*-type and *p*-type silicon are in fig. 1.4 .

In *n*-type extrinsic semiconducting silicon, four of the five electrons in the pentavalent donor atom will form the covalent bonds. The fifth electron is excess and only very weakly bonded: indeed, it requires an energy of only 0.05 eV (0.01 eV for germanium) to dislodge it from the atomic structure and render it available for conduction. For *n*-type silicon, the donor energy level is thus only 0.05 eV below the bottom of the conduction band (fig. 1.5a). Hence, at a temperature of 300 K at which the most probable energy of a particle is 0.026 eV, almost all the 'fifth' electrons of the donor atoms will be raised into the conduction band.

In *p*-type silicon, only three electrons are available in any one trivalent acceptor atom: a hole exists in the covalent bonding which requires four electrons for completion. A very small amount of energy is needed for an electron to leave the valence band and occupy this vacancy. A discrete acceptor energy level is therefore produced just above the top of the valence band (fig. 1.5b). This ready departure of electrons from the valence band to this acceptor level leaves behind excess holes in the valence band which become the chief current carriers.

In extrinsic semiconductor materials the number of thermally generated electrons and holes available for conduction is a small fraction of the total number of free carriers. Most of the current carriers in the *n* and *p* type extrinsic materials are the electrons or holes introduced deliberately as a result of the donor or acceptor additions respectively. These electrons or

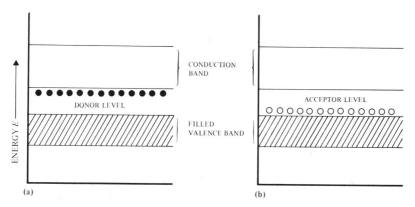

Fig. 1.5 Energy-band structure for (a) *n*-type silicon showing the donor energy level and (b) *p*-type silicon showing the acceptor energy level

holes form the *majority carriers*. In *n*-type semiconductors, the majority carriers are electrons whilst the minority carriers are holes; in *p*-type material, the majority carriers are holes whereas the minority carriers are electrons.

It can be proved that, under thermal equilibrium, the *law of mass action* holds whereby in extrinsic material

$$np = n_i^2 \qquad (1.15)$$

where n = number of electrons per unit volume,
p = number of holes per unit volume,
and n_i = number of minority current carriers per unit volume in the intrinsic material, which depends upon the temperature.

In *n*-type material, almost all the donor atoms lose an electron (the pentavalent donor atom readily loses an electron to the conduction band because only four electrons are needed for the covalent bonding). The donor atoms are hence nearly all positive ions. If the number of donor atoms per unit volume is N_D, they produce N_D positive ions per m³. The total number of positive charges per m³ is therefore $N_D + p$.

In *p*-type material almost every one of the trivalent acceptor atoms will have an electron attached to it: these acceptor atoms are therefore negative ions. If N_A is the number of acceptor atoms per unit volume, there are consequently N_A negative ions per m³. The total number of negative charges per m³ is therefore $N_A + n$.

The semiconductor as a whole in equilibrium is not charged electrically. In a material in which both donors and acceptors exist, then

$$N_D + p = N_A + n \qquad (1.16)$$

In *n*-type material, N_A is zero and there are very few holes so $p \to 0$. From equation (1.16) it follows that

$$N_D = n, \text{ approx.}$$

i.e. the concentration of electrons is approximately the same as the concentration of donor atoms.

Likewise, for *p*-type material, it can be shown that p, the concentration of holes, is approximately the same as N_A, the concentration of acceptor atoms.

EXAMPLE 1.6a. *A sample of n-type extrinsic silicon at 300 K contains 1 donor atom in 10^8 silicon atoms. Given that the electron mobility μ_e at 300 K in silicon is $0.13 m^2 s^{-1} V^{-1}$, calculate the resistivity of this sample. [The Avogadro constant is 6.025×10^{26} molecules/kilomole; the atomic mass of silicon is 28.1 and its density is 2330 kg m^{-3}.]*

28.1 kg of silicon contains 6.025×10^{26} atoms.

The concentration of silicon atoms is therefore

$$\frac{6.025 \times 10^{26} \times 2330}{28.1} \text{ m}^{-3}$$

The concentration of donor atoms (equal to the concentration of electrons) is therefore

$$\frac{6.025 \times 10^{26} \times 2330 \times 10^{-8}}{28.1} \text{ m}^{-3} = 5 \times 10^{20} \text{ m}^{-3}$$

The electrical conductivity σ is given by equation (1.1) in which, in the present case, p is negligible compared with n. Therefore,

$$\sigma = en \mu_e = 1.6 \times 10^{-19} \times 5 \times 10^{20} \times 0.13 = 10.4 \, \Omega^{-1} \text{ m}^{-1}$$

The resistivity is therefore $1/\sigma = 0.096 \, \Omega \text{ m}$.

Note how small this value is compared with the resistivity of intrinsic silicon, which is 2300 Ω m (table 1.3).

EXAMPLE 1.6b. *In n-type silicon, the energy required to release a donor electron may be compared with the energy needed to release an electron from an hydrogen atom. The Bohr theory of the hydrogen atom gives that the ionization energy E_i is $me^4/8\epsilon_o^2 h^2$ where m is the mass and e is the charge of the electron, ϵ_o is the permittivity of free space and h is the Planck constant. The value of the ionization energy of hydrogen is 13.6 eV. Assuming that this expression applies to the donor ionization energy E_D for n-type silicon, calculate E_D on making corrections because the effective mass of the electron in silicon is 0.2 m and the relative permittivity of silicon is 11.7.*

For hydrogen, $E_i = me^4/8 \epsilon_o^2 h^2 = 13.6$ eV.

For *n*-type silicon, $E_D = m^* e^4/8\epsilon_o^2 \epsilon_r^2 h^2$

where $m^* = 0.2\,m$ and $\epsilon_r = 11.7$. Clearly, therefore,

$$E_D = \frac{0.2}{11.7^2}\,E_i = \frac{0.2 \times 13.6}{11.7^2}\;\text{eV}$$

$$= 0.02\;\text{eV}.$$

The result given in example 1.6b for E_D is of the correct order of size. However, the donor ionization energy will depend on the type of pentavalent donor. The actual values of E_D in eV for n-type silicon are 0.045 for phosphorus, 0.049 for arsenic and 0.039 for antimony as the donor.

1.7 The Incidence of Photons on a Semiconductor

In the case of an intrinsic semiconductor, the incidence of a photon of radiation of frequency ν and so energy $h\nu$ (where h is the Planck constant) will be able to cause an electron to leave the valence band and enter the conduction band provided that the energy of this photon exceeds the width E_G of the forbidden energy gap. For silicon at 300 K, $E_G = 1.1$ eV so that the frequency of the radiation required to cause such *intrinsic photo-excitation* must exceed ν_{min} given by

$$h\nu_{min} = 1.1$$

where $h = 4.14 \times 10^{-15}$ eVs

$$\text{Therefore }\nu_{min} = \frac{1.1 \times 10^{15}}{4.14} = 2.66 \times 10^{14}\;\text{Hz}.$$

The maximum wavelength of the radiation is therefore

$$\lambda_{max} = \frac{3 \times 10^8}{2.66 \times 10^{14}} = 1.13 \times 10^{-6}\,\text{m} = 1.13\,\mu\text{m}.$$

This maximum wavelength is in the near infra-red region, so photons in the visible light region are clearly able to create current carriers in pure silicon.

For extrinsic silicon, the frequency of the radiation capable of photo-excitation is ostensibly much lower because the electron has only to go from the donor level (for n-type silicon) into the conduction band, requiring an energy of 0.05 eV, whilst for p-type silicon the incident photon need only have enough energy to raise an electron from the valence band to the acceptor energy level. Apparently, we now have photo-excitation in the far infra-red. In practice, with the usual forms of silicon used in transistor manufacture, such *extrinsic photo-excitation* occurs only to a very limited extent. The photon yield (number of electrons produced per incident photon) is very small because the concentrations of donor and acceptor atoms are tiny compared with that of the intrinsic element. The cross-section for the interaction of a photon with, say, a pentavalent doping atom is exceedingly small.

1.8 The Life-times of Current Carriers

The current carriers (electrons and holes) within a semiconductor are at equilibrium concentrations at a given temperature. Departure from this thermal equilibrium is a consequence of external stimuli. Possible stimuli are (a) the application of an electric field to cause drift of the current carriers; here, in practice, metallic contacts are made to the semiconductor at which, in effect, current carriers are injected; (b) the incidence on the semiconductor of photons or charged particles; (c) a change of temperature; (d) the application of a magnetic flux to cause modifications of current carrier paths.

The creation in a given region of a semiconductor of a departure from the initial equilibrium will be followed by recombination of electrons and holes when the initiating stimulus is removed or reduced. Clearly, the life-times of the current carriers created are of critical importance in relation to the speed at which a semiconductor device can respond to changes of external stimuli.

If a piece of silicon which contains initially per unit volume n_o electrons and p_o holes is irradiated with photons of sufficient energy and this incident radiation is then switched off, the excess electrons and holes generated within the semiconductor by the radiation will subsequently decrease in concentration.

In an introductory treatment, it is reasonable to assume that the rate of change of concentration at time t of the current carriers (created initially by a momentarily applied stimulus over a given volume of the semiconductor) is directly proportional to the excess concentration at time t, and so to the concentration itself at that time because the initial concentration (before application of the stimulus) is assumed to be a constant independent of time. If therefore the concentration of electrons at a given temperature at time t is n and that of holes is p, it follows that,

$$\frac{dn}{dt} = -an \quad (1.17)$$

and

$$\frac{dp}{dt} = -bn \quad (1.18)$$

where a and b are constants, the minus signs being needed because the concentrations decrease with time t.

Integration of equation (1.17) gives

$$\log_e n = -at + \text{const.}$$

At $t = 0$, taken to be the time when the external stimulus is switched on for a vanishingly short time compared with the recombination times of current carriers, $n = n_o$. Therefore,

$$\log_e n = -at + \log_e n_o$$

or $n = n_o \exp(-at)$ (1.19)

Likewise, equation (1.18) integrates to give

$$p = p_o \exp(-bt) \quad (1.20)$$

Thus, the initial concentrations produced decrease exponentially with time.

Referring to equation (1.17)

$$dn = -an\, dt$$

gives the number of electrons, dn, which recombine in the interval of time t to $(t + dt)$. From equation (1.19) put $n = n_o \exp(-at)$, then,

$$dn = -an_o \exp(-at)\, dt$$

The sum of the life-times of all the generated electrons is consequently

$$\int_0^\infty -at\, n_o \exp(-at)\, dt.$$

The mean life-time of an electron is therefore given by

$$\tau_e = -a \int_0^\infty t \exp(-at)\, dt$$

Integration by parts of the r.h.s. of this equation gives $1/a$. Therefore
Therefore $\tau_e = 1/a$.

Hence equations (1.19) and (1.20) can be written

$$n = n_o \exp(-t/\tau_e) \quad (1.21)$$

and

$$p = p_o \exp(-t/\tau_h) \quad (1.22)$$

where τ_e and τ_h are respectively the mean life-times of electrons and holes in the semiconductor.

These results are a simplification inasmuch as it is assumed that the mean life-times are independent of n and p. A full theoretical treatment is beyond the scope needed for the present work.

The recombination of electrons and holes in a semiconductor might at first be thought to be due to electrons from the conduction band returning to the valence band to recombine with holes there. In fact, this is normally improbable. This is because such recombination with a hole means that the electron entering the valence band has to be accommodated within a vacancy in the crystal lattice structure which, within the valence band, implies that it fits in where an electron would normally be in the valence shell structure around one or other of the atomic cores. The chances are small that such an entering electron would have the correct momentum and direction of motion to do this.

Recombination of electrons with holes in semiconductors in fact normally occurs because of the existence of impurities in the bulk or at the surface of the crystal. Single crystals of silicon and germanium inevitably contain imperfections which introduce electron 'traps' or *recombination centres* within the forbidden energy gap. The consequent energy states in the forbidden gap are primarily the locations where generated electrons are accommodated and end their lives.

In many instances, it is essential that carrier life-times be as short as possible so that, after a stimulus causing a change of electron or hole concentration, this concentration returns to the thermal equilibrium value as quickly as possible after the stimulus is removed. The controlled incorporation within silicon of certain specific impurities, e.g. nickel or gold, enables the carrier mean life-times to be substantially reduced. One nickel atom in 10^{18} atoms of silicon will shorten noticeably carrier life-times.

Two life-times are concerned: τ_v the volume life-time and τ_s, the surface life-time. The former is sensitive to lattice imperfections; the latter is sensitive to surface contamination.

Mean life-times range from 1 ns to several μs and from 1 to 2 ms in the case of very pure silicon.

1.9 Diffusion Currents in a Semiconductor

Diffusion currents will exist when the concentration of electrons or holes in a semiconductor is not uniform throughout.

For simplicity, consider one dimension along an x-axis only. Suppose the number of electrons per unit volume around a point x_1 on this axis is n_{x_1} whereas at a different point x_2, on this same axis, the number of electrons per unit volume is n_{x_2}. If $n_{x_1} > n_{x_2}$, electrons will diffuse from x_1 to x_2. Hence, along this x-axis, there will exist at any point x a concentration gradient of dn/dx.

The current which flows as a consequence of diffusion of electrons is in the opposite direction from that of the electric current because, conventionally, a positive current is directed from a positive to a negative region. The diffusion equation for electrons is

$$n v_e = -D_e \, \text{grad} \, n \quad (1.23)$$

where D_e is the diffusion constant (diffusivity) of the material for electrons (unit : $m^2 \, s^{-1}$) and v_e is the diffusion velocity of the electrons.

For an x-direction only, the diffusion current density due to electrons is

$$J_e = n e \, v_e = e D_e \frac{dn}{dx} \quad (1.24)$$

where $-e$ is the electronic charge in coulomb and J_e is in Am^{-2}

In the case of the diffusion of holes, the appropriate equation is

$$J_h = -e D_h \frac{dp}{dx} \quad (1.25)$$

where J_h is the hole diffusion current density, e is the positive hole charge, D_h is the diffusion constant of the material for holes and $\frac{dp}{dx}$ is the hole concentration gradient.

For a semiconductor specimen in thermal equilibrium to which no external influence is applied, the net flow of current must be zero. The separation of charges due to diffusion creates an electric field in a semiconductor. This static electric field must counteract the diffusion flow of current carriers.

The mobility of a charged particle in an electric field and the diffusion constant in a medium are both motions dependent upon the statistics of inter-particle collisions in the medium. Mobility and diffusivity would hence be expected to be inter-dependent. This is indeed the case; the inter-dependence has been expressed by an equation due to Einstein in the form:

$$\frac{\mu}{D} = \frac{q}{kT} \quad (1.26)$$

where for a particle of charge q in a medium at temperature T, μ is the mobility, D is the diffusion constant and k is the Boltzmann constant.

Equation (1.26) may be established in a simple way. Confining attention to an x-axis only, suppose that E is the static electric field created in the x-direction at a point x by diffusion of particles each of charge q. The current density J at x is zero for a material in thermal equilibrium so that

$$J = qD \frac{dn}{dx} + q \mu n E = 0 \quad (1.27)$$

18 Electronics

where n is the number of charges at x.

From the Boltzmann distribution law it is known that

$$n = C \exp(-qEx/kT) \qquad (1.28)$$

where C is a constant of proportionality.

Therefore, $\dfrac{dn}{dx} = C \exp(-qEx/kT)(-qE/kT)$ (1.29)

From equation (1.27)

$$\frac{dn}{ndx} = -\frac{\mu E}{D}$$

Substituting for ndx and dn from equations (1.28) and (1.29) respectively,

$$\frac{dn}{ndx} = \frac{C\exp(-qEx/kT)(-qE/kT)}{C\exp(-qEx/kT)} = -\frac{\mu E}{D}$$

Therefore $q/kT = \mu/D$ which is the Einstein relationship.

For electrons in a semiconductor this becomes

$$-e/kT = \mu_e/D_e$$

where μ_e is the mobility and D_e the diffusion constant of electrons in the material.

For holes, the relationship is

$$e/kT = \mu_h/D_h.$$

EXAMPLE 1.9 *Establish that, at a temperature of 300 K the mobility μ of a particle of charge q in a solid medium is 39 D, where D is the diffusion constant of this particle in the same medium. In silicon at 300 K, the electron mobility is 0.13 $m^2\ s^{-1}\ V^{-1}$. Find the diffusion constant of electrons in silicon at 300 K. [The Boltzmann constant $k = 1.38 \times 10^{-23}\ JK^{-1}$.]*

From equation (1.26)

$$\mu = eD/kT = \frac{1.6 \times 10^{-19}\ D}{1.38 \times 10^{-23} \times 300} = 39\ D$$

For silicon at 300 K, $D_e = \mu/39 = 0.13/39 = 3.33 \times 10^{-3}\ m^2 s^{-1}$.

1.10 Diffusion length

The distance that an electron (or hole) travels in a semiconductor during its mean life-time τ_e (or τ_h) (which is terminated by recombination) is the diffusion length for electrons (or holes) in the material. For electrons, let this diffusion length be L_e. Considering the one-dimensional case of an x-axis only, put

$$n = n_o \exp(-x/L_e) \qquad (1.30)$$

where n is the concentration of electrons at the distance x and $n = n_o$ at $x = 0$.

At a distance x, the diffusion current density due to electrons is given by equation (1.24) to be $eD_e \dfrac{dn}{dx}$. Due to diffusion, electrons consequently arrive at this location at a rate of $D_e \dfrac{d^2 n}{dx^2}$. They are recombining at a rate of $\dfrac{dn}{dt}$, which equals $-an$ or $-n/\tau_e$ (section 1.8). Therefore,

$$D_e \frac{d^2 n}{dx^2} + \frac{n}{\tau_e} = 0. \qquad (1.31)$$

From equation (1.30)

$$\frac{d^2 n}{dx^2} = \frac{-n_o}{L_e^2} \exp(-x/L_e) = \frac{-n}{L_e^2}$$

Substitution in equation (1.31) gives

$$\frac{D_e}{L_e^2} = \frac{1}{\tau_e}$$

or $L_e = \sqrt{D_e \tau_e} \qquad (1.32)$

In the case of holes, the diffusion length is likewise given by

$$L_h = \sqrt{D_h \tau_h} \qquad (1.33)$$

In silicon at 300 K, the diffusivity D_e for electrons is $3.33 \times 10^{-3}\ m^2 s^{-1}$ (*see* example 1.9). An electron with a mean life-time of 10^{-8}s will hence have a diffusion length in silicon at 300 K given by equation (1.32) to be

$L_e = \sqrt{3.33 \times 10^{-3} \times 10^{-8}} = 5.8 \times 10^{-6}\,\text{m} = 5.8\,\mu\text{m}$

1.11 The *p-n* Junction; Open-circuited

Within a single crystal of silicon (or germanium) an excess of holes is formed to one side of a boundary by the addition of a trivalent acceptor and an excess of electrons on the other side by the addition of a pentavalent donor. At the boundary a *p-n* junction is formed. Such a junction is represented schematically by fig. 1.6. Within the main body (excepting the depletion region) of the *p*-type material A almost every one of the acceptor atoms (of which there are N_A per unit volume) will have an electron attached to it, so will be a negative ion, and the number of holes per unit volume, p, will also be N_A approx. (section 1.6). Within the main body (excepting the depletion region) of the *n*-type material B almost every one of the N_D donor atoms in unit volume will have lost an electron so will be a positive ion, and the number of electrons per unit volume, n, is N_D approx.

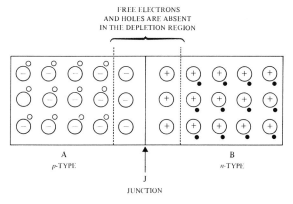

Fig. 1.6 Schematic two-dimensional representation of a *p-n* junction (open-circuited)

Within the *p*-type semiconductor material A, the number of holes per unit volume, p, is large whereas in the *n*-type material, B, it is small. Holes will consequently diffuse across the *p-n* junction from A to B. For similar reasons, electrons will diffuse from B to A. Such diffusion of holes and electrons will take place over the diffusion lengths, terminated by hole recombination in the *n*-type material and electron recombination in the *p*-type. Over a distance of about 0.5 μm extending symmetrically about the junction J, many of the ions (locked in the crystal lattice) will therefore not be neutralised by associated holes or electrons. Thus, the negative acceptor ions within A in the immediate vicinity of J become uncovered because holes in this region have diffused across to B whereas positive donor ions within B near J become uncovered as they have lost electrons by diffusion to A.

Over this distance of some 0.5 μm about the junction there exists therefore a region of space-charge known as the *depletion region*. Within this depletion region, there are no free carriers: the space-charge is dependent upon ions locked in the lattice. This space-charge will have a density ρ which is negative in A near J and positive in B near J; the variation of ρ with distance x from the junction plane is as shown in fig. 1.7a.

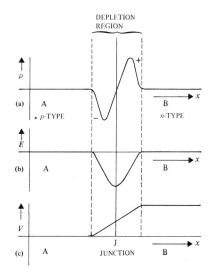

Fig. 1.7 (a) Variation of charge density ρ with distance x around a *p-n* junction; (b) Variation of electric field strength E with distance x; (c) Variation of the potential V with distance x

Choosing an *x*-axis as the axis of symmetry through the semiconductor crystal perpendicular to the plane of the junction (where the dimensions of this plane of cross-section exceed greatly the extent along the *x*-axis of the charge variation), we may write from Poisson's equation in electrostatics:

$$\frac{\partial^2 V}{\partial x^2} = -\frac{\rho}{\epsilon} \qquad (1.34)$$

where ϵ is the permittivity of the medium (silicon or germanium).

Integration of equation (1.34) gives

$$\frac{dV}{dx} = -\int \frac{\rho dx}{\epsilon} = -E \qquad (1.35)$$

where E is the electric field strength.

The manner in which the electric field strength E varies in the immediate vicinity of the junction is therefore as depicted in fig. 1.7b: E is a maximum at the junction itself (where the rate of change of ρ with x is a maximum) and is directed from B to A (opposite to the positive direction of x). At the junction it will attain a value such that it just counteracts the diffusion, giving an equilibrium when no further motion of change takes place across the junction.

Integration of equation (1.35) gives

$$V = -\int E dx$$

where V is the potential. The variation about the junction with distance x is shown in fig. 1.7c. V attains a maximum value of V_o, which is the height of the potential barrier set up at equilibrium against the flow of holes from A to B across the junction. The potential barrier against the flow of electrons from B to A will be of the same height but of opposite sign.

The energy-band diagram for a p-n junction in equilibrium on open-circuit (fig. 1.8) is such that the Fermi energy level E_F is the same in the p-type material as in the n-type. If, in a given volume of the material the numbers of electrons and of holes at any energy level were not the same, current would flow across the junction until these Fermi levels were the same and equilibrium prevailed. With an open-circuited junction this equilibrium is maintained with the energy bands (conduction and valence) of the p-type material above those of the n-type, corresponding to the positive potential of the n-side relative to the p-side, so exhibiting the flow of electrons from n to p and of holes from p to n.

1.12 The p-n Junction; Application of a Bias Voltage

Consider the application across a p-n junction of a potential difference provided by an external source of voltage, e.g. a battery. *Reverse bias* prevails when the negative terminal of this source of supply is connected (via an ohmic contact) to the p-type semiconductor material on one side of the junction and the positive terminal is connected to the n-type material on the opposite side of the junction (fig. 1.9a).

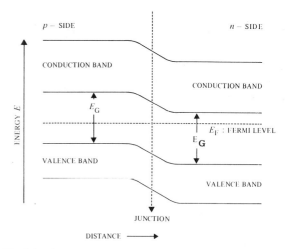

Fig. 1.8 The energy-band diagram for a p-n junction in equilibrium on open circuit

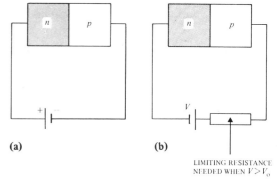

Fig. 1.9 (a) Application of reverse bias across a p-n junction; (b) Application of forward bias across a p-n junction

The application of reverse bias increases the height of the potential barrier against the movement of electrons from the n-type to the p-type material (and also for the oppositely directed movement of holes). Moreover, it causes holes in the p-type material and electrons in the n-type material to be attracted further away from the junction. The width of the depletion layer consequently increases with increase of the reverse bias. The current flow on application of reverse bias is nominally zero. In fact, the current is not zero but of small value I_s because a few electrons are thermally generated as minority carriers in the

p-type material and a few holes in the n-type and these few current carriers will traverse the junction. This small current I_s is independent of the bias value beyond a small amount — it is called the *reverse saturation current* — but it increases with temperature because of the additional thermal generation of electrons and holes.

Forward bias is maintained when the positive terminal of a voltage supply is connected to the p-type semiconductor material and the negative terminal is connected to the n-type material on the opposite side of the junction (fig. 1.9b). Now electrons cross the junction from n-type to p-type under the attraction of the positive connection and holes cross from p-type to n-type. The height of the potential barrier that prevailed across the junction when it was initially on open-circuit is decreased on application of this forward bias. As electrons are thereby moved in the electric field across to the p-type material which is rich in holes, on arrival they are minority carriers: this process is therefore known as *minority carrier injection* and is fundamental in the operation of the p-n junction diode and the bipolar junction transistor. Holes, of course, move the other way during such minority carrier injection. The excess minority carrier concentration in the p-type (or n-type) material brought about by this injection, falls off exponentially with distance from the junction.

If the p-n junction is short-circuited, i.e. the applied bias voltage is zero, the height of the potential barrier is clearly V_o, the same as when the junction is open-circuited.

When the applied forward bias voltage V equals V_o the retarding potential barrier is overcome and the current flow through the p-n junction is restricted only by the bulk resistance of the semiconductor material and any resistance present at the contacts between the supply of voltage and the semiconductor. Hence, with significant values of forward bias, to prevent destruction of the p-n junction, a limiting resistance is essentially required in series with the external voltage supply.

1.13 Semiconductors Based on Compounds

The most widely used element in semiconductor device manufacture (particularly diodes and transistors) is silicon. Compared with its chief rival, germanium, it ensures a lower leakage current in device manufacture, electrical characteristics which are less affected by changes of temperature and enables single components and integrated circuits to be based on the valuable silicon planar technique (*see* Appendix A).

Several other semiconductor materials are used, chiefly for the construction of photo-electric devices and light-emitting (electroluminescent) devices. Examples are the compound semiconductors known as III – V and II – VI semiconductors.

III – V semiconductors are intermetallic compounds such as gallium arsenide (Ga As), gallium phosphide (Ga P) and indium antimonide (In Sb). These are formed from combinations of a Group III element of valence 3 with a group V element of valence 5. In such a compound (e.g. Ga As where Ga is trivalent and As is pentavalent) each atom of the Group V element donates an electron to each atom of the Group III element. As in silicon, there are hence four valence electrons associated with each atom for covalent bonding but where also the trivalent atom is a negative ion whereas the pentavalent atom is a positive ion, so ionic bonding exists in addition.

II – VI semiconductor compounds are formed from a combination of a divalent element from Group II with an element from Group VI of valence 6. Now each Group VI atom donates two electrons (and so becomes a doubly positively charged ion) to each Group II atom (which becomes a doubly negatively charged ion). Each atom in the compound has four electrons in the covalent bonding but there is again also ionic bonding, which is more prevalent than in the case of III – V materials. Examples of II – VI semiconductor compounds are bismuth telluride ($Bi_2 Te_3$), lead selenide (Pb Se), lead sulphide (Pb S) lead telluride (Pb Te) and cadmium sulphide (Cd S).

Donor doping to form n-type semiconductor compounds and acceptor doping to form p-type semiconductor compounds is undertaken. For example, with III – V materials such as gallium arsenide, the donor dopant is a Group VI element such as tellurium of valence 6. The addition in a controlled fashion of such a Group VI element will provide some III – VI combinations within the III – V compound and where every Group VI atom exists in place of a Group V one, a donor electron (giving n-type material) is available. The appropriate acceptor dopant is of a Group II element. Then Group II divalent atoms will be associated with Group V atoms instead of Group III (trivalent) ones; at each such location, a hole will exist, so p-type material is formed.

The width of the forbidden energy gaps in eV in some intrinsic semiconductor compounds are recorded in brackets in the following list:—

III – V semiconductor compounds:
indium antimonide, In Sb (0.17); gallium antimonide, Ga Sb (0.7); indium phosphide, In P (1.17); gallium arsenide, Ga As (1.4).

II – VI semiconductor compounds:
bismuth telluride, $Bi_2 Te_3$ (0.15); lead selenide, Pb Se (0.25); lead sulphide, Pb S, (0.38); lead telluride, Pb Te (0.29); cadmium sulphide, Cd S, (2.5).

All semiconductors have the following main properties: a large negative temperature coefficient of resistance of about $6\% K^{-1}$; sensitivity to photons so are photovoltaic or photoconductive; large values of magnetoresistance and Hall coefficient compared with true pure metals.

1.14 Amorphous Semiconductors

A crystal may be considered as a solid in which there is long range repetition of the positions of atoms so that even at distances over a large number of atomic diameters, there is a regular periodicity in atomic positions. If this long range periodicity does not exist, the material is amorphous or non-crystalline. Amongst amorphous solids, glasses are of interest; by 'glassy' is meant an amorphous solid obtained by supercooling a liquid.

Although all the semiconductor devices widely used in modern electronics are based on crystalline solids, interest in glassy solids was stimulated in 1968 on the discovery by Ovshinsky that certain stable special glasses exhibited fast electrical switching characteristics in that they changed from a low to a high conductivity state when a critical voltage was applied across them. Of the two classes of switch described by Ovshinsky, the first, known as a 'threshold switch' (fig. 1.10a) was based on a glassy solid $Te_{33} As_{27} Ge_{20} S_{20}$ in the form of a film a few μm thick, which is stable and exhibits no tendency to crystallise on heating or slow cooling. In this threshold switch, the conductivity is low below a specific threshold current and high above that current. Operating characteristics displayed are typically: switching voltage 15V, switching time about 10^{-10}s, a conductivity of $10^{-5} \Omega^{-1} m^{-1}$ when voltage and current are below the threshold value whereas above this value, in the 'on' state of the switch, the current could be as much as 100 mA. The second class described, known as a 'memory' switch (fig. 1.10b) was based on a glassy solid $Te_{81} Ge_{15} As_4$ (again in thickness of a few μm) which is different in that the high conductivity state (conductivity about $1 \Omega^{-1} m^{-1}$) is retained even when the applied voltage (about 15V max) is reduced to zero. For this memory switch, the conductivity in the off-state (before any application of voltage) is about $10^{-4} \Omega^{-1} m^{-1}$. The switching time is about 10^{-3}s.

These switching possibilities of glassy films lead to a promise of a versatile set of semiconductor devices based on glassy amorphous solids with the possibility of ready and comparatively cheap manufacture.

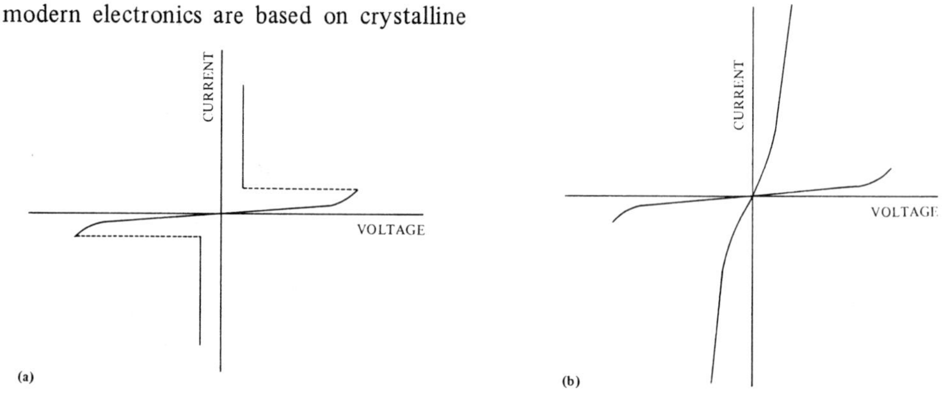

Fig. 1.10 Voltage-current characteristics of **(a)** a threshold switch and **(b)** a memory switch

Exercises I

1. By reference to The Periodic Table of the Elements and the electron-shell structure of the atoms concerned, together with other basic properties, explain why germanium and silicon are the most important intrinsic semiconducting elements in modern electronics.

2. Write explanatory notes, with diagrams as necessary on:
   (a) the forbidden gap (i.e. the band gap) of energy width E_G between the top of the valence band at energy E_V and the bottom of the conduction band at energy E_C in an intrinsic semiconducting element, such as silicon;
   (b) electrical conduction in intrinsic silicon.

3. Explain why the concept of the effective mass of an electron arises in the description of semiconduction in a solid element and derive an equation which relates the effective mass m^* of an electron to its energy, where the Planck constant h and $k = 2\pi/\lambda$ are also involved, λ being the de Broglie wavelength of the electron.

4. On the basis of a simple model together with the relevant mathematical theory, show that the Fermi level in an intrinsic semiconductor lies midway in the energy gap.

5. State the Fermi-Dirac distribution function. Choosing appropriate axes, draw graphs of this function at 0 K and T K, and explain briefly the significance of each graph.

6. Calculate the electrical conductivity of intrinsic germanium at 300 K assuming that the density of states in the conduction band is $4.83 \times 10^{21} T^{3/2}$ m^{-3}, the energy gap width is 0.72 eV, the electron mobility is 0.38 m^2 V^{-1} s^{-1} and the hole mobility is 0.18 m^2 V^{-1} s^{-1}.

7. Explain with the aid of suitable diagrams the formation of extrinsic (a) n-type and (b) p-type silicon by means of doping with suitable donor and acceptor elements respectively.
   Draw and comment on the energy-band diagrams for (c) n-type silicon showing the donor energy level and (d) for p-type silicon showing the acceptor energy level.

8. A sample of n-type germanium at 300 K contains 1 donor atom in 10^8 germanium atoms. If the electron mobility at 300 K in germanium is 0.38 m^2 V^{-1} s^{-1}, calculate the resistivity of this sample, where the necessary physical constants involved are taken from suitable tables or given in this text.

9. Explain how the donor ionisation energy for n-type silicon may be calculated approximately on the assumption that the Bohr theory of the hydrogen atom is applied with a suitable correction for the relative permittivity of silicon.

10. In the case of a semiconductor, write notes on the following, giving relevant mathematical detail, as required:
    (a) recombination centres;
    (b) the life-times of current carriers;
    (c) diffusion currents.

11. Establish from first principles the equation

    $q/kT = \mu/D$

    where for a current carrier in a semiconductor q, μ and D are respectively the electrical charge, the mobility and the diffusion constant, T is the absolute temperature and k is the Boltzmann constant.
    Establish that, at $T = 300$ K, the diffusion constant of electrons in silicon is 3.33×10^{-3} m^2 s^{-1}, given that the electron mobility is 0.13 m^2 s^{-1} V^{-1} and the Boltzmann constant is 1.38×10^{-23} JK^{-1}.

12. Write a short essay on the p-n junction dealing with the open-circuit case, the case where a bias voltage (both forward and reverse) is applied across the junction, the depletion region and the reverse saturation current.

13. Explain what is meant by the III – V intermetallic compound semiconductors and the II – VI semiconductor compounds, giving examples in each case.

14. Write an account of the so-called 'glassy' semiconductors, including a speculation on the possible future of these materials in semiconductor devices.

2·Semiconductor Diodes

2.1 The p-n junction diode

Provided that the single crystalline structure is preserved throughout, on arranging a junction between n-type and p-type semiconductor material a diode with rectifying properties may be formed. When forward bias is applied (the p-type material is made positive in potential with respect to the n-type) current flows readily through the junction with electrons leaving the n-type material to enter the p-type and holes moving in the opposite direction: minority carrier injection (section 1.2) occurs. On the application of reverse bias (the p-type material being made negative with respect to the n-type) the current flow is very small.

The basic equation which applies to the current flow through a p-n junction (derived by Shockley in 1949) is

$$I = I_s \left[\exp(eV/\eta kT) - 1\right] \quad (2.1)$$

where I is the current, V is the voltage applied across the junction itself, I_s is the reverse saturation current (this saturation current is obtained when adequate reverse bias is applied), k is the Boltzmann constant, T is the absolute temperature and η is a constant which, in the majority of cases, may be taken to be unity.

Equation (2.1) is the result of the addition of currents given by

$$I_h = I_{hs} \left[\exp(eV/\eta kT) - 1\right] \quad (2.2)$$

where I_h is the net hole current through the p-n junction and I_{hs} the reverse saturation current due to holes, together with

$$I_e = I_{es} \left[\exp(eV/\eta kT) - 1\right] \quad (2.3)$$

where I_e and I_{es} refer to electrons.

At 300K, $kT = 300/11600 = 0.026 = 1/39$ eV

so that equation (2.1) becomes

$$I = I_s \left[\exp(39V) - 1\right] \quad (2.4)$$

In equation (2.3) the magnitude of I_{es} depends on the modest concentration of electrons in the p-type material whereas in equation (2.2) I_{ps} depends on the modest concentration of holes in the n-type material.

In 1949 it was shown by Shockley that

$$I_{hs} = ep_h D_p \Big/ \sqrt{D_h \tau_h} \quad (2.5)$$

where e is the electronic charge, p_h is the concentration of holes in the n-region, D_h is the diffusion constant of the material for holes and τ_h is the mean life-time of a hole in the n-region. A similar equation holds for I_{es}.

As the mass action law (equation 1.15) holds for a semiconductor in equilibrium, it follows that n-type material with a high concentration of electrons will have a small concentration of holes, i.e. p_h is small so that I_{ph} is small, as is seen from equation (2.5). This indicates one requirement for obtaining very small reverse saturation currents. Another necessity is that there be a very small leakage current through the surface in the neighbourhood of the p-n junction. Modern encapsulation methods for silicon junction diodes (the most frequently used) are so excellent that, for example, the 1 N 3064 silicon epitaxial diode has $I_s = 10^{-7}$ A at 50V. Germanium diodes have rather larger reverse saturation currents but these are not encountered to any extent in the forms which make use of a p-n junction 'plane'; the majority of germanium diodes are point contact and gold-bonded types.

If the forward voltage V applied across the

junction itself is considerably greater than kT (which is equivalent to 0.026eV at 300 K) equation (2.1) simplifies to become

$$I = I_s \exp(V/kT) \qquad (2.6)$$

or

$$\log_e I = \log_e I_s + V/kT \qquad (2.7)$$

on taking η to be unity.

Consequently, except for values of V less than V_γ, the 'cut-in voltage' (which is about 0.6V for silicon) with forward bias (V positive) the current through the silicon junction diode increases exponentially with voltage.

Clearly, when $V_\gamma > V > 0$, the diode current is very small. The threshold voltage (positive) below which the diode current is insignificant is V_γ.

When a reverse bias of magnitude significantly in excess of kT is applied, equation (2.1) reduces to

$$I = -I_s \text{ approx,}$$

so I_s, the reverse saturation current, is almost constant in magnitude independently of the magnitude of the reverse bias above small values.

The static characteristic of the diode current I against the applied voltage V is therefore typically as shown in fig. 2.1. Note that the reverse currents are insignificantly small ($< 1\mu A$ for a silicon diode) whereas the forward currents are in mA when $V > V_\gamma$. The sudden increase of the reverse current above $V = -V_Z$ (the Zener voltage) is considered in section 2.6. The Mullard silicon junction diode O A 200 (a general purpose diode for applications at high temperature and low reverse currents) has the typical forward characteristics at temperatures of 25°C and 125°C shown in fig. 2.2, whilst the reverse saturation current at 25°C is 0.02 μA at a reverse voltage of 50, which reverse current increases to 1.0μA at 125°C.

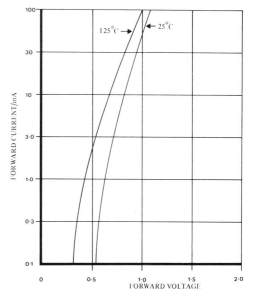

Fig. 2.2 Forward characteristics of the Mullard silicon junction diode OA200

The incremental resistance r of the diode is the reciprocal of the slope of the $V - I$ characteristic at a given point. It is not a constant but depends on the operating voltage.

Thus $r = \dfrac{\partial V}{\partial I}$

and the incremental conductance $g = \dfrac{\partial I}{\partial V}$.

From equation (2.1) with $\eta = 1$:

$$g = \frac{\partial I}{\partial V} = \frac{e}{kT} I_s \exp(eV/kT) = (I + I_s)e/kT$$

With a forward bias in excess of V_γ, $I_s \ll I$ so that

$$g = Ie/kT$$

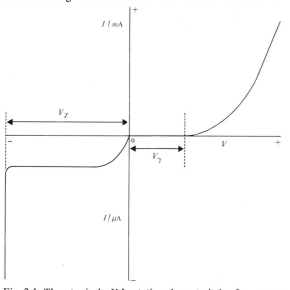

Fig. 2.1 The typical *V-I* static characteristic for a *p-n* junction diode

26 Electronics

At 300 K, $e/kT = 1.6 \times 10^{-19}/1.38 \times 10^{-23} \times 300$

$$= 38.6 \text{ V}^{-1}$$

Therefore, $g = (38.6\, I)\, \Omega^{-1}$

and $r = (0.026\, I^{-1})\, \Omega$

or $r = \dfrac{26}{I_{mA}}\, \Omega$

where I_{mA} is I in mA.

2.2 The basic diode circuit

The basic circuit associated with a single diode (fig. 2.3) comprises an input voltage v_i across the diode D and a load resistance R_L in series, where the output voltage across the load is v_o.

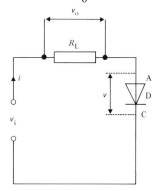

Fig. 2.3 The basic diode circuit

The voltage v across the anode (A) and cathode (C) of the diode is clearly related to v_i and iR_L (the voltage drop across the load where i is the current in the circuit) by

$$v = v_i - iR_L = v_i - v_o \qquad (2.8)$$

The way in which v varies with i is known from the static characteristic (the characteristic of the diode without any load present) which is known from the manufacturer's data and is given as typical in fig. 2.4a. What is required in practice is the current i in terms of the input voltage v_i; then the output voltage v_o is given by iR_L. This plot of the current against the input voltage is the *dynamic characteristic* (fig. 2.4b) which includes the effect of the resistor R_L and is obviously of smaller slope and more nearly linear than the static characteristic.

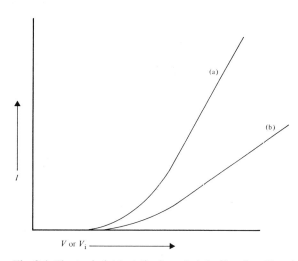

Fig. 2.4 The typical **(a)** static characteristic (I against V) and **(b)** dynamic characteristic (I against V_i) of a silicon junction diode

A plot of the output voltage v_o against the input voltage v_i is the *transfer characteristic* of the circuit; as $v_o = iR_L$, where R_L is a constant, this curve is clearly of the same shape as the dynamic characteristic (i against v_i).

2.3 The junction diode as a rectifier

Silicon diodes are much used in circuits for rectifying an alternating current supply. The instantaneous input voltage v_i at time t to the rectifier circuit is very frequently of sinusoidal wave-form and is represented by

$$v_i = V_P \sin \omega t$$

where V_P is the peak value of this input alternating voltage of frequency $f = \omega/2\pi$.

The basic half-wave rectifier circuit (fig. 2.5a) comprises usually the ac supply to a transformer across the secondary of which are connected the diode D and the load resistance R_L in series.

In an introductory theory which is satisfactory for most practical purposes it may be assumed for simplicity that the dynamic characteristic of the diode is linear (i.e. the diode forward resistance R_f is a constant irrespective of voltage), that the resistance of the diode when reverse bias is applied across it (i.e. the reverse resistance) is so large that reverse currents are negligibly small and that the positive threshold voltage V_γ, beyond which the diode conducts

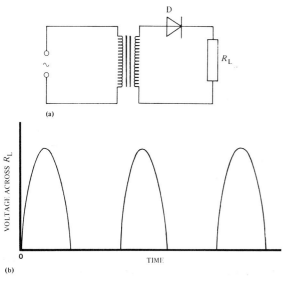

Fig. 2.5 (a) The basic half-wave rectifier circuit. (b) The unidirectional voltage output across the load resistance for a sinusoidally varying input voltage

significantly in the forward direction, is small enough compared with the peak input voltage V_P as to be negligible. In practice V_γ is about 0.6V for a silicon diode.

The output current i_L from the rectifier through the load resistance R_L during positive-going input half-cycles is then given by

$$i_L = \frac{V_P \sin \omega t}{R_L + R_f} \quad (2.9)$$

where V_P is the peak voltage across the transformer secondary and R_f is the forward resistance of the diode.

The instantaneous output voltage v_o is iR_L, and hence is given by

$$v_o = \left(\frac{V_P \sin \omega t}{R_L + R_f}\right) R_L \quad (2.10)$$

During those half-cycle periods when the input voltage is negative across D and R_L, $v_o = 0$.

Thus v_o is given by equation (2.10) during any one cycle when $0 \leqslant \omega t \leqslant 2\pi$ whereas v_o is zero when $\pi \leqslant \omega t \leqslant 2\pi$. The output voltage provided by a half-wave rectifier circuit is shown in fig. 2.5b.

Putting

$$i_L = I_{PL} \sin \omega t$$

where I_{PL} is the peak load current, the average value I_{av} (or I_{dc}, the current recorded by a moving-coil ammeter) of the current during any one half-cycle is given by

$$I_{av} \frac{T}{2} = \int_0^{\frac{T}{2}} I_{PL} \sin \omega t \, dt = \frac{I_{PL}}{\omega} \int_0^{\pi} \sin \omega t \, d(\omega t)$$

where T is the period.

$$= -\frac{I_P}{\omega} \left[\cos \omega t\right]_0^{\pi} = -\frac{I_P}{\omega} \left[-1 - 1\right] = 2 I_P / \omega$$

Therefore,

$$I_{av} = 4I_P / \omega T = 2I_P/\pi = 0.637 \, I_P.$$

In half-wave rectification, during every other half-cycle the current is zero, hence

$$I_{av} = I_P / \pi = 0.319 \, I_P \quad (2.11)$$

and the average voltage across the load resistance is given by

$$V_{avL} = \frac{I_{PL} R_L}{\pi} = \frac{V_P R_L}{\pi (R_f + R_L)} \quad (2.12)$$

2.4 Full-wave rectifier circuits

Two frequently employed full-wave rectifier circuits are fig. 2.6a, in which use is made of a transformer with a centre-tapped secondary winding and fig. 2.6b, the bridge rectifier circuit.

The use of a centre-tapped transformer secondary winding enables voltages at the secondary winding extremities A and B relative to the centre-tap at C to be in anti-phase. Hence, when A is positive with respect to C, diode D_1 is conducting and the direction of the conventional current flow (opposite to that of electron flow) through the load resistance R_L is as shown by the arrow (fig. 2.6a) whereas B is negative so that the current through the diode D_2 is zero. During the immediately succeeding half-cycle of the sinusoidal input, A is negative with respect to C whereas B is now positive so that diode D_1 is

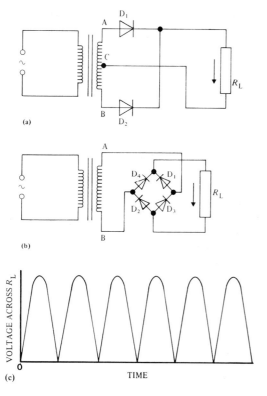

$$\frac{v}{2} = \frac{V_P \sin \omega t}{2}$$

In full-wave rectification the average current is (section 2.3)

$$I_{av} = 2I_P/\pi$$

and the average voltage across the load resistance R_L is given by

$$V_{avL} = 2I_{PL}R_L/\pi = V_P R_L/[\pi(R_f + R_L)] \quad (2.13)$$

which is the same as given by equation (2.12), so the average voltage across the load is the same in half-wave rectification as in full-wave, presuming that the same diodes are used in both cases and also the same transformer voltages. This arises from the fact that, although the unidirectional current persists during both the positive and negative half-cycles of the input in full-wave rectification, the input voltage is halved because of the division into two parts by the centre-tap of the transformer secondary.

In the case of the full-wave bridge rectifier circuit (fig. 2.6b) the output voltage is twice that obtained in the centre-tapped transformer case except that four diodes are used instead of two so that R_f in equation (2.12) becomes $2R_f$ because there are two diodes in series with R_L during conduction.

Fig. 2.6 Full wave rectifier circuits (a) using a centre-tapped transformer (b) the bridge configuration (c) the unidirectional output voltage across the load resistance in the full-wave rectifier circuit for a sinusoidal input voltage

non-conducting whereas diode D_2 conducts. The direction of the conventional current flow through R_L is readily seen to be the same as before. Consequently, a unidirectional voltage appears across R_L during both the positive and the negative half-cycles of the input (fig. 2.6c): full-wave rectification is achieved.

In the full-wave rectifier bridge circuit (fig. 2.6b) diodes D_1 and D_2 conduct during those half-cycles when A is positive with respect to B whereas diodes D_3 and D_4 conduct during those half-cycles when A is negative with respect to B. The current through the load resistance R_L passed by diodes D_1 and D_2 is in the same direction as that passed by D_3 and D_4, so again a unidirectional voltage is established across R_L during both the positive and negative half-cycles of the input.

For full-wave rectification where the transformer secondary winding is centre-tapped, the input voltage across one rectifier and the load resistance R_L is

2.5 Smoothing filters for rectifier circuits

To obtain from a half- or full-wave rectifier a steady output voltage, the supply provided by the rectifier circuit itself must be 'smoothed'. This means that the output voltage across the load is, as far as possible, rendered constant independently of time. This involves using the unidirectional voltage pulses from the rectifier to charge up a capacitor. Moreover, this capacitor must discharge through the load R_L at a considerably smaller rate than it charges.

The use of a simple capacitor filter in a half-wave rectifier circuit (fig. 2.7a) involves a 'smoothing' or 'reservoir' capacitor C across the load resistance R_L with a series safety resistor R_S between the diode and C to limit the charging current. If R_S were omitted the diode would become damaged by large current surges. The rate at which C is charged by rectified current pulses depends on the current-carrying capability of the diode and the resistor R_S.

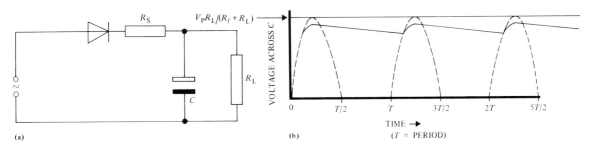

Fig. 2.7 (a) A simple capacitor filter within a half-wave rectifier circuit. (b) Graphical representation of the action of this smoothing capacitor

If the time constant CR_L for the discharge of C through R_L is considerably greater than T the period of the ac supply, C will charge up nearly to the peak voltage $V_P R_L/(R_f + R_L)$ during the first conducting half-cycle (fig. 2.7b). During the immediately following period T, the voltage across C will decrease exponentially at a rate decided by $1/CR_L$ which, being small, will mean that this voltage only falls by a small fraction of V_P before a second re-charging occurs as a consequence of the next positive-going current pulse from the rectifier. Reasoning on these lines shows that the voltage across R_L is now as represented by the full-line curve in fig. 2.7c, which is reproduced in somewhat greater detail in fig. 2.8.

Suppose the forward resistance R_f of the diode plus the safety resistance R_S is much less than the load resistance R_L. To a close approximation, the maximum voltage to which the smoothing capacitor C is charged will then be V_P, the peak voltage across the transformer secondary. Let the load current be I_L, then in the time T of one period, the charge which capacitor C loses is $I_L T$, so the voltage across C falls by $I_L T/C$, taken to be much less than V_P. The peak voltage across C is therefore V_P whereas the minimum voltage across C is $(V_P - I_L T/C)$ (fig. 2.8). The average voltage across C, i.e. the average output voltage from this power supply, is therefore $V_P (1 - I_L T/2C)$. This output voltage is subject to a *ripple voltage* taken to be half the decrease of the capacitor voltage during the period T, i.e. it is $I_L T/2C$. Hence, the percentage ripple voltage

$$= \frac{I_L T/2C}{V_P (1 - I_L T/2C)} \times 100 \qquad (2.14)$$

In terms of the load resistance R_L instead of the load current I_L put $I_L = V_P/R_L$ approximately, because $I_L T/2C \ll V_P$, then the percentage ripple voltage

$$= \frac{100 (V_P T/2CR_L)}{V_P(1 - V_P T/2CR_L)} = \frac{100 T/2CR_L}{1 - V_P T/2CR_L} \qquad (2.15)$$

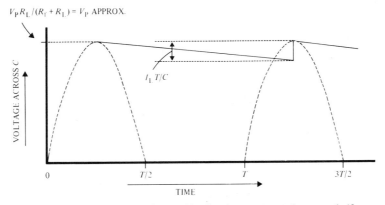

Fig. 2.8 Variation of voltage across a smoothing capacitor and load resistor connected across a half-wave rectifier

EXAMPLE 2.5 *A half-wave rectifier circuit is to be smoothed by a simple capacitor filter. If the transformer secondary voltage is 40V rms at 50Hz and the average load current supplied is 10mA, calculate the minimum capacitance of the smoothing capacitor C needed to ensure a percentage ripple voltage below 5.*

The peak voltage across the transformer secondary = V_p = 40 $\sqrt{2}$ = 56V. The loss of charge by the capacitor C during the period of 0.02s is 0.02 x 10^{-2} coulomb so the decrease of voltage across C will be 0.02 x $10^{-2}/C$, which is the peak-to-peak ripple voltage. The ripple voltage is half this value, i.e. $10^{-4}/C$

For the percentage ripple not to exceed 5,

$$\frac{10^{-4}/C}{56(1 - 10^{-4}/C)} = \frac{5}{100}$$

Therefore, $10^{-2}/C = 280 - 280 \times 10^{-4}/C$

Therefore, $\dfrac{3.8 \times 10^{-2}}{C} = 280$

Therefore, $C = \dfrac{3.8 \times 10^{-2}}{280}$ F = $\dfrac{3.8 \times 10^{-4}}{2.8}$ F

The capacitor must therefore have a capacitance of at least 136 μF.

When the half-wave rectifier circuit is first switched on, the capacitor C is uncharged. If the resistance of the secondary winding of the transformer is very low, only the safety resistance R_S will prevent rapid charging of C. The maximum charging current supplied to C will be V_p/R_S. R_S must be chosen so that this current does not exceed the maximum instantaneous current rating of the diode used.

The simple capacitor smoothing filter is widely used. If a full-wave rectifier circuit is employed instead of a half-wave one, the time over which the capacitor loses charge is reduced to $T/2$ so the percentage ripple voltage will be halved.

2.6 The Zener diode

The plot of the current through a *p-n* junction against the voltage V across it is typically that the reverse current due to surface leakage and thermally generated electrons and holes is essentially small and fairly constant until the reverse voltage across the junction reaches a breakdown value V_Z. For reverse voltage of magnitude exceeding V_Z, the inverse current increases greatly (fig. 2.9). The dynamic slope resistance R_Z changes from high values below V_Z to low values above V_Z.

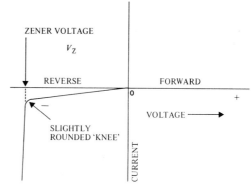

Fig. 2.9 The current-voltage characteristic of a typical *p-n* junction showing the onset of a large reverse current beyond the breakdown voltage

This breakdown is due to two phenomena (a) the *Zener effect*, named after C. Zener who developed in 1934 the theory of breakdown in the case of dielectrics, long before the advent of the Zener diode, and (b) the *avalanche effect*, put forward for semiconductors by McKay in 1953 and developed a year later by Wolff to explain serious discrepancies between the theoretical predictions based on the Zener effect and experimental observations.

In the Zener effect the internal electric field existing across the junction, and so within the depletion layer, is large enough (reaches values of the order of 3×10^7 V m^{-1}) to rupture directly a covalent bond within the silicon so that an electron is released (the phenomena has thereby some comparability with field electron emission from electrodes) and a hole is created. This production of an electron-hole pair increases the reverse current because the electrons so produced on the *p*-side of the junction are attracted across to the *n*-side (which is positive w.r.t. the *p*-side for reverse bias) whereas the holes move in the opposite direction. As the reverse voltage across the junction is increased, the electric intensity will tend to rise but also the width of the depletion layer will increase. The width of this depletion region is also inversely proportional to the concentration of the doping elements. Hence heavily doped silicon (i.e. of low resistivity) will result in a narrow depletion region, favouring high electric field intensity for modest values of the applied reverse voltage. For heavily doped *p* and *n* materials forming the diode junction, the values of the reverse voltage V_Z at which there is an onset of the Zener effect are between 3V and 5V but values exceeding 7V are needed as the doping concentration is reduced and the depletion region width is greater.

Whereas the Zener effect involves the creation of electron-hole pairs by electric field disruption, the *avalanche mechanism* is due to acceleration across the reverse-biased junction of minority carriers already present (electrons in the *p*-side and holes in the *n*-side semiconductor materials). When the reverse voltage across the *p-n* junction is high enough, electrons (or holes) acquire energies exceeding the ionization energy needed to release electrons from the covalent bonding in the silicon. Each such collision results in an additional electron and hole. These new current carriers are also accelerated in the electric field and are able to cause ionization. The process is therefore cumulative — an essential feature of an avalanche mechanism.

In general, the Zener effect predominates with thin junctions (narrow depletion regions) in heavily doped materials at reverse voltages across the junction of below about 5V whereas the avalanche effect predominates in lightly doped materials at voltages exceeding about 7V. However, there is no clear distinction in the case of a particular junction as to where the Zener effect ends and the avalanche effect takes over. Indeed, the development of the avalanche mechanism was to explain serious discrepancies: as an example, a Zener diode made with *n*-type silicon of resistivity 10^{-4} ohm m on one side and *p*-type silicon of negligible resistivity on the other side was predicted theoretically to have $V_Z = 3.9V$ in accordance with Zener theory but in fact was found to have experimentally a V_Z of 15V.

For silicon diodes with $V_Z < 5V$, the magnitude of V_Z decreases with increase of temperature T, so dV_Z/dT is negative. This negative temperature coefficient is consistent with decrease of the forbidden energy gap E_G for silicon with increase of temperature (for silicon, E_G is 1.21 eV at 0K and 1.1 eV at 300 K) so Zener effect disruption of covalent bonding is more probable.

However, silicon diodes for which $V_Z > 6V$ have a positive temperature coefficient. This is consistent with predominance of the avalanche mechanism because carrier mobilities decrease with temperature rise so that higher voltages are needed to ionize the silicon.

Although theory predicts that breakdown is sharp, there being an abrupt discontinuity in the reverse voltage against reverse current characteristic, in practice, the characteristic in this breakdown region exhibits a slightly rounded 'knee', which is particularly noticeable at low breakdown voltages. This roundness is due to surface phenomena, thermal effects and resistivity of the bulk material.

The breakdown voltage V_Z for Zener diodes (so-called even if the breakdown mechanism is not primarily due to the Zener effect) can be controlled by careful adjustment of the grading of the doping concentrations near the junction. Zener diodes (also known as *voltage regulator diodes*) over a wide range of values of V_Z from as low as 2.4V up to 200V are commercially available, with power ratings of 0.25, 0.4, 0.75, 1.0, 1.5, 10 and 50W. The tolerances of the nominal Zener voltages are within some ± 10%.

Voltage reference diodes are also available such as the Mullard BZX48, 49 and 50 set which have very low temperature coefficients by careful control of manufacture whereby the negative temperature coefficient associated with the Zener effect is balanced against the positive temperature coefficient characteristic of the avalanche mechanism. These voltage reference diodes are capable of providing nominal reference voltages as accurately as standard cells and, in many applications, can replace such cells. For example, the BZX48 provides at I_Z = 2.0mA a reference voltage of 6.5 ± 5% with a temperature coefficient as low as 0.001% per °C over the temperature range from 0 to 70°C.

The value of the Zener diode rests on the fact that, when the voltage across it exceeds slightly the breakdown value V_Z, this voltage is independent of the current through it. It therefore lends itself admirably to the development of voltage regulator circuits, typical of which is that shown in fig. 2.10. This consists of a supply source of voltage V_i across the resistor R in series with the Zener diode (note the circuit symbol for the Zener diode). The load resistance R_L [across which a stable voltage V_L (= V_Z) is to be maintained] is in parallel with this Zener diode. The resistor R is an essential requirement: if it were absent, any slight increase of the supply voltage V_i above V_Z would cause the reverse diode current to rise inordinately (as is seen from the characteristic, fig. 2.9) and burn out the junction.

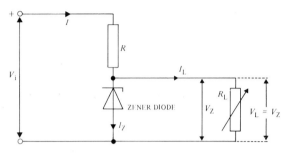

Fig. 2.10 A simple voltage regulator circuit based on a Zener diode

Applying Kirchhoff's laws to the circuit (fig. 2.10) it follows that:

$$V_i = V_Z + IR \qquad (2.16)$$

where the Zener voltage $V_Z = V_L$, the voltage across R_L, and I is the current from the source.

Note that V_i must exceed V_Z. It is usually chosen to be about $1.2 V_Z$.

The current I through R is the sum of I_Z the current through the Zener diode, and I_L, the load current, hence

$$I = I_Z + I_L \qquad (2.17)$$

V_i (often obtained from a silicon diode rectifier power unit with capacitor smoothing) may well vary because of variations in the ac mains supply voltage to the rectifier power unit or for other reasons. However, V_Z and so V_L remain constant. If V_i increases, V_Z remains the same but I increases. It is seen from equations (2.16) and (2.17) that this increase of I is accommodated by increase of I_Z without change of V_Z; as I increases, IR goes up to counteract any tendency for increase of $V_Z = V_L$. Likewise, if V_i decreases, $V_Z = V_L$ remains constant because the voltage IR across R goes down, the decrease of I being due to fall-off of I_Z. Moreover, this constant output voltage is also maintained if the load current I_L increases or decreases because the current through the Zener diode will decrease or increase correspondingly.

This simple circuit therefore maintains constant the voltage across the load irrespective of changes of the input voltage or of the load current.

When I_L is a maximum at $I_{L\,max}$, the current I_Z through the Zener diode is a minimum in a given circuit (equation 2.17). I_Z must not be allowed to decrease too much as it is essential to utilise that part of the characteristic beyond the 'rounded knee' (fig. 2.9). Hence I_Z must exceed a certain minimum value $I_{Z\,min}$. The usual empirical rule adopted is that $I_{Z\,min}$ should not be less than $0.1\,I_{L\,max}$.

From equations (2.16) and (2.17) it follows that

$$V_i = R\,(I_Z + I_L) + V_Z \qquad (2.18)$$

The value of R required is therefore calculated from

$$R = \frac{V_i - V_Z}{I_Z + I_L} \qquad (2.19)$$

Furthermore, it must be ensured that the power P_Z dissipated in the Zener diode does not exceed a maximum permissible value $P_{Z\,max}$. For silicon p-n junction diodes this is related to the maximum permissible junction temperature of typically 175°C in continuous operation. P_Z is given by

$$P_Z = I_Z V_Z \qquad (2.20)$$

and, since from equation (2.18),

$$I_Z = \frac{V_i - V_Z}{R} - I_L$$

therefore

$$P_Z = \left(\frac{V_i - V_Z}{R} - I_L\right) V_Z \qquad (2.21)$$

To arrive at the best value for the limiting resistance R, it is first clear that I_Z is kept as small as possible to minimise P_Z (equation 2.20, in which V_Z is constant). However I_Z must not be allowed to decrease below $0.1 I_{L\,max}$. Therefore equation (2.19) is written

$$R = \frac{V_i - V_Z}{1.1 I_L} \quad (2.22)$$

where I_L becomes $I_{L\,max}$ in the case where I_L varies because R_L is variable.

In the general case where both V_i and I_L vary, R must be small enough to enable the current demand of $1.1 I_{L\,max}$ to be supplied when V_i is at its minimum value $V_{i\,min}$. Equation (2.22) then becomes

$$R = \frac{V_{i\,min} - V_Z}{1.1 I_{L\,max}} \quad (2.23)$$

The power dissipated in the Zener diode is a maximum of $P_{Z\,max}$ when V_i reaches its maximum value. Equation (2.21) then becomes

$$P_{Z\,max} = \left(\frac{V_{i\,max} - V_Z}{R} - I_{L\,min}\right) V_Z \quad (2.24)$$

Equations (2.23) and (2.24) are the design equations for the simple Zener diode regulator circuit of fig. 2.10.

EXAMPLE 2.6 *A Zener diode has a Zener voltage of 5.8V and a maximum power rating of 300mW. It is to be used in a simple regulator circuit to provide a maximum load current of 10mA where the voltage supply input varies from 10V to 20% above the Zener voltage. Calculate the value of the series resistance required, and show that the choice of a resistance selected nearest this value of R from amongst those commercially available still ensures that the maximum power rating of the Zener diode is not exceeded.*

The value of R required is given by equation (2.21)

$$R = \frac{V_{i\,min} - V_Z}{1.1 I_{L\,max}}$$

on choosing empirically that I_Z is not allowed to decrease below $0.1 I_{L\,max}$. In this equation, $V_{i\,min}$ is 20% above $5.8V = 1.2 \times 5.8V = 7V$, and $I_{L\,max} = 10mA = 0.01A$. Therefore,

$$R = \frac{7 - 5.8}{0.011} = \frac{1.2}{0.011} = 109 \text{ ohm}.$$

The nearest preferred value of R from amongst the resistor values supplied by manufacturers is 110 ohm. The power dissipation will then be given by equation (2.21) to be

$$P_Z = \left(\frac{10 - 5.8}{110} - 0.01\right) 5.8 \text{ watt}$$

$= 162mW$, which is much less than the maximum power rating of 300mW.

The use of Zener diodes in series (fig. 2.11a) enables a number of different stable output voltages to be obtained. Two Zener diodes in parallel (each provided with series resistors) (fig. 2.11b) gives even greater stability and is useful for suppressing ripple on a voltage supply. Fig. 2.11c shows how the circuit of fig. 2.11b is readily developed to provide a voltage supply which can be varied by adjustment of the variable resistor R_2 but where at each setting the output voltage is constant.

2.7 Voltage reference diodes

Zener diodes with very small temperature coefficients have already been mentioned. Fig. 2.12 shows for a silicon Zener diode having V_Z of nominally 5.5V the reverse current against reverse voltage curves at two different temperatures of 25°C and 150°C. It is seen that for a particular value of the reverse current (45mA in this example) the temperature coefficient is zero. On this basis, packaged series assemblies are available. An example is the 1N1735 series of Zener reference elements which covers the voltage range from 6.2 to 49.6V in 6.2V steps.

Another possibility is exemplified by the 1N3157 assembly which consists of three series connected Zener diodes, two of which have positive temperature coefficients and the third a negative temperature coefficient. This provides thermal stability to within ± 0.001% per °C over the temperature range from −55°C to 100°C.

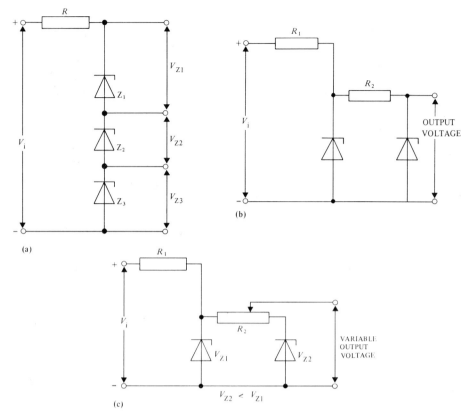

Fig. 2.11 (a) Zener diodes in series (b) Zener diodes in parallel (c) Obtaining by the use of two Zener diodes in parallel, a supply voltage which can be varied

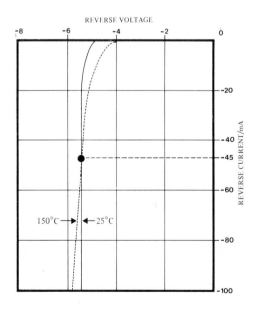

Fig. 2.12 Current against voltage characteristics of a silicon Zener diode at two different temperatures

2.8 Other applications of Zener diodes

Fig. 2.13 shows a number of circuit applications of Zener diodes:

(a) for a unidirectional pulse, a clipper or squarer;
(b) a limiter or clipper for an alternating voltage of sinusoidal wave-form;
(c) a means of providing by a difference method a regulated voltage of lower value than the minimum V_Z values;
(d) an arrangement to protect a voltmeter from overload

The versatile Zener diode has many other applications, particularly in conjunction with bipolar transistors, other semiconductor devices, and in integrated circuits. Some of these are considered, as appropriate, later in this text.

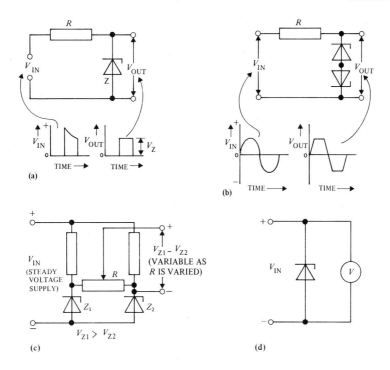

Fig. 2.13 Some further applications of Zener diodes: (a) a clipper for unidirectional pulses; (b) a clipper for an alternating voltage; (c) a difference circuit for providing a stable low voltage; (d) protection of a voltmeter from overload

2.9 Capacitance of *p-n* junction

There are three capacitative effects associated with *p-n* junctions: (a) the depletion region or transition capacitance C_T concerned when a reverse bias voltage is applied across the junction; (b) the diffusion capacitance C_D involved when a forward bias voltage is applied; and (c) the dynamic diffusion capacitance C_D' concerned when the applied forward voltage varies with time.

(a) *The depletion region capacitance* C_T: when a reverse bias voltage is applied across a *p-n* junction, the movement of majority carriers away from the junction itself results in a depletion region of width which increases with the reverse voltage V applied. This increase of width uncovers stationary charges (due to acceptor ions in the *p*-side and donor ions in the *n*-side) in the depletion region. Thus, a change of voltage from V to $(V + dV)$ – both of which are negative – causes a change of charge of dQ, so that an incremental depletion-region capacitance C_T is defined by

$$C_T = \frac{dQ}{dV}$$

Moreover, if the alteration dV occurs in a time dt, a current i will result given by

$$i = \frac{dQ}{dt} = C_T \frac{dV}{dT}$$

Two main types of *p-n* junction are involved in practice: step-graded junctions, as in alloy junctions (appendix A) in which acceptor ions in the *p*-side change abruptly at the junction to donor ions in the *n*-side; and graded junctions (typically in an integrated circuit transistor) in which there is not an abrupt change from acceptor to donor ions but a grading in that the number of such ions per unit volume varies gradually on progressing from the *p*-side to the *n*-side.

(b) *The diffusion capacitance* C_D: when a forward bias is applied across a *p-n* junction, injected charge stored near the junction results in a capacitance which is considerably greater than the depletion region capacitance C_T. The time constant associated with this diffusion capacitance is not usually large because the dynamic forward resistance of the diode is small. Indeed, this time constant is equal to the mean life-time of the minority carriers which, depending on the *p-n* junction doping and geometry, may be from 1ns up to a few hundred μs.

(c) *The dynamic diffusion capacitance* C_D': depends on how the applied voltage varies with time. For a sinusoidally alternating applied voltage, C_D' is a function of frequency.

2.10 Varactor diodes

Also known as the *variable capacitance diode* or the *parametric diode*, the varactor diode offers a capacitance which varies with the reverse voltage applied across a *p-n* junction. As the depletion layer width increases with reverse voltage, the capacitance decreases. In fig. 2.14, (a) shows the circuit symbol for the varactor diode, (b) the typical variation of the capacitance with applied voltage and (c) the equivalent circuit of the varactor diode. This equivalent circuit comprises the variable capacitance C_T (typically 10pF max.), the series ohmic resistance R_S which is a few ohm, and a parallel resistance R_r of a few megohm, being the resistance associated with the small reverse current.

The varactor diode is useful in a number of applications. A few which may be mentioned are (i) in an *LC* circuit for high frequency signals, tuning to resonance is achievable by applied voltage variation; (ii) an ac bridge circuit which includes capacitance may be balanced by voltage variation; (iii) varactor diodes with graded *p-n* junctions providing a non-linear capacitance variation with voltage are available which operate in the microwave region up to 100GHz. Across a coaxial transmission line or a wave-guide, such a varactor diode will provide a low impedance (short-circuit) when a small forward bias is applied and the capacitance is large and high impedance (open-circuit) when the bias is adequately reversed and the capacitance is small. This technique leads to a method of switching power in the microwave region whereby a control signal of a few milliwatt can switch an output power of several kilowatt.

For use at frequencies below about 100MHz, varactor diodes are made from silicon. Above this frequency, gallium arsenide is often used.

2.11 Tunnel diodes

The usual *p-n* junction diode has normally an impurity concentration of about 1 part in 10^8, and the width of the depletion layer at small voltages is about 1μm. This depletion layer inhibits the movement of majority carriers from one side to the other of the junction.

As already mentioned (section 2.6) on increase of the impurity concentrations, the width of the depletion region and the reverse breakdown voltage both decrease. The limit to the doping concentrations

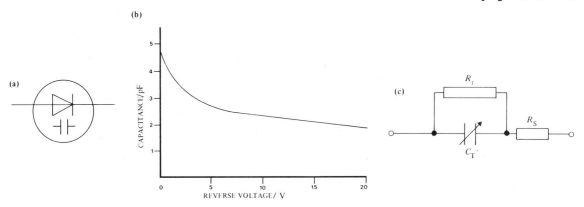

Fig. 2.14 The varactor diode : **(a)** the circuit symbol; **(b)** the typical variation of capacitance with applied voltage; **(c)** the equivalent circuit

possible is set by the solubility of the donor (or the acceptor) atoms in the host semiconductor material. Nevertheless, impurity concentrations as high as 1 part in 10^3 may be achieved with corresponding junction barrier widths at zero applied bias as small as $10^{-2}\,\mu\text{m}$. This width is small compared with the de Broglie wavelength λ ($\lambda = h/mv = h/\sqrt{2Vem}$) of electrons of velocity v and energy Ve. Consequently, in accordance with the quantum mechanics of the 'tunnel effect' whereby a particle of energy E has a definite probability of penetrating a thin reverse potential energy barrier of height V even though $E < V$, electrons (and holes in the opposite direction) are able to penetrate the junction barrier potential.

For a conventional p-n junction the energy-band structure is with the Fermi level in the forbidden gap (fig. 1.8). In the tunnel diode, for which the donor concentration in the n-side material exceeds $10^{25}\,\text{m}^{-3}$ (corresponding to about 1 impurity atom in 10^3 host atoms), there is such a large number of electrons that the Fermi level is within the conduction band. Consequently, many electrons exist in energy states between E_F and E_{Cn}, the bottom of the conduction band for the n-side material, where E_F is above E_{Cn} (the shaded area on the n-side in fig. 2.15a represents partial filling of the conduction band by electrons). Moreover, the excessively high concentration of acceptor atoms in the p-side material (and so large number of holes per unit volume) results in the Fermi level (which must be the same on the two sides of the junction when the applied voltage is zero) also being within the valence band on the p-side. There are hence a number of holes in the energy states between E_F and E_{Vp}, the top of the valence band in the p-side, where E_{Vp} is above E_F.

Consider a voltage V to be applied across the junction, giving rise to a corresponding energy of eV in electron-volt. If V is negative (reverse bias) the Fermi level on the n-side will be depressed relative to the Fermi level on the p-side by an amount eV. Hence E_F will become E_{Fn} on the n-side where E_{Fn} is below E_F. Now there will be some energy states near the top of the filled part of the valence band on the p-side which are at the same energy levels as vacant energy states in the conduction band on the n-side. Electrons will therefore be able to penetrate by tunnel effect from the p-side to the n-side. This will cause a reverse current to flow which will clearly increase as this reverse bias is increased (fig. 2.15b).

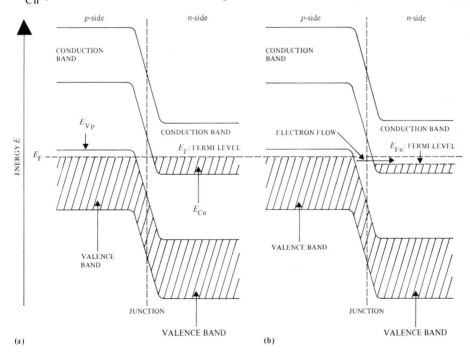

Fig. 2.15 The energy-band structure characteristic of a tunnel diode at room temperature: (a) applied voltage across junction is zero: (b) applied voltage across junction is V as reverse bias

If V is positive (forward bias), the Fermi level on the n-side will be raised. Now there are energy states in the conduction band on the n-side at the same levels as vacancies in the valence band on the p-side. Electrons can therefore tunnel from the n-side to the p-side, causing a forward tunnel effect current to flow in addition to the usual drift and diffusion currents. As V is increased positively, this forward current will increase. However, when V reaches a value of V_p, the forward current will attain a peak value of I_p. As V is increased beyond V_p, the Fermi level within the conduction band in the n-side will become raised to such an extent that the bottom of the n-side conduction band at energy E_{Cn} will eventually be above the top of the valence band at energy E_{Vp} in the p-side. When $V = V_V$, there will no longer be any vacant sites on one side of the junction at the same energy levels as occupied sites on the other side so the tunnelling current will be zero and only the smaller drift and diffusion currents will flow.

The voltage against current characteristic of the tunnel diode (fig. 2.16) will therefore exhibit a negative resistance region from $V = V_p$ to $V = V_V$. For values of V greater than V_V, the characteristic will be as for the corresponding normal diode injection current due to the drift and diffusion of current carriers. The minimum current I_V is known as the valley current. Beyond the valley voltage V_V, the injection current increases with voltage.

Tunnel diodes are made from the semiconducting elements (suitably doped to provide p and n type materials) germanium, gallium arsenide or silicon. The ratio of the peak current to the valley current (I_p/I_V) for tunnel-diodes made from germanium can be between 10 and 30, as much as 60 for gallium arsenide but for silicon (which is not much used for tunnel diode construction) this ratio is restricted to maximum values of about 5. The voltage swing V_S between the two points on the curve where the current is the peak value I_p (fig. 2.16) is 0.45V for tunnel diodes based on germanium, 0.7V for silicon and about 1.0V for gallium arsenide.

The two chief applications of tunnel diodes are (i) in very high-frequency oscillator circuits where advantage is taken of the negative resistance region of the $V - I$ characteristic and (ii) in very high speed switch circuits where advantage is gained from the extremely rapid rate at which current carriers will tunnel through the very thin junction barrier region.

When operated over the voltage region from V_p to V_V as a negative resistance device of minimum value R_{min} at the point of inflection of the curve between V_p and V_V, the equivalent circuit of the tunnel diode (the circuit symbol is shown in fig. 2.17a and the equivalent circuit in fig. 2.17b) consists of the ohmic resistance R_S (typically about 1 ohm) of the semiconducting material, a small series inductance L_S (typically about 5nH) determined by the con-

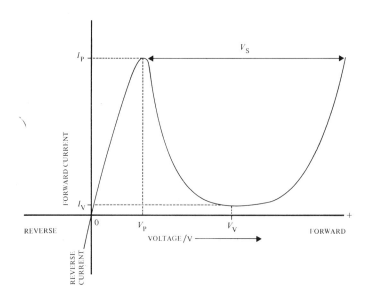

Fig. 2.16 The typical voltage-current characteristic of a tunnel diode

struction of the device, the junction capacitance C (about 10pF) and R_{min} of about 50 ohm.

A simple oscillator based on a tunnel diode (fig. 2.17c) comprises the appropriate supply voltage to operate the diode in the negative resistance region and with the tunnel diode in series with an external inductor of inductance L.

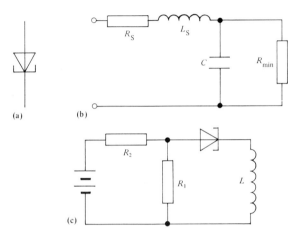

Fig. 2.17 (a) Conventional circuit symbol for a tunnel diode (b) Equivalent electrical circuit of a tunnel diode (c) A simple series oscillator circuit utilising a tunnel diode

Oscillations are produced when

$$R |R_d| = L/C$$

at a frequency f given by

$$f = \frac{1}{2\pi} [(1 - R/R_d)/LC^{1/2}]$$

where

$$R = R_1 R_2 / (R_1 + R_2) + R_S + R_L,$$

R_L being the ohmic resistance of the inductor L, and R_d is the resistance of the diode in the negative resistance region.

In such a simple circuit, the frequency and stability are not good because they depend on R_d which is not constant but varies with voltage and temperature. A preferable simple tunnel-diode oscillator circuit is that shown in fig. 2.18 where an inductor-capacitor (LC) circuit is used which is more readily made stable, and where operation at frequencies of 100MHz or more are possible if sufficiently small values of L and C are available.

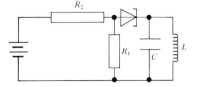

Fig. 2.18 A simple series-parallel oscillator circuit utilising a tunnel diode

As a switch, the tunnel diode can be very fast acting in times as short as 1ns or less. The promise of extremely fast switching circuits for digital computers has not been fully realised, however, because of subsequent advances in integrated circuit technology based on developments from conventional diodes and transistors. The basic tunnel diode switch circuit (fig. 2.19a) operates with a load resistance R_L which exceeds R_{min}. The load line is then as in Fig. 2.19b: it intersects the tunnel diode characteristic at three points A, B and C. When a trigger voltage is applied in series with the supply voltage E and the tunnel diode, the diode switches rapidly in a time less than 1ns from A to C on the characteristic, giving corresponding diode current changes and hence voltages across R_L.

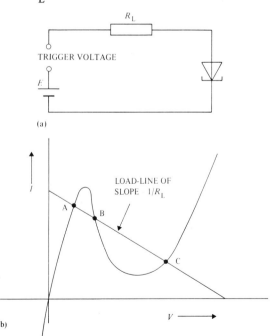

Fig. 2.19 The tunnel diode as a switch (a) the basic circuit; (b) the load-line involved

The chief disadvantages of the tunnel diode are the low voltage output swing possible and the fact that isolation between its input and output is not possible because it is a two-terminal device.

The tunnel diode was first introduced in 1948 by Esaki.

2.12 Other semiconductor diodes utilised in microwave oscillators

The *IMPATT* (IMPact Avalanche Transit Time) *diode* is based on a specially made *p-n* junction across which the reverse voltage is varied rapidly above and below the avalanche breakdown value. There is then a phase lag of the current through the diode on the alternating voltage applied. This arises for two reasons: the build-up and subsidence of the avalanche effect occupies a specific short time period; there is a transit time for the current carriers through the semiconductor. The IMPATT diode characteristic also has a negative resistance region. Used in conjunction with an appropriate inductance-capacitance (*LC*) circuit, oscillations are sustained if the *L* and *C* values in relation to those of the diode (which depend on its physical dimensions) are chosen to be resonant in the frequency range where the phase lag (which can be arranged at very high frequency to be between 90° and 270°) is appropriate. These oscillations can have wavelengths of a centimetre or less, i.e. are in the microwave and millimetre wavelength regions. IMPATT diode oscillators are available commercially which produce 100mW at 10^{10} Hz (10GHz).

The IMPATT diode was introduced in 1958 by Read as a *p-n-i-n* device, i.e. with alternate layers (separated by junctions in a single crystal) of *p*-type and *n*-type, intrinsic and *n*-type material, usually silicon. It was difficult to make so the simpler *p-n* type was introduced in 1964 by Johnston, De Loach and Cohen.

Gunn diodes, introduced in 1964 by Gunn, are made from *n*-type gallium arsenide sandwiched between metal electrodes in the form of deposited metallic films on opposing faces of a slab of the gallium arsenide about 0.2 mm thick. On applying voltage pulses across these electrodes to give rise to pulsed electric fields of intensities of several thousand volts per cm in the gallium arsenide, oscillations are sustained of period near the transit time of the current carriers between the electrodes.

2.13 Light-emitting diodes

Light-emitting diodes (LEDs) or electroluminescent diodes make use of a forward-biased *p-n* junction based upon III – V semiconductor compounds designed to enhance the electron component of the current. They are described in Chapter 8.

Exercise 2

1. The current *I* which flows through a *p-n* junction at an absolute temperature *T* across which there is maintained a pd of *V* is given by the equation

   $I = I_s [\exp(eV/\eta kT) - 1]$

   where I_s is the reverse saturation current, e is the electronic charge, *k* is the Boltzmann constant and η is a constant.
   Describe a simple experiment (based on the use of this equation) with a germanium *p-n* junction diode which would enable η to be determined.

2. Explain with appropriate diagrams and mathematics the function of a *p-n* junction diode as a half-wave rectifier for a sinusoidally varying input. Discuss the effects of any non-linearity of the dynamic characteristic of the diode and that a small positive threshold voltage, V_γ, is involved.

3. Draw the circuit diagram and explain the operation of a power unit which supplies a steady voltage output from an ac sinusoidal input in which a bridge rectifier configuration is used and capacitor smoothing.
   State the advantages and disadvantages of this circuit arrangement compared with a configuration which makes use of a centre-tapped transformer.

4. A full-wave rectifier circuit is to be smoothed by a capacitor filter. If the voltage across the whole of the centre-tapped secondary of the trans-

former employed is 50V rms at 50Hz and the average load current supplied is 15mA, calculate the minimum capacitance of the smoothing capacitor required to ensure that the percentage ripple voltage is not greater than 3.

5. Write explanatory notes on (a) the *Zener effect* and (b) the *avalanche effect* in relation to the breakdown experienced when a reverse voltage of sufficient magnitude is maintained across a *p-n* junction diode.
Explain the importance of this effect in the development of the Zener diode.

6. Give an account of voltage reference diodes and their uses.

7. Explain, with appropriate diagrams and theory, the action of a simple voltage regulator circuit based on a Zener diode.

8. A Zener diode has a Zener voltage of 5.2V and a maximum power rating of 250mW. It is used in a simple regulator circuit to provide a maximum load current of 12mA where the voltage supply input to this circuit varies from 9V to 11V. Calculate the value of the series resistance R required in this circuit.

9. Give circuit diagrams and explanations of the use of a Zener diode to:

   (a) clip a unidirectional input voltage pulse;
   (b) limit an alternating input voltage of sinusoidal wave-form to a given amount less than the peak value;
   (c) provide a regulated voltage V where $V < V_Z$, the Zener voltage;
   (d) protect a voltmeter from overload.

10. Define and explain the terms:

    (a) depletion region capacitance;
    (b) diffusion capacitance;
    and
    (c) dynamic diffusion capacitance
    as applied to a *p-n* junction.

11. Write an account of the varactor diode, including some of its applications.

12. With the aid of a diagram of the appropriate energy-band structure, explain the action of a tunnel diode. What are the main uses of the tunnel diode? Why has its early promise as a fast-acting switch for use in computer circuitry not led to its widespread application?

13. Write explanatory notes on

    (a) the IMPATT diode;
    (b) the Gunn diode
    and indicate in outline the applications of these devices.

14. Referring to an appropriate manufacturer's catalogue, design the circuit of a power unit which provides a steady voltage output 15V–0– –15V and where the input to this power unit is from the ac mains at 240V rms 50Hz.

3 · Field Effect Transistors

3.1 Introduction

As the name suggests, the field effect transistor (fet) is an active device in which the current flowing between two electrodes through a silicon channel is controlled by the application of an electric field set up within the conducting region by a voltage applied to a third electrode called the gate G. Early development work on field effect devices was no doubt influenced by attempting to make a solid-state device comparable with the triode valve. In this three-electrode vacuum tube the current of electrons from a thermionic cathode to an anode is controlled by a voltage maintained on a grid between the cathode and anode. This voltage on the grid is normally negative with respect to the cathode so it does not attract electrons — the grid current is negligibly small. Yet, being close to the cathode, the electric field due to the grid voltage has a marked influence on the electron space charge and so controls the anode current. As this control in a triode valve by an electric field demands virtually zero current from the input signal source, the triode has a very high input resistance, as compared with the bipolar junction transistor (Chapter 4) which has a relatively small input resistance and was developed before the field-effect transistor.

Early attempts to produce an electric field to control the current within a semiconductor by maintaining a voltage across insulated electrodes at or near the surface of the semiconductor were unsuccessful. In all cases, the surfaces appeared to screen electrically the interior of the semiconductor. Indeed, it was while investigating these surface effects (and making a solid-state counterpart of the triode valve) that Shockley and his associates Bardeen and Brattain discovered transistor action in 1948. This work resulted in the point-contact transistor, quickly to be replaced by the bipolar junction transistor with its low input resistance. Two developments led to the state at the present time when two families of field effect devices are available. The first provided the junction-gate field effect transistor (jgfet); the second led to the insulated-gate field effect transistor (igfet).

The jgfet (often referred to simply as the 'fet' without distinguishing it from the igfet) was due to Shockley in 1952, who realised that a controlling electric field could be created within the body of a semiconductor by making use of a reverse-biased p-n junction. The action depends on the creation and control of a depletion region produced at a reverse-biased junction where the width of this region is related to the width of the conducting channel.

The igfet depends on the ability to produce thin stable layers of insulating material (often of silicon dioxide, SiO_2) on the surface of a slice of silicon. The most common igfet is the metal-oxide semiconductor field effect transistor (the mosfet). The name describes the structure which consists of a metal electrode in intimate contact with the insulating oxide film (the metal being often aluminium deposited on to SiO_2) the voltage on which electrode is able to influence the carrier concentration in the semiconductor layer immediately below the oxide.

Each member of the fet families has a high input resistance of value about $10^9\,\Omega$ for the jgfet (the resistance of a reverse-biased silicon junction) and approximately $10^{12}\,\Omega$ for the igfet. The use of a field effect transistor thus enables good electrical insulation to be achieved between the controlling circuit and the controlled circuit.

There are six basic types of fet: two of these are jgfets and four are igfets. They are listed in fig. 3.1, with the appropriate circuit symbols.

3.2 The junction-gate field effect transistor (jgfet)

A jgfet consists essentially of a semiconducting current channel, the resistance of which is controlled by an electric field applied at right angles to the

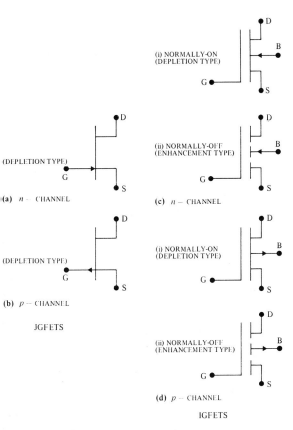

Fig 3.1 The six basic types of field effect transistor (S:source; G:gate; D:drain; terminal B is attached to the bulk substrate and is generally connected internally to the source S)

direction of current flow. Unlike the bipolar junction transistor (Chapter 4) in which both types of current carrier (holes and electrons) are involved, the jgfet is a *unipolar device,* so-called because only one type of carrier is involved: electrons in an *n*-channel jgfet and holes in a *p*-channel jgfet. The simplified diagram of an *n*-channel fet (fig. 3.2) shows ohmic contacts at the ends of a rod of *n*-type silicon with heavily doped layers of *p*-type silicon on the sides of the rod. Electrons enter the channel at the source S and are collected at the drain D. The control electrode, the gate G, is normally operated at a negative potential with respect to the source. As the voltage on the gate is made more negative with respect to the source, the depletion layer at the reverse-biased *p-n* junction extends into the channel, reducing its effective width and so increasing the channel resistance. Therefore, the channel current, normally called the drain current I_D is a maximum when the applied control signal voltage $V_{GS} = 0$.

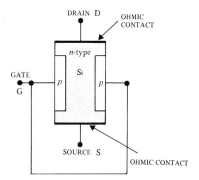

Fig. 3.2 Structure of an *n*-channel fet (schematic)

Immediately the *p-n* junction is created during manufacture, a depletion region forms in the silicon crystal as illustrated in fig. 3.3a. The effect of increasing V_{GS} (where the gate is negative with respect to the source) is shown in figs. 3.3b and c. At some voltage, labelled V_2 in fig. 3.3c, the depletion region extends across the whole rod so that the conducting channel width is reduced to zero. These diagrams are for the case where the voltage V_{DS} applied between the drain and the source is zero, and hence $I_D = 0$.

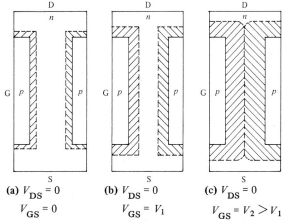

Fig 3.3 Depletion region width increases with V_{GS} where the gate is negative with respect to the source, $V_{DS} = 0$.

As V_{DS} is increased (with the drain positive with respect to the source) free electrons flow in the channel. There is now a potential gradient set up along the silicon rod which produces a depletion region shaped as in fig. 3.4b where the channel is most restricted in the region near the drain electrode because, there, the reverse bias at the *p-n* junction is a maximum. Note that, because no current flows in the heavily doped *p*-region, it may be considered as an equipotential region. Once again the width of the conducting channel is reduced, this time as the drain-source voltage V_{DS} is increased. In general, both forms of bias (V_{GS} and V_{DS}) are applied together resulting in the characteristic shown in fig. 3.8.

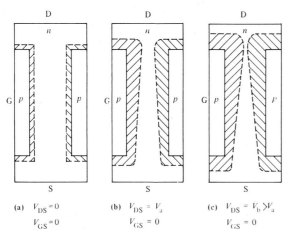

Fig 3.4 Variation of the depletion region extent with the drain-source voltage V_{DS}; $V_{GS} = 0$

A structure much more like that of the actual manufactured *n*-channel fet is represented in fig. 3.5. A + sign marked against *n* or *p* simply indicates that the region is heavily doped. Such regions are created where metal contacts are to be made because it is easy to make an ohmic contact between a heavily-doped semiconductor and a metal.

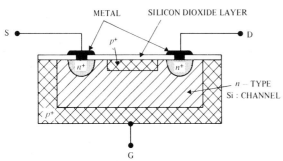

Fig. 3.5 An *n*-channel fet

With a channel of lightly-doped *p*-type silicon and heavily-doped *n*-type silicon providing the control electrode, a *p*-channel fet is produced. Both *n* and *p* channel types are available. As the mobility of electrons in silicon is approximately twice that of holes, *n*-channel fets give superior performance and are more widely used: the discussion here will therefore concentrate on *n*-channel fets.

3.3 The *n*-channel fet as a voltage-controlled resistor and as a switch

If an ohmmeter having a low internal voltage supply is connected across the drain-source electrodes of an *n*-channel jgfet of which the gate is set at a negative potential with respect to the source (fig. 3.6a), a good indication is obtained of the growth of the depletion region, and the corresponding increase in channel resistance, with increase of this negative potential.

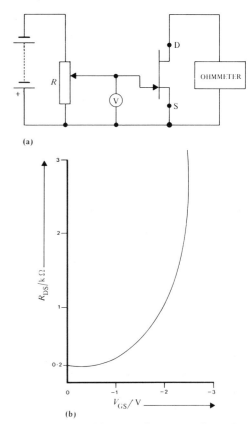

Fig. 3.6 The *n*-channel fet as a voltage-controlled resistance

Note that a digital ohmmeter with its built-in constant-current source is unsuitable for this circuit; furthermore, when using a universal meter as an ohmmeter, care should be taken to ensure that the positive voltage available from the universal meter is connected to the drain electrode of the fet.

The characteristic of drain-source resistance (R_{DS}) against gate-source voltage (V_{GS}) is typically of the form shown in fig. 3.6b. With $V_{GS} = 0$, R_{DS} may be of the order of 200Ω rising as V_{GS} is increased negatively, to several megohms.

Thus, if the fet is to be used as a switch it can be operated by a voltage signal: its resistance will be very high in the open state (ideal for a switch) but in the closed state its resistance will still be appreciable (approximately 200Ω for the fet used to obtain the characteristic shown in fig. 3.6b). In many applications this high closed resistance can be tolerated especially when, on the credit side, the fast switching action inherent in the jgfet is present.

3.4 The characteristics of an *n*-channel jgfet

The drain characteristics are obtained by plotting the drain current I_D against the drain-source voltage V_{DS} for a set of chosen values of the gate-source voltage V_{GS} (where V_{GS} is such that the gate is negative with respect to the source). Fig. 3.7 shows a circuit suitable for this purpose.

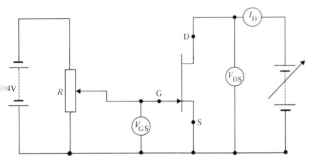

Fig 3.7 Determination of the drain characteristics of an *n*-channel jgfet

The characteristics (fig. 3.8) exhibit two distinct regions:

(a) A voltage-controlled resistance region in which the drain current I_D is proportional to V_{GS} at a constant value of V_{DS}.
Within this region for any chosen value of V_{DS}, the drain current depends on V_{GS} and the fet is acting as a voltage-controlled resistor. Field effect transistors designed especially to be used in this way are called field effect resistors (fer) or voltage dependent resistors (vdr).

(b) The constant current region. Beyond the knee in the characteristic, the drain current is virtually independent of the drain-source voltage V_{DS}.
It is in this region that the fet is normally operated when used as an amplifier. From the reasoning put forward in relation to figs. 3.3 and 3.4, it is to be expected that the constant current is reached at lower values of V_{DS} for larger values of V_{GS}.

Some authorities refer to the region to the left of the knee as 'pre-pinch-off' and to the constant current region as 'pinched-off'. These terms arise from the channel widths involved. In this text, the term 'pinched-off' is to be retained for a well-defined voltage V_p on the transfer characteristic (fig. 3.9).

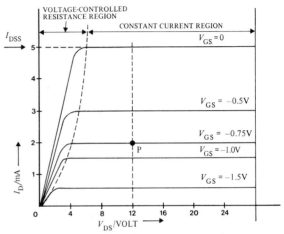

Fig. 3.8 The drain characteristics of an *n*-channel jgfet

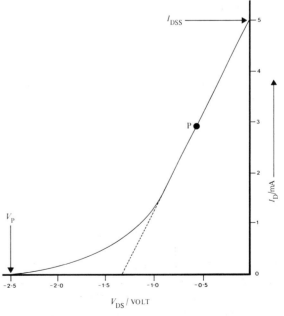

Fig. 3.9 The transfer characteristic of an *n*-channel jgfet

The transfer characteristic is the plot of the drain current I_D against the gate-source voltage V_{GS} for a given value of the drain-source voltage V_{DS} (fig. 3.9). This transfer characteristic is part of a parabola having a nearly linear portion over a considerable part of the curve. Two points on the transfer characteristic are important:

(a) I_{DSS}: the drain current for zero bias, i.e. $V_{GS} = 0$.
(b) V_P: the gate-source voltage for zero drain current. V_P is called the *pinch-off voltage*. The value of V_P can only be obtained from such a characteristic as that of fig. 3.9 by extrapolation to $I_D = 0$. Manufacturers usually quote V_P for $I_D = 1$nA.

The parabolic transfer characteristic is represented by an equation:

$$I_D = I_{DSS} (1 - V_{GS}/V_P)^2 \quad (3.1)$$

An operating point on the transfer characteristic would be chosen near the centre of the linear portion, say point P in fig. 3.9. The gradient $(\partial I_D/\partial V_{GS})$ of the curve at this point is called the mutual conductance of the fet, denoted by g_m and expressed in mA/V.

An expression for g_m can be obtained by differentiating equation (3.1) with respect to V_{GS}:

$$g_m = \partial I_D/\partial V_{GS} = -2(I_{DSS}/V_P)(1 - V_{GS}/V_P) \quad (3.2)$$

$$= g_{mo}(1 - V_{GS}/V_P) \quad (3.3)$$

where g_{mo} is the value of g_m when $V_{GS} = 0$, i.e.

$$g_{mo} = -2I_{DSS}/V_P.$$

Equation (3.3) gives the mutual conductance g_m at any value of I_D and may be written

$$g_m = g_{mo} \sqrt{I_D/I_{DSS}} \quad (3.4)$$

3.5 Parameters of a jgfet

The constant current region of the drain characteristic (fig. 3.8) is most important in the design of a fet amplifier. To express the behaviour in this region in mathematical terms, the effect of the gate-source voltage V_{GS} and the drain-source voltage V_{DS} are considered as independent variables whereas the drain current I_D is the dependent variable. Then,

$$\delta I_D = \left(\frac{\partial I_D}{\partial V_{GS}}\right)_{V_{DS}} \delta V_{GS} + \left(\frac{\partial I_D}{\partial V_{DS}}\right)_{V_{GS}} \delta V_{DS}$$

$$\quad (3.5)$$

This equation expresses the manner in which I_D will change when small changes occur in V_{GS} and V_{DS} about some operating point such as P in fig. 3.8. It is customary to write lower case letters to denote a small change: hence $i_d = \delta I_D$ and $v_{gs} = \delta V_{gs}$. The term $(\partial I_D/\partial V_{GS}) = g_m$ is a constant at a particular point and its value is usually obtained from the slope of the transfer characteristic at point P (fig. 3.9). The value of g_m, the mutual conductance, at $V_{DS} = 12$V, is about 3mA/V.

The term $(\partial I_D/\partial V_{DS})$ has the dimensions of conductance so is denoted by g_d and called the *output conductance* or *drain conductance*. Often g_d is written as $1/r_d$ where r_d is the *output* or *drain resistance*. From the characteristics of fig. 3.8 it is seen that g_d will have a very low value, say $10^{-5}\Omega^{-1}$ so that $r_d = 10^5 \Omega$. Equation (3.5) can now be written:

$$i_d = g_m v_{gs} + (1/r_d) v_{ds} \quad (3.6)$$

Because the term $(1/r_d) v_{ds}$ is small, the behaviour of the jgfet in the constant current region is represented approximately by

$$i_d = g_m v_{gs} \quad (3.7)$$

This leads to the statement that the jgfet is a voltage controlled device because equation (3.7) indicates that any change in I_D (represented by i_d) is due entirely to variations occurring in V_{GS} (represented by v_{gs}). Hence basic amplifier design with a jgfet is particularly straightforward for small signal behaviour (i.e. input signal voltage variations are small) and over the frequency region in which equation (3.7) is valid.

An *amplification factor* (μ) for the jgfet may be defined as

$$\mu = -\left(\frac{\partial V_{DS}}{\partial V_{GS}}\right)_{I_D} = -(v_{ds}/v_{gs})_{i_d = 0} \quad (3.8)$$

Note that, from equations (3.7) and (3.8) it follows that

$$\mu = g_m r_d \quad (3.9)$$

3.6 A source resistor to provide gate-source bias

The circuit of fig. 3.10 shows how the gate may be maintained at a negative potential with respect to the source without the need for a second voltage supply. This circuit arrangement is particularly valuable (providing that the applied voltage V does not drop below the knee voltage) because the circuit acts as a constant current source in which the magnitude of the constant current I_D is determined by the value of R_S.

It is a worthwhile experiment to use the circuit of fig. 3.10 to investigate how (a) the current in the load resistor R_L varies with the applied voltage V for a fixed value of R_S and (b) the current in the load R_L varies with the magnitude of R_S while the applied voltage V is kept constant.

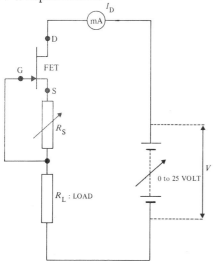

Fig. 3.10 A source resistor R_S provides gate-source bias by virtue of the pd across it; the jgfet is a constant current source

3.7 The common-source amplifier

The circuit of a common-source (CS) amplifier making use of an *n*-channel jgfet (fig. 3.11a) planned to operate at mid-band frequencies (i.e. at audio-frequencies over the range from 1kHz to 10kHz) has the equivalent circuit shown in fig. 3.11b.

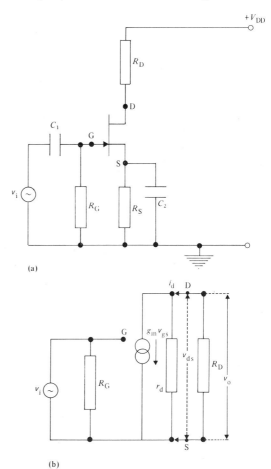

Fig 3.11 (a) The common-source amplifier; (b) the equivalent circuit

The steady bias voltage on the gate is provided by the source resistor R_S; this determines the operating point. This steady bias voltage is a consequence of the drain current which flows through R_S and so develops across it a voltage drop such that the source is at a positive steady bias with respect to earth. At the same time, the mean steady voltage (about which the input signal voltage v_i alternates) on the gate is zero because the gate is earthed by the resistor R_G. Hence, the steady voltage on the gate G is negative with respect to the source S, i.e. V_{GS} is negative.

The equivalent circuit (fig. 3.11b) is concerned only with the ac performance of the amplifier so the steady voltage supplies are absent; it incorporates a number of points of special interest:

(a) The by-pass capacitor C_2 (fig. 3.11a) serves to provide a very low impedance path across R_S for ac so that the steady bias voltage V_{GS} is not significantly varied by the alternating input. Hence neither C_2 nor R_S appear in fig 3.11b.
(b) No dc supply voltage (V_{DD}) is shown. Although this supply is essential for operation of the amplifier and it is the origin of the increased power made available at the amplifier output terminals, as regards ac only the internal impedance (at the ac operating frequency) of the voltage supply source V_{DD} connects one end of R_D (the load resistor in the fet drain circuit) to the source S. As this internal impedance is ensured to be negligible, it is not included in the equivalent circuit.
(c) The intersecting circles symbol used in fig. 3.11b is the conventional one representing an ideal current generator, i.e. one with no internal impedance. This generator in the present case provides an alternating current of $g_m v_{gs}$ on the basis that it is equivalent to the jgfet of mutual conductance g_m to which the input signal voltage is v_{gs} (equation 3.7). This current is independent of the load resistance connected across it because the jgfet is being used in the constant current region.
(d) The drain resistance r_d may be neglected if the load resistance R_D in the drain circuit is much less than r_d (see table 3.1).
(e) The resistance of the reverse-biased junction, which may be represented by r_{gs} or r_{gd}, is very large so the gate is considered to be electrically isolated from the channel (see table 3.1).
(f) By the nature of the jgfet structure (fig. 3.2) a small capacitance (C_{gd}) must exist between the gate and the drain and also one (C_{gs}) between the gate and the source. Over the mid-band frequency range, the effects of C_{gd} and C_{gs} have been neglected (see table 3.1).

The voltage gain of the amplifier of fig. 3.11 can be obtained from the equivalent circuit. Assuming $R_D \ll r_d$,

the output voltage $v_o = -g_m v_{gs} R_D$ *(3.10)*

v_{gs} = the input voltage v_i

the voltage gain A_v, defined as v_o/v_i is therefore given by

$$A_v = v_o/v_i = -g_m R_D \qquad (3.11)$$

The negative sign indicates that inversion occurs, i.e. the output is 180° out-of-phase with the input voltage. Voltage gains of about 10 are typical for a simple CS amplifier.

Note that if r_d cannot be neglected in the equivalent circuit (fig. 3.11b) the current source is connected across the parallel combination of r_d and R_D. This combination may be represented by a resistor R'_D, where

$$R'_D = \frac{R_D r_d}{R_D + r_d} \qquad (3.12)$$

Therefore,

$$v_o = -g_m v_{gs} R'_D$$

and $A_v = \dfrac{v_o}{v_i} = \dfrac{v_o}{v_{gs}} = -g_m R'_D$

$$= \frac{-g_m R_D r_d}{R_D + r_d} \qquad (3.13)$$

Table 3.1

Typical parameter values for a jgfet

Parameter:	g_m	r_d	C_{ds}	C_{gs}	C_{gd}	$r_{gs} = r_{gd}$
Value:	0.4 to 8 mA/V	300 kΩ	0.4 pF	3 pF	3 pF	10^{10} Ω

If $r_d \gg R_D$, equation (3.13) reduces to equation (3.11) as would be expected.

3.8 Design features of a CS amplifier

The required value of R_S, the source resistor used to establish the chosen operating point, can be found either by analysis or graphically.

EXAMPLE 3.8 *In a typical CS amplifier stage which makes use of an n-channel jgfet (fig. 3.11(a)) for which $V_p = -2.0V$ and $I_{DSS} = 1.4mA$, it is planned to bias the circuit at $1_D = 0.7mA$ with $V_{DD} = 20V$. Assuming that $r_d \gg R_D$ and that R_S is by-passed with a very large capacitor C_2, find (a) V_{GS} (b) g_m, (c) R_S, and (d) R_D, so as to provide a voltage gain of at least 20dB. Choose suitable values for C_1 and R_G.*

From equation (3.1)

$$0.7 = 1.4 (1 + V_{GS}/2)^2$$

Therefore, $V_{GS} = -0.59V$

The mutual conductance at zero bias $g_{mo} =$

$$\frac{-2 I_{DSS}}{V_P}$$

Therefore, $g_{mo} = -2 \times 1.4/2 = 1.4 \text{mA/V}$

At $I_D = 0.7\text{mA}$, $g_m = 1.4\sqrt{\frac{0.7}{1.4}}$ (from equation 3.4)

Therefore, $g_m = 1.0 \text{mA/V}$.

The source resistor $R_S = -V_{GS}/I_D = (0.59/0.7)\text{k}\Omega$
$= 840\Omega$

As 20dB corresponds to a voltage gain $A_v = 10$,

$A_v = g_m R_D = 10$

Therefore, $R_D = (10/1.0)\text{k}\Omega$

Hence, R_D must be 10kΩ or greater.

The capacitor C_1 serves to block any dc voltage from the source which would affect the operating point. A value of 0.1μF is suitable bearing in mind that R_G would have a value between 2MΩ and 10MΩ.

R_G is included to maintain the gate at earth potential; it is, in fact connected across the input terminals of the amplifier stage. Hence, in this mid-frequency range (1kHz to 10kHz), R_G represents the input resistance of the amplifier. A value of $R_G = 10\text{M}\Omega$ is suitable because the gate leakage current through R_G is negligible and so would not affect the operating point. If an input resistance in excess of 10MΩ is needed, a different circuit configuration is used (Section 3.12).

Fig. 3.12 illustrates the graphical method of determining R_S by making use of the static transfer characteristic.

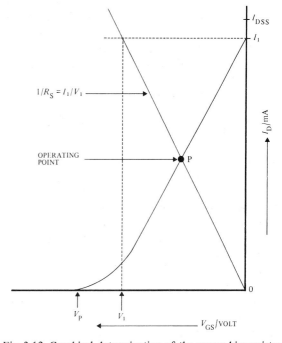

Fig. 3.12 Graphical determination of the source bias resistor R_S

A bias load-line (which is straight because it refers to an ohmic resistance) is drawn from the origin through the chosen bias point P (fig. 3.12). Because no gate-to-source voltage V_{GS} can be developed when $I_D = 0$, the self-bias load-line must pass through the origin. The required value of R_S is the reciprocal of the slope of this line.

Equally useful is the ability to determine the bias point resulting from the combination of a given jgfet and source resistor. A line of slope $1/R_S$ (R_S now being given) is drawn through the origin of the axes on which is drawn the static transfer characteristics which intersects this characteristic at the bias point.

Although one should really use the dynamic

transfer characteristic for the given fet–load resistor combination, in fact the dynamic (i.e. I_D vs V_{GS} curve with R_D in series with the drain) and static (R_D omitted) characteristics are virtually identical when taken at a value of V_{DS} near the desired operating point, so that catalogue static transfer characteristics can be used for this purpose with negligible error.

3.9 The operating point and the load line

In choosing the quiescent or dc operating point P (fig. 3.13) we attempt to obtain:

(a) Maximum voltage gain.
(b) Linear behaviour in that whatever the magnitude or sign of the input voltage v_i, the output voltage $v_o = -A_v v_i$ is such that A_v is a constant.
(c) Stable operation: the performance of the amplifier will not vary with temperature changes or with small changes in the applied voltage V_{DD}.

On the drain or output characteristics for a jgfet is drawn a load line AB which indicates how the operating point moves when a signal is applied to the input (fig. 3.13). A load line is represented by the equation

$$v_{DS} = V_{DD} - i_D R_D - V_S \qquad (3.14)$$

where v_{DS} is the instantaneous voltage between drain and source, i_D is the total drain current at any instant and made up from the steady current I_D and the varying signal component i_d, V_S is the steady voltage across the source resistor R_S (the ac component is by-passed by the capacitor C_S so V_S is steady – which is why it is expressed as a capital letter) and R_D is the drain or load resistor.

To draw a load line it is most convenient to re-arrange equation (3.14) to indicate the two end-points of this line. Thus,

$$i_D = \frac{V_{DD} - V_S}{R_D} - \frac{v_{DS}}{R_D} \qquad (3.15)$$

and the two points are when

$$i_D = 0, v_{DS} = V_{DD} - V_S$$

and when

$$v_{DS} = 0, i_D = (V_{DD} - V_S)/R_D.$$

The slope of this load line is $1/R_D$ because equation (3.15) has the form $y = mx + c$ where m is the slope. The load resistor R_D corresponding to the line in fig. 3.13 is 6250 Ω.

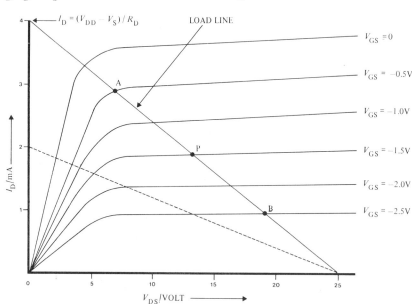

Fig. 3.13 A load-line drawn on typical drain (I_D vs. V_{DS}) characteristics

Note that the resistor in the drain circuit has been labelled R_D; the symbol R_L is to be reserved for the load imposed on the amplifier stage when it is coupled, usually by an RC network, to a subsequent stage.

A load line corresponding to a larger value of R_D is shown dotted in fig. 3.13. This larger value of R_D could provide for larger output voltage swings and hence a larger value for A_v. However, some distortion would then have to be tolerated because of the non-uniform spacing of the intercepts on this load line by neighbouring characteristic curves.

In practice, a more linear behaviour is achieved either by using a much smaller input signal or by employing negative feedback whereby the amplification, although reduced, can be made independent of variation in the jgfet parameters.

3.10 A stable CS amplifier

The circuit of a simple but stable CS amplifier providing a voltage gain of approximately 10 (fig. 3.14) has the gate of the jgfet maintained at its correct dc operating point by the combination of a potential divider network across the voltage supply V_{DD} and the drain current I_D producing a voltage drop across the source resistor R_S. To obtain a high input resistance, the 10MΩ resistor is connected between the divider and the gate electrode. Because the gate leakage current is negligible, this large resistance does not affect the steady bias and yet maintains, at medium frequencies, a resistance of at least 10MΩ across the source terminals. This stage was designed with a jgfet having the following parameters: $V_P = -2.0$V and $I_{DSS} = 0.5$mA. It was planned to bias the circuit at $I_D = 0.2$mA with $V_{DD} = 25$V.

As $I_D = I_{DSS}(1 - V_{GS}/V_P)^2$

$0.2 = 0.5(1 - V_{GS}/2.0)^2$

Therefore, $V_{GS} = -0.74$V

With the potential divider network shown in fig. 3.14, the point A is at +10V. Hence the value of R_S required to provide the correct gate-source bias is given by

$10.74 = 0.2 \times 10^{-3} R_S$

Therefore, $R_S \simeq 56$kΩ.

The mutual conductance at zero bias, $g_{mo} = -2I_{DSS}/V_P$

Therefore, $g_{mo} = \dfrac{2 \times 0.5}{2.0} = 0.5$ mA/V.

At $I_D = 0.2$mA, $g_m = g_{mo}\sqrt{I_D/I_{DSS}}$

Therefore, $g_m = 0.5\sqrt{0.2/0.5} = 0.32$ mA/V

Because r_d is inversely proportional to I_D, the low drain current (0.2mA) enables a large value of R_D to

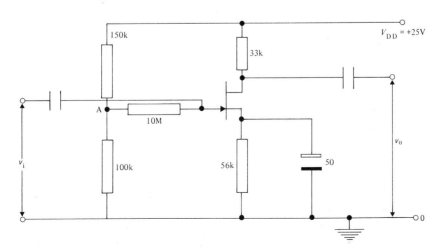

Fig. 3.14 A stable CS amplifier stage

be chosen without violating the assumptions of section 3.7. For a voltage gain of 10, the value of R_D required is given by

$$-10 = -g_m R_D, \text{ where } g_m = 0.32 \text{mA/V}.$$

Therefore, $R_D = 31 \text{k}\Omega$.
The proposed value of 33kΩ for R_D is therefore suitable.

The dc stability of the amplifier is provided by the large value of R_S which is possible only because the potential divider network has been used. Any small change in I_D due, say, to a temperature change, would alter the steady bias in such a manner as to compensate for the change. If I_D increases, V_{GS} increases and tends to reduce I_D. This stability is thus achieved utilising *dc negative feedback*.

EXAMPLE 3.10 *If the jgfet used in the CS amplifier of fig. 3.14 has a drain (or output) resistance r_d of 300kΩ at $I_D = 0.5$mA, calculate the error in the voltage gain A_v involved in neglecting r_d.*

$r_d \propto 1/I_D$. Hence, r_d at $I_D = 0.2$mA $= \dfrac{300 \times 0.5}{0.2}$

$= 750 \text{k}\Omega$.

But, $A_v = \dfrac{-g_m r_d R_D}{r_d + R_D} = -\dfrac{0.32 \times 750 \times 33}{783} = -10.1$

Using $A_v = -g_m R_D$,

$A_v = -0.32 \times 33 = -10.6$.

The error involved is hence about 5%, which is negligible unless close tolerance components are selected.

The above worked example has been included because a resistor R_D as large as 33kΩ is not common and it emphasises the need to examine the equivalent circuit of fig. 3.10b and assess the effect of r_d before using the simple relationship $A_v = -g_m R_D$.

3.11 The CS amplifier without a by-pass capacitor across R_S

Two methods will be examined which yield the voltage gain A_v of a CS amplifier in which the source resistor R_S is not by-passed. The first assumes that $r_d \gg R_D$ and provides a quick method of calculating A_v. The second method is more rigorous in that it examines the amplifier in a perfectly general way and yields all the amplifier equations used so far.

Method 1. Looking in at the source terminal, the mutual conductance g_m of the jgfet may be represented by a resistor of magnitude $(1/g_m)$kΩ connected between the gate and the source. The gate voltage therefore appears across the resistors $1/g_m$ and R_S connected in series (fig. 3.15).

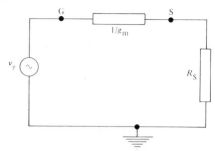

Fig. 3.15 Regarded from the signal source, the mutual conductance is represented as a resistor

The gate-source voltage, $v_{gs} = \dfrac{v_g (1/g_m)}{(1/g_m) + R_S}$ (3.16)

The current generator provides a current of magnitude

$$g_m v_{gs} = v_g / [(1/g_m) + R_S] \quad (3.17)$$

The output voltage

$$v_o = -g_m v_{gs} R_D$$

$$= \dfrac{-v_g R_D}{(1/g_m + R_S)}$$

The voltage gain, $A_v = v_o/v_g = -R_D/[(1/g_m) + R_S]$
.... (3.18)

The voltage gain, A_v, can be written as $-g'_m R_D$

where $g'_m = 1/[(1/g_m) + R_S]$ (3.19)

It is apparent from equation (3.18) and immediately obvious experimentally, that if the by-pass capacitor is removed, the voltage gain of the stage drops appreciably. Without a by-pass capacitor across R_S, which provides virtually a short-circuit to ac, an alternating voltage of magnitude $i_d R_S$ appears across R_S. This voltage is in-phase with the input signal voltage v_i. Hence, the control voltage v_{gs} is no longer equal to v_i but is given by

$$v_{gs} = v_i - i_d R_S \qquad (3.20)$$

Because v_{gs} is thus reduced, so also is the output voltage v_o and the voltage gain $A_v = v_o/v_i$ is considerably lower. The decrease in gain is due to ac negative feedback (see Appendix B).

Method 2. The small − signal equivalent circuit for a jgfet in CS connection and operating in the mid-frequency band (1kHz to 10kHz) is shown in fig. 3.16.

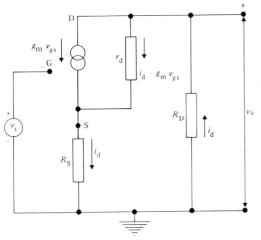

Fig. 3.16 Small signal equivalent circuit for a jgfet in CS connection

Kirchhoff's second law applied to the output circuit, yields

$$i_d R_D + (i_d - g_m v_{gs}) r_d + i_d R_S = 0 \qquad (3.21)$$

The voltage between gate and source, v_{gs}, is given by equation (3.20):

$$v_{gs} = v_i - i_d R_S$$

Combining equations (3.20) and (3.21) and using relationship $\mu = g_m r_d$, gives

$$i_d R_D + i_d r_d - \mu v_i + \mu i_d R_S + i_d R_S = 0$$

Hence,

$$i_d = \frac{\mu v_i}{R_D + r_d + (\mu + 1) R_S} \qquad (3.22)$$

The output voltage $v_o = -i_d R_D$ so

$$v_o = \frac{-\mu v_i R_D}{R_D + r_d + (\mu + 1) R_S} \qquad (3.23)$$

and $A_v = \dfrac{-\mu R_D}{R_D + r_d + (\mu + 1) R_S}$

If, as is usually the case, both R_S and R_D are very small compared with r_d, equation (3.23) can be written:

$$A_v = \frac{-\mu R_D}{r_d + r_d g_m R_S} = \frac{-R_D}{1/g_m + R_S} \qquad (3.24)$$

which is the same as equation (3.18) derived by Method 1.

Note that if $R_S = 0$ because it is effectively by-passed for ac, and r_d is large compared with R_D, equation (3.23) becomes

$$A_v = -\mu R_D/r_d = \frac{-r_d g_m R_D}{r_d} = -g_m R_D$$

in accordance with equation (3.11).

3.12 The common drain (CD) amplifier

Having derived the results for a CS amplifier with a source resistor R_S which is not by-passed, it is useful to consider the behaviour if $R_D = 0$ and the output signal is across R_S (fig. 3.17).

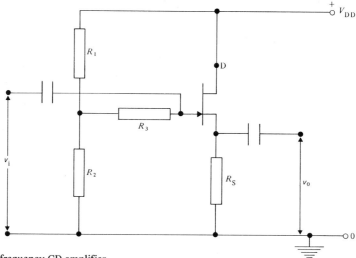

Fig. 3.17 A basic low-frequency CD amplifier

The jgfet is said to be connected in the common-drain (CD) configuration which is used when the highest possible input impedance is required and voltage gains greater than unity are not essential. The role of a CD amplifier is strictly as an impedance changer: it is sometimes called a *buffer amplifier* because it is used between a very high impedance signal source and a conventional amplifier stage.

Bearing in mind the equivalent circuit of fig. 3.15, the output circuit of fig. 3.16 can be used to calculate the output voltage v_o across the source resistance R_S.

$$v_o = \frac{v_g R_S}{1/g_m + R_S} \qquad (3.25)$$

and $A_v = v_o/v_g = g_m R_S/(1 + g_m R_S)$

Notice that the output voltage is in phase with the input signal voltage and also that $A_v \simeq 1$ for $g_m R_S \gg 1$. A voltage gain of unity means that the output voltage (taken from the source) follows the input voltage variations (at the gate). Hence, the CD configuration is called a *source follower*.

Typical component values for the amplifier of fig. 3.17 might be $V_{DD} = 20V$, $R_1 = 330k\Omega$, $R_2 = 100k\Omega$, $R_3 = 500M\Omega$ and $R_S = 2.2k\Omega$. With $C = 1nF$, this amplifier would have a voltage gain of about 0.7 and an input resistance of 500MΩ down to a frequency of 1Hz.

Recalling the expression for i_d derived from fig. 3.16,

$$i_d = \frac{\mu v_i}{R_D + r_d + (\mu + 1) R_S}$$

The output voltage across R_S with $R_D = 0$ can be written

$$v_o = i_d R_S = \mu v_i R_S/[r_d + (\mu + 1) R_S]$$

If $(\mu + 1) R_S \gg r_d$,

$$A_v = \mu R_S/[(\mu + 1) R_S] = \mu/(\mu + 1) \qquad (3.26)$$

Again, it is apparent that $A_v \to 1$ for $\mu \gg 1$.

It must be remembered that all the circuits discussed so far have been designed to operate at frequencies for which the capacitance between the gate and the drain (C_{gd}) and the gate and source (C_{gs}) can be ignored. This is certainly valid for frequencies up to 50kHz.

The low voltage gain of between 0.7 and 0.9 for a source follower stage must not be regarded as a serious disadvantage because the signal amplitude can readily be increased in subsequent stages. The important factor is that the use of the CD configuration compared with the CS one enables an

Table 3.2

Features of CS and CD amplifier stages

Configuration	Nature	Voltage gain (A_v)	Input resistance (R_i)	Output resistance (R_o)
CS	Inverting	$\dfrac{-g_m r_d R_D}{r_d + R_D}$ or $-g_m R_D$	R_G, say 10 MΩ	R_D
CD	Non-inverting	$\dfrac{g_m R_S}{1 + g_m R_S}$ or $\dfrac{\mu}{\mu + 1}$	500 MΩ	$1/g_m$

amplifier stage to be made with (a) a much higher input resistance, (b) a low output resistance which is virtually independent of R_S, and (c) a much more linear behaviour (*see* equation 3.26) in that changes in device parameters have virtually no effect on the gain.

In table 3.2 the features of a CS and a CD amplifier stage (using a jgfet in the frequency range 1kHz to 50kHz) are compared. Expressions for the input and the output resistance have not been derived in this text: they are merely quoted here for comparison.

3.13 Behaviour of the jgfet at higher frequencies

At higher frequencies, say above 50kHz, the effects of the gate-drain capacitance (C_{gd}) and of the gate-source capacitance (C_{gs}) must be considered. Fig. 3.18 shows the important features for the ac operation of a CS amplifier stage. Note that the resistance of the signal source (denoted by R_{SS}) is included. Whereas R_{SS} was of no consequence in the previous work discussed where no current was flowing into the device across the reverse-biased junction, both C_{gd} and C_{gs} will offer a conducting path at the higher frequencies so that a voltage drop will occur across R_{SS}.

The input circuit (fig. 3.19) comprises the input signal connected across a divider network consisting of the signal source resistance R_{SS} in series with the lumped input capacitance, C_i. At low frequencies, the impedance of C_i is very large so that $v_{gs} = v_i$, as used in the earlier analysis here.

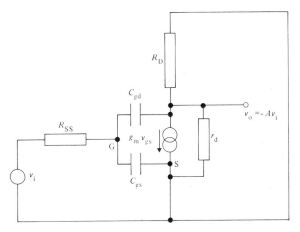

Fig. 3.18 Junction capacitances C_{gd} and C_{gs} govern high frequency performance

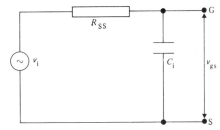

Fig. 3.19 The input circuit of a CS stage with lumped capacitance C_i

At the higher frequencies, we must write

$$v_{gs} = \frac{v_i(-jX_C)}{R_{SS} - jX_C} \qquad (3.27)$$

where X_C is the reactance of the capacitance C.

Therefore, $v_{gs} = \dfrac{v_i}{1 + j2\pi f R_{SS} C_i}$ (3.28)

Defining a frequency $f_1 = 1/2\pi R_{SS} C_i$ (3.29)

$v_{gs} = \dfrac{v_i}{1 + jf/f_1}$ (3.30)

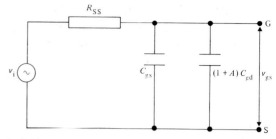

Fig. 3.20 Input circuit to emphasise the importance of Miller effect

The voltage gain is now seen to be frequency-dependent with the gain A at a frequency f given by

$A = \dfrac{A_L}{\sqrt{1 + (f/f_1)^2}}$ (3.31)

where A_L is the voltage gain at low frequency (1kHz→10kHz) and $f \ll f_1$.

The frequency f_1 is called the *upper 3dB point*. It is the upper frequency at which the voltage gain has dropped by 3dB of its mid-frequency value A_L. This is manifest from equation (3.31) because, at $f = f_1$

A dB $= A_L$ dB $- 10 \log_{10} 2$

$\quad = A_L$ dB $- 3$ dB

The lower 3dB point of such a stage would be determined primarily by the capacitance C_S connected across the source resistance R_S which provides the steady bias. The resistor R_S is not shown in fig. 3.18 because it is considered to be effectively by-passed and hence short-circuited to ac.

What factors determine the magnitude of C_i? Essentially, this input capacitance consists of three components:

(i) the gate-source capacitance, C_{gs};

(ii) the effective capacitance connected across the input terminals due to the Miller effect (Appendix C); this is of magnitude $(1 + A_L) C_{gd}$ where A_L is the mid-frequency voltage gain;

(iii) stray capacitance due to the wiring and the circuit layout. We shall assume careful wiring and neglect this component.

The input circuit is drawn as in fig. 3.20 to show the importance of the Miller effect on the high frequency performance of a jgfet. According to the Miller effect (Appendix C) any impedance Z existing between the input and output terminals of an amplifier, so acting as a feedback element, may be considered as an impedance of magnitude $Z/(1 - A_L)$ connected across the input terminals. Because of the inverting nature of the CS amplifier A_L is negative and the input capacitance is therefore much larger than may be supposed from looking at the magnitudes of C_{gs} and C_{gd} alone.

EXAMPLE 3.13 *A jgfet used in a CS amplifier stage has the following parameters: $C_{gd} = 5pF$, $C_{gs} = 5pF$; $g_m = 2mA/V$; and $r_d = 100k\Omega$. If the resistor R_D in the drain lead has a value of $25k\Omega$, calculate the mid-band gain and the upper 3dB point when the signal source has a resistance $R_{SS} = 25k\Omega$.*

The mid-band gain denoted by A_L is given by

$A_L = \dfrac{-g_m r_d R_D}{r_d + R_D} = \dfrac{-2.0 \times 100 \times 25}{125}$

$\quad = -40.$

$C_i = C_{gs} + (1 + A_L) C_{gd} = 5 + (41 \times 5) = 210 pF.$

The upper 3dB point f_1 is given by

$f_1 = \dfrac{1}{2\pi R_{SS} C_i} = \dfrac{10^{12}}{2\pi \times 25 \times 10^3 \times 210} = 30$ kHz.

3.14 Designs to reduce the input capacitance

It is evident from section 3.13 that the term $(1+A) C_{gd}$ is the largest of those contributing to the input capacitance C_i. Three circuit designs are examined each of which has a much smaller value of the input capacitance, C_i.

(a) **The CD amplifier.** The basic common drain or source follower is shown in fig. 3.17. The input capacitance C_i is given by

$$C_i = C_{gs} + (1 - A_v) C_{gd}$$

However, because A_v is positive (there is no phase inversion) and less than 1, the input capacitance is, in fact, less than $C_{gs} + C_{ds}$. Hence, the CD amplifier will have a good high frequency performance, although its input resistance is still limited by R_3 (fig. 3.17) which may be of the order of 500MΩ.

(b) **The technique of 'bootstrapping'.** This is not confined to jgfet circuits; it is incorporated into many types of amplifier circuit to achieve a very high input resistance. The principle involved is that an ac signal, approximately equal in magnitude to the input voltage and in-phase with it, is applied to the end of the resistor R_G remote from the gate. Applied to a source follower circuit, it enables the good high frequency behaviour to be associated with very high input resistance.

The voltage gain of the source follower is close to unity and the output voltage is in-phase with the input voltage. The resulting signal voltage appearing across R_G is very much less than v_i and hence the equivalent input resistance due to R_G is very much increased. For the circuit of fig. 3.21, the input resistance R_i is given by

$$R_i = R_G (1 + g_m R_S) \qquad (3.32)$$

With $R_G = 500\text{M}\Omega$, $g_m = 2\text{mA/V}$ and $R_S = 3\text{k}\Omega$ an input resistance of $3.5 \times 10^9 \, \Omega$ is obtained.

(c) **The cascode* amplifier.** This arrangement enables the large input capacitance resulting from the Miller effect to be largely eliminated. Two identical jgfets (usually manufactured simultaneously on the same silicon slice) are connected as shown in fig. 3.22. Because we are more concerned with good high-frequency performance than with very high input resistance, jgfet T_1 is connected in CS configuration with jgfet T_2 as the load. The gate of T_2 is maintained at a constant potential, positive with respect to earth but negative with respect to its source. When the input signal v_i is applied, a common signal current i flows in both fets. Then,

$$i = g_m v_i \qquad (3.33)$$

if the voltage at the drain of T_1 changes by v,

$$i = -g_m v \qquad (3.34)$$

because, with the gate voltage of T_2 fixed, the control voltage $v_{gs} = v$ (g_m is the same for T_1 and T_2). Hence, the voltage gain A_v of the stage is given by:

$$A_v = v_o/v_i = -iR_D/v_i = -g_m R_D \qquad (3.35)$$

which is the same as that of any CS amplifier with $r_d \gg R_D$.

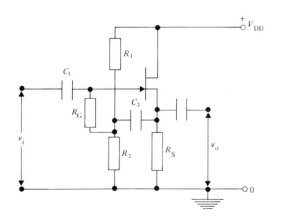

Fig. 3.21 Application of the 'boot strapping' technique to a source follower

---

*The term 'cascode' is used in the case where an active device serves as the load for a similar active device.

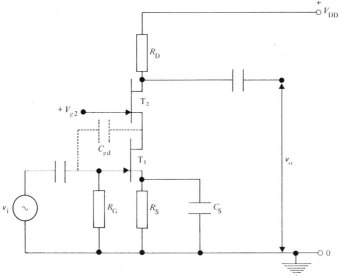

Fig. 3.22 A cascode amplifier in which two jgfets are used

From equations (3.33) and (3.34)

$$v_i = -v$$

When the gate voltage of T_1 increases by v_i, the drain voltage drops by a similar amount and the total voltage across C_{gd} of T_1 always remains equal to $2v_i$. The size of C_{gd} has therefore effectively been increased by a factor of two instead of $(1 + A_L)$ as it is in the conventional CS amplifier. Hence the cascode arrangement of fig. 3.22 provides the same voltage gain as a conventional CS amplifier but without ruination of the high-frequency performance by the Miller effect.

If the amplifier is designed to operate over a very wide frequency range, C_S may consist of two capacitors in parallel: a high value electrolytic capacitor to provide good low frequency behaviour and a high quality low value capacitor for the high frequency. This is desirable because the performance of electrolytic capacitors is often poor at high frequencies in that they have an appreciable impedance and would not effectively by-pass R_S.

3.15. Low frequency response of an amplifier based on jgfets

In a single-stage amplifier based on the CS configuration, to provide a good low frequency response, a capacitor C_S of large capacitance is connected across R_S effectively to short-circuit this source resistor even for low-frequency ac. The gain at medium and low frequencies is then virtually constant.

To join two such CS amplifier stages together to provide a two-stage amplifier in which the output voltage (across the load resistor of a fet T_1) is made the input to a second stage based on fet T_2, resistance – capacitance (RC) coupling is widely used in the case where the input signal voltage to stage 1 is alternating (fig. 3.23). This RC coupling requires the capacitor C_C and the resistor R_{G2}. The capacitor C_C is needed to isolate the steady voltage on the drain of fet T_1 from the gate of fet T_2. At the same time, C_C and R_{G2} are in series across the load R_D of T_1 and serve as an alternating potential divider (fig. 3.24) so that the alternating voltage which appears across R_{G2} is the input between the gate of fet T_2 and earth. R_{G2} is also the gate resistor for T_2 and would have a value of several megohms.

The voltage across R_{G2}, which is the input voltage to the second stage (based on T_2) is denoted by v_{i2}. The drain resistance r_d of the first stage fet (T_1) is very much larger than R_D (the load resistor in

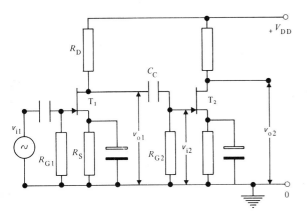

Fig. 3.23 A two-stage resistance-capacitance coupled amplifier based on jgfets

the drain circuit of T_1) so has been omitted in fig. 3.24b. It follows that

$$v_{i2} = \frac{-g_m v_{i1} R_D R_{G2}}{R_D + X_C + R_{G2}} \quad (3.36)$$

where v_{i1} is the input signal voltage to the first stage and X_C is the impedance of the capacitor C_C.

The voltage gain of the first stage (including the RC network) at low frequencies is denoted by

$$A_L = v_{i2}/v_{i1}$$

and

$$A_L = \frac{-g_m R_D R_{G2}}{R_D + Z_C + R_{G2}} = \frac{-g_m R_D}{1 + \dfrac{R_D}{R_{G2}} + \dfrac{1}{j\omega R_{G2} C_C}}$$

Field Effect Transistors 59

Neglecting R_D/R_{G2}, therefore,

$$A_L = -A / [1 + (1/j\omega R_{G2} C_C)] \quad (3.37)$$

where A is the voltage gain at mid-band frequencies (1kHz to 10kHz). Hence, the magnitude of the voltage gain drops and becomes equal to $A/\sqrt{2}$ when

$$\omega = 1/R_{G2} C_C$$

i.e. $f_L = 1/2\pi R_{G2} C_C \quad (3.38)$

The frequency f_L is termed the low 3dB frequency. For good low frequency performance, f_L should be small, so that both R_{G1} and C_C should be large. For a low 3dB frequency of 10Hz with $R_{G2} = 2.2\text{M}\Omega$, the capacitance of $C_C = 0.007\mu\text{F}$.

Note that not only has the voltage gain dropped by $1/\sqrt{2}$, there is also a phase-change of $45°$ at $f = f_L$. This phase-change will introduce waveform distortion when the input signal is not of sinusoidal waveform. Unless it is necessary to have very low and even zero frequency behaviour, in which case the two stages have to be directly coupled, it is often good policy to use the RC coupling network. This is because it rejects unwanted low frequency signals such as those due to ac mains pick-up.

The low-frequency performance of the amplifier in fig. 3.23 may be examined by considering the RC coupling as a high-pass filter (fig. 3.25). Then,

$$v_o = v_i R/(R - jX_C) = v_i/[1 - (j/2\pi fRC)]$$

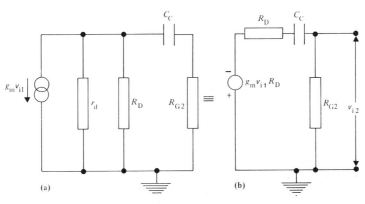

Fig. 3.24 Equivalent circuits for the amplifier of Fig. 3.23

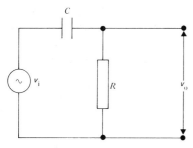

Fig. 3.25 A high-pass *RC* network as an inter-stage coupling

If we define a frequency $f_L = 1/2\pi RC$,

$$v_o = v_i / [1 - j(f_L/f)]$$

The voltage transfer function (VTF) = v_o/v_i is of magnitude

$$\text{VTF} = 1/\sqrt{1 + (f_L/f)^2} \qquad (3.39)$$

With $f \gg f_L$, VTF = 1

with $f = f_L$, VTF = $1/\sqrt{2}$.

In the amplifier circuit of fig. 3.23. as R_{G2} is large (say, 2.2 MΩ) the *RC* coupling network provides no shunt across R_D. Also, because the input resistance of fet T_2 is very high, there is no shunt across R_{G2}. The network consisting of C_C and R_{G2} may now be considered as a frequency-dependent potential divider.

The voltage input to the second stage will have dropped to $1/\sqrt{2}$ of its mid-band value at a frequency $f_L = 1/2\pi R_{G2} C_C$, in accordance with equation (3.38)

3.16. A differential amplifier based on jgfets

The very useful, multi-stage differential or difference amplifiers based on fets are much utilised in operational amplifiers (Chapter 5), particularly in integrated-circuit form. A basic, symmetrical differential amplifier in which the jgfets are assumed to be identical (fig. 3.26) has the unusual feature of having two input signals v_{i1} and v_{i2} and two outputs v_{o1} and v_{o2}.

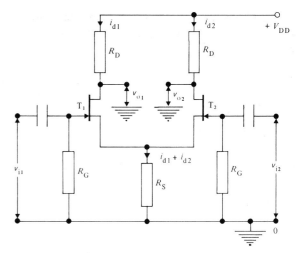

Fig. 3.26 A basic symmetrical jgfet differential amplifier

Considering the small signal behaviour of jgfet T_1,

$$i_{d1} = g_m [v_{i1} - (i_{d1} + i_{d2})R_S] \qquad (3.40)$$

or

$$i_{d1}(1 + g_m R_S) + i_{d2} g_m R_S = g_m v_{i1}.$$

Similarly for jgfet T_2,

$$i_{d1} g_m R_S + i_{d2}(1 + g_m R_S) = g_m v_{i2} \qquad (3.41)$$

where the mutual conductance g_m is the same for each fet.

From equations (3.40) and (3.41)

$$i_{d1}(2 + 1/g_m R_S) = g_m v_{i1}(1 + 1/g_m R_S) - g_m v_{i2}. \qquad (3.42)$$

If the source resistor R_S is very large so that $1/g_m R_S \ll 1$, equation (3.42) becomes

$$i_{d1} \simeq (g_m/2)(v_{i1} - v_{i2}) \qquad (3.43)$$

Similarly,

$$i_{d2} \simeq (g_m/2)(v_{i2} - v_{i1}) \qquad (3.44)$$

Hence,

$$i_{d2} = -i_{d1} \qquad (3.45)$$

Listed below are several important features of differential amplifiers apparent from this simple analysis:

(a) The current change in either jgfet is proportional to the difference between the voltage applied to the gates.
(b) The output signal voltage from fet $T_1 = v_{o1} = -i_{d1}R_D$
Therefore $v_{o1} = -(g_m R_D/2)(v_{i1} - v_{i2})$.
The output signal from fet $T_2 = v_{o2} = -i_{d2}R_D$.
Hence,
$v_{o2} = -(g_m R_D/2)(v_{i2} - v_{i1}) = (g_m R_D/2)(v_{i1} - v_{i2})$.
The two output signals, v_{o1} and v_{o2}, are therefore equal in magnitude but in anti-phase. Note that v_{o1} is 180° out-of-phase with v_{i1} whereas v_{o2} is in-phase with v_{i1}.

(c) If $v_{i2} = 0$, that is the gate of fet T_2 is connected to earth, there are still two output voltages of equal magnitude and 180° out-of-phase because

$v_{o1} = -(g_m v_{i1} R_D/2)$ and $v_{o2} = g_m v_{i1} R_D/2$.

The amplitude of these output voltages is one-half the output $g_m v_{i1} R_D$ which is obtained from a single CS stage having R_S by-passed. An amplifier stage producing these balanced or 'push-pull' outputs is called a *paraphase amplifier*.

(d) If $v_{i1} = v_{i2}$, the output voltage from either T_1 or T_2 is zero. At either drain, the output voltage produced by one input cancels exactly the output due to the other input.

This ideal behaviour of the differential amplifier is dependent on the following two assumptions:

(i) the differential amplifier is perfectly symmetrical;
(ii) the resistor R_S is very large compared with $1/g_m$.

Integrated circuit technology (Appendix A) has assisted in the attainment of good symmetry because two jgfets can be produced with virtually identical characteristics in very close proximity on the same silicon slice. Many manufacturers market these dual jgfets within the same package.

From equation (3.42) it is seen that the performance of the differential amplifier approaches the ideal as the value of R_S is increased.

If we define A_D as the *differential voltage gain*, and $v_{i2} = 0$,

$$A_D = v_{o2}/v_{i1} = -g_m R_D/2 \qquad (3.46)$$

The *common-mode gain* A_C is defined by

$$A_C = v_{o2}/v_C$$

where

$$v_C = \frac{1}{2}(v_{i1} + v_{i2}).$$

If $v_{i1} = v_{i2}$,

$$A_C = v_{o2}/v_{i1}$$

and, although this term (v_{o2}/v_{i1}) has not been derived in the simple analysis, it has a value of $-R_D/2R_S$ (fig. 3.26).
Hence,

$$A_C = -R_D/2R_S \qquad (3.47)$$

A figure of merit for a differential amplifier is the common-mode-rejection ratio (CMRR) denoted by C and defined by the equation

$$C = |A_D/A_C| \qquad (3.48)$$

For the circuit of fig. 3.26, therefore,

$$C = |A_D/A_C| = (g_m R_D/2)/(R_D/2R_S)$$
$$= g_m R_S \qquad (3.49)$$

The significance of the CMRR will be examined in more detail in section 3.17. At this point it suffices to give a simple example. If $v_{i1} = +50$mV and $v_{i2} = -50$mV, the difference signal voltage $v_d = v_{i1} - v_{i2} = 100$mV whereas the average, or common-mode voltage $v_C = (v_{i1} + v_{i2})/2 = 0$. If then, the signal voltages are changed such that $v_{i1} = +850$mV and $v_2 = +750$mV, the difference voltage $v_d = v_{i1} - v_{i2} = 100$mV, while the average, or common-mode signal voltage $v_C = (v_{i1} + v_{i2})/2 = 800$mV.

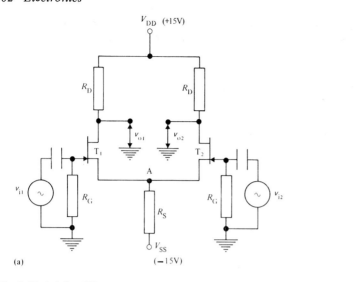
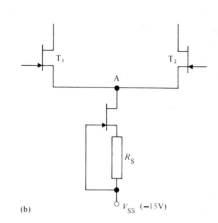

Fig. 3.27 A jgfet differential amplifier: (a) with the input gate at earth potential; (b) with R_S replaced by a constant current source

An ideal differential amplifier would give the same output voltage for each of the above cases because the difference signal voltage is 100mV in each case, but a practical amplifier would not. How well the amplifier can ignore or reject the common-mode signal is expressed in this CMRR term. The larger the value of C, the more closely will the amplifier approach the ideal. From equation (3.49) it is seen that a large value of R_S is essential for C to be large.

Fig. 3.27a shows a typical differential stage having two stabilised voltage supplies: $V_{DD} = +15V$ and $V_{SS} = -15V$. This results in a convenient arrangement with each gate maintained at earth potential (0V). If R_S is increased, V_{SS} must also be increased. A more common solution (fig. 3.27b), particularly for operational amplifiers (Chapter 5) for which a high CMRR is required, is to replace R_S by a constant current source (section 3.6). The constant current is determined by the self-bias resistor R_S. The output impedance of this constant current source may be between 1MΩ and 10MΩ — a value unobtainable if a passive component is used with realistic values of V_{SS}.

This amplifier configuration which makes use of two matched active devices (fets or other active components) with a large common source (or emitter in the case of a bipolar transistor) resistor is often called a *long-tailed pair*.

One noteworthy feature of this arrangement is if the resistor in the drain circuit of jgfet T_1 (fig. 3.26)

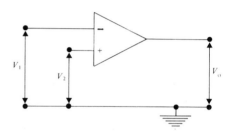

Fig. 3.28 A general differential amplifier

is equal to zero and no signal is applied to the gate of T_2 which is assumed to remain at earth potential. Because there is no change in the drain potential of T_1 when the input to its gate is altered, the input capacitance C_{gd} of T_1 is not enhanced by Miller effect (Appendix C). The output voltage, across the load in the drain circuit of T_2 is in phase with the input signal voltage v_{i1}, but the input capacitance of the stage is low, resulting in good high frequency performance.

3.17. The general case of the differential amplifier

Because the term 'differential amplifier' refers to a class of amplifier rather than a particular circuit, the general case is examined. The amplifier, represented by the arrow head (triangle) in fig. 3.28 has two input signal voltages V_1* and V_2 and an output voltage

---

*Capital letters are used here because V may be steady or varying.

$V_o \propto (V_1 - V_2)$ in the ideal case. Input V_1 is applied to the inverting input terminal identified by the negative sign (−) whereas V_2 is applied to the non-inverting terminal identified by the positive sign (+). The amplifier corresponds with the particular case of the jgfet differential amplifier (fig. 3.26) in which the output is taken from the drain of jgfet T_2.

The output voltage from the amplifier can be written

$$V_o = A_1 V_1 + A_2 V_2 \qquad (3.50)$$

where A_1 is the voltage gain when $V_2 = 0$, and A_2 is the voltage gain when $V_1 = 0$.

The input signal can be considered as made up from two parts: a difference mode signal, V_D where

$$V_D = V_1 - V_2 \qquad (3.51)$$

and a common-mode signal V_C equal to the average value of the two input signal voltages, i.e.

$$V_C = \frac{1}{2}(V_1 + V_2) \qquad (3.52)$$

From equations (3.51) and (3.52) we can write

$$V_1 = V_C + V_D/2 \text{ and } V_2 = V_C - V_D/2$$

Substituting these values in equation (3.50) gives

$$V_o = A_D V_D + A_C V_C \qquad (3.53)$$

where

$$A_D = \frac{1}{2}(A_1 + A_2) \text{ and } A_C = A_1 + A_2.$$

The common-mode rejection ratio C, which serves as a figure of merit for a differential amplifier, is defined by

$$C = |A_D/A_C|$$

Hence, equation (3.53) can be written

$$V_o = A_D V_D (1 + A_C V_C/A_D V_D)$$

$$= A_D V_D (1 + V_C/C V_D) \qquad (3.54)$$

Equation (3.52) is deduced and used again in Chapter 5 where operational amplifiers are described.

It is clear from equation (3.52) that, as $C \to \infty$, the amplifier output is determined by the difference signal voltage only. With good design, C can be as great as 10^5.

EXAMPLE 3.17. *A differential amplifier has a differential gain A_D of 100 and a common-mode rejection ratio C of 1000. If the input signals are initially (1) $V_1 = +5.0mV$, $V_2 = -5.0mV$ and later (2) are $V_1 = +105mV$ and $V_2 = +95mV$, calculate the output voltage in each case.*

Case 1

$$V_D = V_1 - V_2 = 10\text{mV}$$

$$V_C = \frac{1}{2}(V_1 + V_2) = 0$$

From equation (3.52) therefore,

$$V_o = 100 \times 10 = 1000\text{mV}.$$

Case 2

$$V_D = V_1 - V_2 = 10\text{mV}$$

$$V_C = \frac{1}{2}(V_1 + V_2) = 100\text{mV}$$

$$V_o = 100 \times 10 \left(1 + \frac{100}{1000 \times 10}\right) = 1010\text{mV}.$$

The difference between the output voltages in the two cases is 1%, or the error is 1% in this case; this is reduced as C is increased.

3.18. Insulated gate field effect transistors (igfets)

The jgfet suffers from the disadvantage that the input resistance is very high (approximately $10^{10} \Omega$) only when the gate-channel junction is reverse-biased. If, for any reason, the gate voltage is reversed so that the junction becomes forward-biased, the resistance falls to a very low value (tens of ohms) and current carriers are injected into the channel. Junction damage can result from such a change of voltage.

A field-effect transistor which retains its very high input resistance irrespective of the polarity of the gate voltage with respect to the source is the *insulated gate field-effect transistor* (the igfet). The most common form of igfet makes use of a silicon oxide (SiO_2) film to isolate electrically the gate electrode from the underlying semiconductor; it is frequently known as a *mosfet* – a word indicating the structure sequence of metal-oxide-semiconductor in the fet.

The high quality of the silicon oxide (SiO_2) as an insulator enables mosfets to be manufactured having an input resistance in excess of $10^{12} \Omega$. Values as high as $10^{15} \Omega$ are possible.

The structure of an enhancement type *p*-channel mosfet (fig. 3.29) comprises a substrate of *n*-type silicon with two heavily doped *p*-type regions (designated p^+ to indicate the very high trivalent impurity concentration) diffused into the surface to serve as the source S and the drain D. A thin insulating layer of SiO_2 is grown over the gap between the source and drain areas and aluminium electrodes are deposited on the source, on the drain and as a thin film on the oxide layer. The aluminium film on the oxide layer between the source and the drain forms the gate electrode (G).

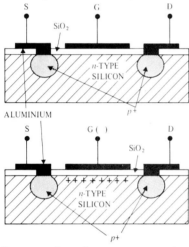

Fig 3.29 Structure of an enhancement type *p*-channel mosfet (Thickness of SiO_2 greatly exaggerated)

In normal use, the drain is made positive in potential with respect to the source. The layer of *n*-type silicon between the drain and source regions forms the channel. With the gate-source pd $V_{GS} = 0$ the drain-to-channel junction is reverse-biased and no drain current flows, i.e. $I_D = 0$. Both *p*-channel and *n*-channel enhancement mosfets behave in this way with no drain current flowing when $V_{GS} = 0$. They are sometimes called 'normally off' mosfets. This behaviour is quite unlike that of a jgfet, which is a depletion device and provides a maximum drain current I_{DSS} when $V_{GS} = 0$.

If the gate potential is made negative with respect to the source, positive holes are attracted into the region immediately below the oxide layer and form a narrow conducting channel between the source and the drain (fig. 3.29b). The conductivity of this channel increases (is enhanced) as the magnitude of V_G is increased (with G negative with respect to S). The conducting channel is sometimes called an *inversion layer* because its impurity type appears to invert from *n*-type to *p*-type as holes are attracted to this region.

An *n*-channel enhancement mosfet has a *p*-type substrate and n^+ regions constituting source and drain: it also is a 'normally-off' device but now conduction is initiated when the gate potential is made positive with respect to that of the source.

In the circuit symbols for the *n* and *p* channel enhancement mosfets (fig. 3.30) note that the normally-off type has a broken (three-section) channel line. Against each symbol is shown the transfer characteristic in order to emphasise the normally-off feature.

The circuit of fig. 3.31 enables the output and transfer characteristics of a *p*-channel mosfet to be

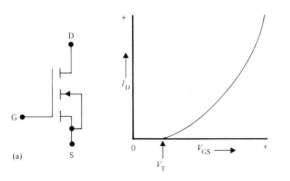 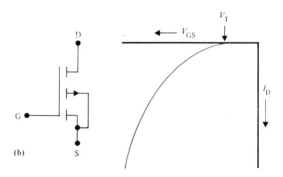

Fig. 3.30 (a) *n*-channel enhancement mosfet and its 'normally-off' feature (b) *p*-channel enhancement mosfet and its 'normally-off' feature

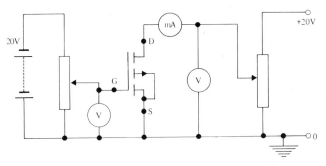

Fig. 3.31 Circuit to obtain the characteristics of a *p*-channel enhancement type mosfet

obtained. The value of V_{GS} at which the drain current I_D begins to flow is called the threshold voltage V_T. In practice V_T is taken as the value of V_{GS} at which $I_D = 10\,\mu A$.

The drain and transfer characteristics of a *p*-channel enhancement type mosfet (fig. 3.32) are similar to those for a jgfet but one marked difference is the larger values of V_{GS} required for control of the mosfet. The value of the threshold voltage V_T of $-4V$ (fig. 3.32) is typical. Recent developments have enabled devices with lower threshold voltages to be produced.

The structure of an *n*-channel depletion type mosfet is shown in fig. 3.33a. In this case, a channel of *n*-type silicon is created by diffusion between the drain and source regions. In common with a jgfet — also a depletion device — current will flow through the channel with $V_{GS} = 0$. This 'normally-on' mosfet has a circuit symbol with a continuous channel line (fig. 3.33b) and a transfer characteristic indicating that it can be used with V_{GS} positive or negative.

With the gate negative with respect to the source, positive charges are induced in the *n*-type channel below the oxide layer so making it less conducting because the re-distribution of charge in this channel causes a depletion of majority carriers. Hence, the drain current is reduced as the negative gate potential is increased, which leads to the designation of depletion-type mosfet.

Because of the voltage drop along the channel, the region near the drain is more depleted than that near the source. This wedge-shaped depletion region is

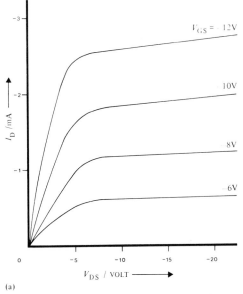

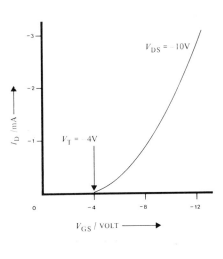

Fig. 3.32 For a *p*-channel enhancement type mosfet: **(a)** the drain characteristics (I_D against V_{DS} with V_{GS} as parameter); **(b)** the transfer characteristic (I_D against V_{GS} for a given value of V_{DS})

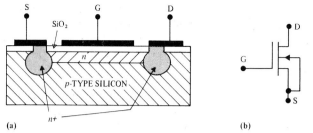 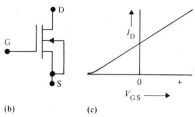

Fig. 3.33 (a) Structure of the *n*-channel depletion type mosfet (b) Circuit symbol (c) Transfer characteristic

similar to that produced in a jgfet except that, in the igfet case, the depletion is created by induced charges instead of a reverse-biased *p-n* junction.

With the gate potential positive with respect to the source, the device of fig. 3.33 can be operated in an enhancement mode. The conductivity of the channel is increased as more electrons move into the channel.

The characteristics of the *n*-channel depletion type mosfet (fig. 3.34) are similar to those of a jgfet; the value of V_{GS} corresponding to negligible I_D (I_D = 1nA) being called the *gate-source cut-off voltage* instead of the term 'pinch-off voltage' used with jgfets.

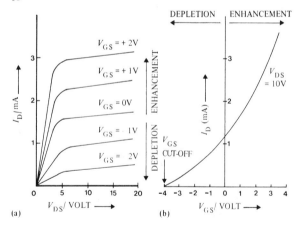

Fig 3.34 For an *n*-channel depletion type mosfet: (a) the drain characteristic (b) the transfer characteristic

As with jgfets, the mosfets display similar well-defined regions of operation. For low values of V_{DS}, the mosfet behaves as a voltage-controlled resistance R where R is approximately inversely proportional to V_{GS}. For the *n*-channel depletion mosfet, the drain current is given by

$$I_D = I_{DSS}(1 - V_{GS}/V^*_{GS})^2$$

where V^*_{GS} is the value of V_{GS} at cut-off, i.e. with I_D = 1nA.

The output resistances and the inter-electrode capacitances for mosfets are similar to those encountered with jgfets. Because the equivalent circuits are the same, any analysis presented earlier in this chapter for jgfets can readily be applied to mosfets.

The individual mosfet on a silicon slice occupies only about 5% of the space required for an equivalent bipolar device. Furthermore, the mosfet structure lends itself to integrated circuit construction (Appendix A). The mosfet family is being used extensively in large-scale integration (LSI) to produce complex logic circuits and semiconductor memory devices.

3.19. Protecting the gate of a mosfet

Special care is needed to protect the mosfet gate. The oxide layer which separates the gate from the underlying semiconductor is extremely thin — of the order of 100nm. Electric charge will accumulate readily on an open-circuit gate and, because the capacitance is low (a few picofarad) the voltage developed across the oxide film can become excessive and damage this insulating layer. Manufacturers often incorporate a Zener diode into the structure between the gate and the substrate. In normal operation, this reverse-biased Zener diode does not affect the behaviour of the mosfet, but the voltage on the gate can never exceed the Zener breakdown voltage.

When transporting or handling a mosfet, it is usual to connect the three leads together by means of a small conducting ring or clip. Within a circuit based on mosfets, a switching facility should be arranged so that the gate electrode of any mosfet is never open-circuit.

Exercises 3

1. 'The jgfet is a unipolar device which operates by the creation and control of the depletion layer at a reverse-biased *p-n* junction'.
   Explain, with the aid of diagrams, the meaning of this statement.

2. Draw a diagram of a circuit to obtain the drain and transfer characteristics of an *n*-channel jgfet. Sketch typical drain and transfer characteristic curves and use them to:
   (a) explain the use of a jgfet as a voltage controlled resistor;
   (b) define and indicate the magnitude of (i) the mutual conductance and (ii) the output resistance.

3. How can a jgfet be used as a constant current source?

4. The drain current I_D of a jgfet is related to the gate-source voltage V_{GS} by the equation

   $$I_D = I_{DSS}\,[1 - (V_{GS}/V_P)]^2$$

   where I_{DSS} and V_P are constants.
   Define these constants and use this equation to deduce an expression for the mutual conductance of the fet in terms of V_{GS}.

5. Draw the circuit diagram of a CS amplifier stage in which the dc bias is obtained by means of a source resistor R_S.
   Draw an equivalent circuit for this amplifier for the frequency range 1kHz to 10kHz and use it to deduce an expression for the voltage gain of the stage.

6. Draw the circuit diagram of a source follower (CD) stage which utilises a jgfet, and in which a potential divider network is arranged to fix the operating point and for the input resistance to be 500MΩ.
   Under what circumstances would such an amplifier stage be used?

7. Draw the equivalent circuit for a CS amplifier stage in which the source resistor R_S is not by-passed.
   Deduce an expression for the voltage gain of such a stage and simplify this expression by assuming that the source resistor R_S and the drain resistor R_D are both small compared with the output resistance r_d.

8. Fig. 3.35 shows a typical *n*-channel jgfet CS amplifier stage for operation in the audio-frequency range. For this fet, I_{DSS} = 2.0mA and V_P = −3.0V. It is planned to bias the circuit at I_D = 1mA. Assuming that $r_d \gg R_D$ and that R_S is by-passed by a very large capacitance C_S, calculate the values of (a) V_{GS}, (b) g_m, (c) R_S and (d) R_D to provide a voltage gain of at least 5.

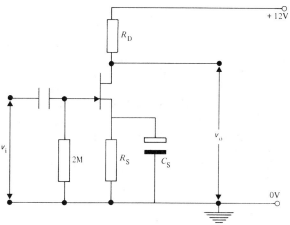

Fig. 3.35 Concerning exercise 8

9. Comment on the effect of the junction capacitance on the behaviour of a jgfet used in a CS amplifier stage at high frequencies.
   A jgfet used in a CS amplifier stage has the following parameters: C_{gd} = 10pF, C_{gs} = 10pF, g_m = 2mA/V and r_d = 180kΩ. If the drain resistor R_D = 20kΩ, calculate the mid-band voltage gain and the upper 3dB point, if the signal source has a resistance of 20kΩ.

10. Discuss circuits designed to reduce the effective input capacitance of a jgfet.

11. What would be the effect of a temperature rise of, say, 30°C on the performance of a jgfet? Explain why thermal runaway is not a hazard with a fet as it is with a bipolar transistor (*see*, also, Chapter 4).

12. Draw a circuit diagram of a differential amplifier based on jgfets. Discuss the special features of a differential stage, dealing, in particular, with its role as a paraphase amplifier.

13. Discuss the use of jgfets in a cascode amplifier.

14. What is meant by the common-mode rejection ratio (CMRR) of a differential amplifier? Derive the expression:

$$V_o = A_D V_D [1 - (V_D/CV_C)]$$

where V_o is the output voltage, C is the CMRR, A_D is the voltage gain for the difference signal voltage V_D, and V_C is the common-mode voltage.

15. Describe the structure and characteristics of a *p*-channel enhancement type mosfet.

16. Compare the structure and characteristics of an *n*-channel jgfet with those of an *n*-channel depletion-type mosfet.

17. In the circuit of fig. 3.36, the two resistors R_1 and R_2 both have negligible temperature coefficients of resistance. They are used in conjunction with R_{DS}, the resistance of the drain-source channel, to provide an effective resistance R_E which can be voltage controlled and yet possess a smaller temperature coefficient of resistance than R_{DS} used alone. Write down the appropriate expression for R_E, deduce $\partial R_E / \partial R_{DS}$, and calculate its value if $R_1 = R_{DS}$ and $R_2 = 2R_1$.

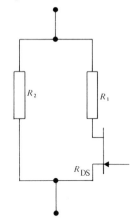

Fig. 3.36 Concerning exercise 17

4 · Junction or Bipolar Transistor

4.1 Introduction

The junction transistor is, without doubt, the most important active semiconductor device. Though integrated circuit methods enable several hundred transistors to be manufactured within a single silicon slice, the physical principles governing transistor operation remain the same, and circuit design based on the integrated circuit procedures or with discrete devices can only be aided by a knowledge of the fundamental processes occurring within the discrete active device and its immediately associated passive components.

A prime concern in electronics is with amplification. Field-effect devices were studied first (Chapter 3) because in an amplifier stage based on an fet there is much greater electrical isolation between the control (input) and the controlled (output) circuits than in an amplifier stage based on a bipolar transistor. The fet amplifier, with little (often negligible) interaction between the input and output circuits, therefore yields to simpler circuit analysis.

The term *bipolar transistor* is synonymous with 'junction transistor' and is used because both electrons and holes contribute to the current, to distinguish this device from the field-effect transistor or unipolar transistor in which only one current carrier is involved.

4.2 Structure of the junction transistor

The junction transistor consists of two *p-n* junctions positioned very close together within a single crystal slice. The region common to the two junctions, called the *base* may be either of *n*-type semiconductor material or of *p*-type. In the first case, the *n*-type base is flanked on each side with *p*-type material, forming a *p-n-p* junction transistor. In the second case, the *p*-type region is flanked on each side by *n*-type material forming an *n-p-n* junction transistor. The thin central region – the *base* B – is of high resistivity material sandwiched between the more heavily doped (lower resistivity) flanking material comprising the *emitter* E on one side and the collector C on the other side. Both *n-p-n* and *p-n-p* devices are available in germanium and in silicon. In general, silicon devices give better performance and are thermally more stable than the germanium equivalents.

The two types of structure are illustrated in fig. 4.1a with the circuit symbol and current and voltage conventions for each in fig. 4.1b. The arrow head on the emitter specifies the direction of conventional current flow (opposite to the direction of electron flow) when the emitter-base junction is forward-biased. The two junctions formed are the emitter-base junction, J_{EB}, and the collector-base junction, J_{CB}.

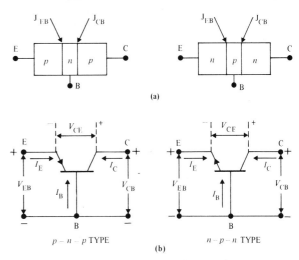

Fig. 4.1 The *p-n-p* and the *n-p-n* junction transistors

Note that in each case I_E, I_C and I_B representing the emitter current, collector current and base current, respectively, are considered positive when flowing into the transistor. Similarly, the voltages across the junctions have a sign convention expressed in fig. 4.1b.

Each junction behaves in a similar way to the *p-n* junction already described in Chapter 2. However, provided that the junctions are correctly biased (i.e. have small voltages of the correct polarity across them) additional and most useful behaviour results from the junctions being so close together, separated only by the thin base region.

To obtain transistor action, the emitter-base junction must be forward-biased and the collector-base junction reverse-biased. Consider the operation of an *n-p-n* silicon transistor. (That of a *p-n-p* type will become apparent, bearing in mind that the applied voltage will be of opposite polarity and the majority carriers will be of opposite sign). For forward bias to prevail across the emitter-base junction, the base must be made positive with respect to the emitter so that electrons — the majority carriers in the *n*-type emitter — will flow readily to the *p*-type base. This causes 'minority carrier injection' into the *p*-type base. (Note the word 'minority' is used here because, when they enter the *p*-type base, the electrons are minority carriers within the base). Movement of electrons from the emitter causes more electrons to move into it (from the external circuit) to maintain equilibrium.

Once in the base region, the electrons do not have far to travel by diffusion through the thin base before they are accelerated towards the positive collector electrode. During such transit by diffusion through the base, some of the electrons will combine with holes in the *p*-type base. This loss of holes will be compensated by more holes moving into the base which constitute the base current I_B.

As the collector is positive with respect to the base, the base is negative with respect to the collector: a reverse-bias exists across the collector-base junction. As the collector voltage is increased positively, the reverse voltage (as regards the inhibition of electron motion back to the base) across the collector-base junction will increase and so the depletion region will extend further into the *p*-type base. This effectively decreases the base width so the electrons moving towards the collector have a shorter distance to travel in the base and consequently the number of electrons lost in transit is reduced. Increasing positively the collector-voltage consequently causes the collector current to increase.

For a silicon *p-n* junction, very little current flows in the forward-biased junction until $V_{EB} \simeq 0.6V$ ($\simeq 0.2V$ for germanium) and then the current rises very rapidly (section 2.1). Because the input signal will always be applied via a forward-biased junction, the input resistance will be low. On the other hand, the output signal is obtained via a reverse-biased junction, the resistance of which is high. This transfer of signal from a low resistance to a high resistance is the origin of the word transistor or 'transfer-resistor'. This feature of resistance transfer identifies the device as an amplifier. Even if no current gain is achieved, i.e. the current flowing into the device is equal to that flowing out, a power gain is still achieved by virtue of the output resistance being higher than the input resistance.

4.3 Basic junction transistor connections

A junction transistor may be used to provide an amplifier stage in one of three ways: common-base (CB), common-emitter (CE) or common-collector (CC). The electrode stated to be common is that one which is common to both the input and the output circuit.

The characteristics of an *n-p-n* transistor in CB connection can be examined by means of the circuit of fig. 4.2. The static characteristic curves (fig. 4.3) show the steady state relationships among the input

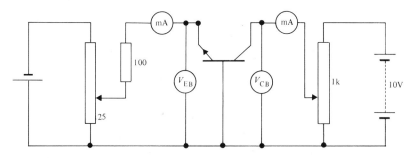

Fig. 4.2 Circuit for obtaining the static characteristics of an *n-p-n* junction transistor in CB connection

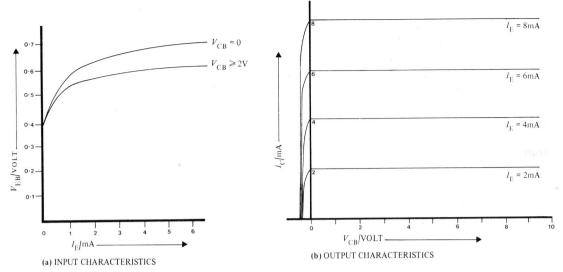

Fig. 4.3 The static characteristic curves of an *n-p-n* transistor in CB connection

and output currents and voltages for a silicon *n-p-n* junction transistor. Note that in the CB connection of fig. 4.2, the input signal is applied between the emitter and base while the output signal is obtained between the collector and base, i.e. the base is common to both input and output circuits.

Some special features of these characteristic curves are noted:

(a) *The input characteristics:* V_{EB} against I_E with V_{CB} as parameter. Very little current flows across the emitter-base junction, even though it is forward-biased, until $V_{EB} \simeq 0.6\text{V}$ (for silicon).

The gradient of the input characteristic $\left(\dfrac{\Delta V_{BE}}{\Delta I_E}\right)$ (taken at a specific value of I_E, say 4 mA) has the dimensions of resistance, is termed the input resistance of the transistor and is denoted by h_{ib}, where the subscript i refers to input and b to common-base connection. Lower case letters are used for the subscripts because the interest is in small alternating signal behaviour and the input resistance is an incremental resistance. Its value for a junction transistor in CB connection depends markedly on I_E but is low, of the order of 50Ω.

(b) *The output characteristics:* I_C against V_{CB} with I_E as parameter. The collector current I_C varies very little with the collector-base voltage V_{CB}: the characteristics are virtually straight lines parallel to the V_{CB} axis. The transistor is behaving as a constant current source. The slope of a line at a particular value of I_C, i.e. $\left(\dfrac{\Delta I_C}{\Delta V_{CB}}\right)$ is the output conductance of the transistor. This output conductance, denoted by h_{ob}, is very low. The reciprocal of h_{ob} is the output resistance. Although the characteristic curves obtained by experiment do not enable an accurate value for the output resistance to be obtained, it is very high, certainly $> 0.5\text{M}\Omega$. This very low input resistance and very high output resistance means that coupling between stages is difficult, i.e. it is not readily possible to make the output of a first stage the input to a second stage, so that CB connection is rarely used in amplifier design.

The *forward current transfer ratio* h_{fb}, also called the forward current gain, is defined by the relationship

$$h_{fb} = \dfrac{\Delta I_C}{\Delta I_E} \text{ at constant } V_{CB}.$$

The value of h_{fb} cannot be obtained with sufficient accuracy from the experimental curves of fig. 4.3b: they indicate a value of unity whereas its value must be slightly less than unity, perhaps 0.99.

The *current transfer characteristic* (I_C plotted against I_E for a fixed value of V_{CB}) is a straight line passing through the origin with a slope of nearly unity, $\left(\dfrac{\Delta I_C}{\Delta I_E}\right) = 1$

The value of the direct current gain, $h_{FE} = I_C/I_E$, must be slightly less than unity.

The behaviour of an *n-p-n* junction transistor in CE connection can be examined by use of the circuit of fig. 4.4. Note that the emitter is common to both the input and output circuits and that the input is via a forward-biased junction, J_{BE}, while the output is from a reverse-biased junction, J_{CB}. Typical static characteristic curves for an *n-p-n* silicon transistor are shown in fig. 4.5.

Some special features of these characteristics are noted:

(a) *The input characteristics* (V_{BE} against I_B with V_{CE} as parameter). Again, very little current flows across the forward-biased base-emitter silicon junction until V_{BE} exceeds about 0.6V. The gradient of the input characteristic $\left(\dfrac{\Delta V_{BE}}{\Delta I_B}\right)$ at a given value of I_B, say $20\mu A$, has the dimensions of resistance and is denoted by h_{ie}.

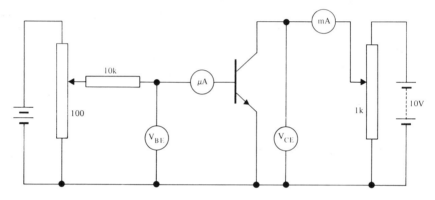

Fig. 4.4 Circuit for obtaining the static characteristics of an *n-p-n* junction transistor in CE connection

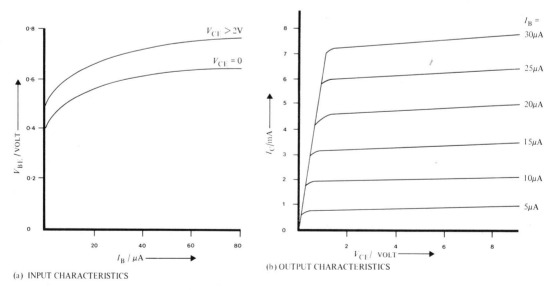

(a) INPUT CHARACTERISTICS

(b) OUTPUT CHARACTERISTICS

Fig. 4.5 The static characteristic curves of an *n-p-n* transistor in CE connection

The input resistance of a junction transistor in CE connection is typically a few thousand ohms.

(b) *The output characteristics* (I_C against V_{CE} with I_B as parameter) are uniformly spaced lines with a gradient $\frac{\Delta I_C}{\Delta V_{CE}}$ which is the output conductance of the junction transistor. The output resistance, $1/h_{oe}$, may be of the order of $10^5 \Omega$. The current gain h_{fe}, is defined as

$$h_{fe} = \frac{\Delta I_C}{\Delta I_B}$$

at some chosen point on the output characteristics. It can be obtained from the characteristics of fig. 4.5b. Values of h_{fe} vary between approximately 200 and 500 depending on the junction transistor and the value of I_C at which it is specified. The *dc current gain* h_{FE}, given by

$$h_{FE} = I_C / I_B$$

may be obtained from the transfer characteristic obtained by plotting I_C against I_B for a given value of V_{CE}.

4.4 The relationship between h_{FE} and h_{FB}

Assuming that the leakage current across a *p-n* junction is negligible — valid in the case of silicon — the relationship between h_{FE} and h_{FB} may be deduced by considering the conventional currents in a *p-n-p* junction transistor (fig. 4.6).

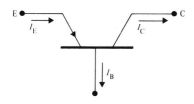

Fig. 4.6 Conventional currents in a *p-n-p* junction transistor

From Kirchhoff's first law

$$I_E = I_C + I_B$$

Substituting $I_E = I_C / h_{FB}$ gives

$$I_C / h_{FB} = I_C + I_B$$

Therefore $\dfrac{1}{h_{FB}} = 1 + 1/h_{FE}$

or $$h_{FE} = \frac{h_{FB}}{1 - h_{FB}} \qquad (4.1)$$

The value of h_{FB} is typically 0.995. Hence, from equation (4.1),

$$h_{FE} = \frac{0.995}{1 - 0.995} \simeq 200.$$

4.5 Leakage currents and thermal runaway

If the collector-base junction of a junction transistor is reverse-biased (the applied test voltage is usually 6V) and the emitter is open-circuit, a leakage current, denoted by I_{CBO}, will flow between collector and base. For a low power germanium transistor I_{CBO} may have a value of $4\mu A$ at $20°C$. Fig. 4.7a shows the

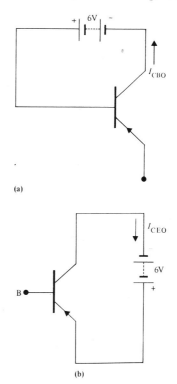

Fig. 4.7 Circuit configurations used to define (a) I_{CBO} and (b) I_{CEO} for a *p-n-p* junction transistor

circuit configuration used to define I_{CBO} for a *p-n-p* junction transistor.

With 6V applied between the collector and the emitter with the collector junction reverse-biased and the base open-circuit, a leakage current I_{CEO} flows in the collector lead (fig. 4.7b). The value of I_{CEO} may be 400µA for a germanium transistor at 20°C. This leakage current doubles approximately for a 10°C rise in temperature. The value of I_{CEO} for a silicon transistor will not exceed a few µA at 20°C and will scarcely change for a 10°C rise in temperature. Manufacturers usually quote the value of I_{CBO} at 20°C; from this value, that for I_{CEO} may be calculated.

The leakage currents are represented in fig. 4.8 from which one can readily determine the relationship between I_{CEO} and I_{CBO}. Thus, the total leakage current $I_{CEO} = (1 + h_{FE})I_{CBO}$

or $I_{CEO} \simeq h_{FE} I_{CBO}$

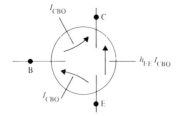

Fig. 4.8 Relationship between the leakage currents I_{CBO} and I_{CEO}

There are two effects of leakage current:

(a) A rise in temperature will increase the leakage current and shift the operating point, possibly causing distortion of the signal during amplification.

(b) The leakage current causes a temperature rise which results in increased leakage current and further temperature rise. This cumulative process, called *thermal runaway* can cause destruction of the transistor.

Leakage currents are particularly troublesome with germanium junction transistors, so care should be taken in the provision of adequate heat sinks and in the choice of a biasing circuit to ensure thermal stability.

With silicon devices, the leakage currents are usually negligible providing the devices are made to operate at reasonable temperatures.

In any laboratory concerned with semiconductor devices, it is essential to have available a unit with which one can check quickly whether a low-power junction transistor or a diode is useful or useless. For a junction transistor, answers to three basic questions are required: (i) are both junctions intact? (ii) what is the leakage current I_{CEO} at room temperature? and (iii) what is the current gain (h_{FE}) of the transistor?

A unit based on a simple circuit (fig. 4.9) is arranged to provide the answers immediately. It is such that the transistor cannot be damaged by plugging it in the wrong way round, the meter is protected and the power dissipated in the transistor can never damage the junctions. Much more complex transistor testers are available but this simple circuit can be readily constructed and is invaluable for making rapid checks on transistors before looking elsewhere for circuit faults.

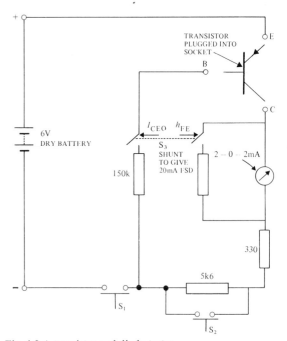

Fig. 4.9 A transistor and diode tester

The 6V dry battery is shown connected to test a *p-n-p* junction transistor. A toggle switch can be included to reverse the battery connections for testing *n-p-n* devices. A transistor socket can be readily arranged to accept TO 18, TO 5 or other transistor outlines (TO) with which one must become familiar. With a *p-n-p* transistor in the socket and switch S_3 in the position I_{CEO}, switch S_1 is closed (a press-button switch is ideal). If the meter reads 1mA or more, the transistor is useless because the E and C terminals must be short-circuited, i.e. the collector-base junction is damaged. If the collector-base junction is intact, only a small leakage current will be recorded when S_1 is pressed and the true leakage current I_{CEO} is measured when S_2 is closed as well. One would expect I_{CEO} for a germanium transistor to be less than 600µA at 20°C whereas for a silicon transistor it should be undetectable.

A finger placed on the metal can of the germanium transistor shows immediately the marked increase of I_{CEO} with temperature.

If the leakage test has shown that the transistor is satisfactory, switch S_3 is set to the position marked h_{FE}. The meter is now shunted to give a full-scale deflection of 20mA and current flows in the base of the transistor. The base current $I_B = 6/(150 \times 10^3)$A $= 40\mu$A, if one neglects the voltage drop across the forward-biased emitter-base junction. If S_1 and S_2 are then closed, the meter records I_C and h_{FE} can be calculated from I_C/I_B. A collector current I_C of 10mA will indicate a value of $h_{FE} = $ 10mA/40µA = 250.

A diode can be tested or its electrodes identified by connecting it between the terminals marked E and C, the test being carried out with switch S_3 in the position I_{CEO}.

4.6 Bias circuits and stability

For good transistor amplifier performance, the quiescent or operating point Q (fig. 4.10b) must be correctly chosen and maintained despite temperature fluctuations. We shall only consider the biasing and stability of a junction transistor in the most common amplifier configuration: namely, CE.

Selecting an operating point, i.e. selecting the steady current that flows through the transistor in the absence of an input signal, is not difficult: one merely requires to choose a point on the load-line (fig. 4.10b) which will provide the most linear performance. However, leakage currents are a problem because they are so temperature sensitive. The drift of the operating point which occurs with temperature change is very marked with high-leakage germanium transistors but very much less so with silicon. A biasing circuit should therefore be chosen which not only stabilises the working point but also prevents thermal runaway.

(a) *The fixed bias circuit.* A resistance R_B is connected between the supply voltage terminal $(-V_{CC})$ and the base (fig. 4.10a). The basic current I_B required to establish the best working

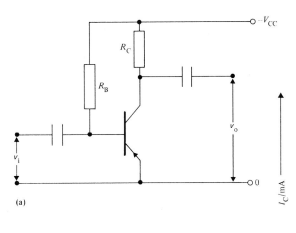
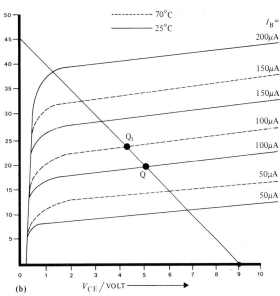

Fig. 4.10 (a) The fixed bias circuit (b) The load-line on the CE output characteristics which are full-line for 25°C and dashed-line for 70°C

point Q is known from the characteristics (fig. 4.10b). It is then a simple matter to calculate R_B because

$$I_B = \frac{V_{CC} - V_{EB}}{R_B} \quad (4.2)$$

where V_{EB} is 0.2V for a germanium transistor and 0.6V for a silicon one.

This circuit is thermally unstable: any increase in temperature which causes the leakage current I_{CEO} to increase will increase the collector current. Even if thermal runaway does not occur, distortion is likely at elevated temperatures.

The load-line given by the equation

$$V_{CE} = V_{CC} - I_C R_C$$

is drawn on the output characteristics of a germanium transistor in CE connection (fig. 4.10b). This load-line corresponds to $R_C = 200\Omega$ and $V_{CC} = 9V$. If the operating point is chosen at Q on the 25°C characteristic, it can be seen that the position of Q will drift with temperature, being located at Q_1 at 70°C. The collector current has changed as a result of this drift from $I_C = 19\text{mA}$ to $I_C = 23\text{mA}$. Thermal instability of this order must be avoided so the fixed bias circuit is only occasionally used with silicon devices.

(b) *Collector-to-base bias.* An improvement in stability is obtained if the resistor R_B is connected between the collector and base (fig. 4.11). If the collector current tends to increase, a greater voltage drop occurs across R_C and the base current I_B falls tending to bring I_C back to its original value. This is a case of dc negative feedback (*see* Appendix B). The value of R_B required is obtained from:

$$I_B = \frac{V_{CC} - I_C R_C - V_{EB}}{R_C + R_B} \quad (4.3)$$

A fundamental disadvantage of this circuit is that part of the alternating output signal which is 180° out-of-phase with the input signal is fed back via R_B to the input circuit. This ac negative feedback reduces the gain (*see* Appendix B) whereas the dc negative feedback improves the thermal stability.

(c) *Collector-to-base bias with the bias resistor decoupled.* The reduction in gain resulting from ac negative feedback may be eliminated by dividing R_B (fig. 4.11) into two equal resistors and connecting the junction between these resistors to earth via a capacitor which has negligible reactance at the signal frequency. Collector-to-base bias is often used, especially with silicon transistors in which the leakage currents are low.

(d) *Potential divider and emitter resistor stabilising bias.* The necessary steady voltage at the base is provided by a potential divider network consisting of resistors R_1 and R_2 (fig. 4.12). Furthermore, a resistor R_E is connected into the emitter lead. This self-bias or emitter bias is the

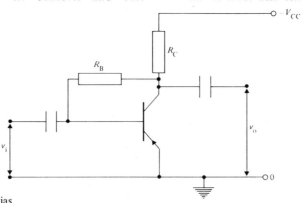

Fig. 4.11 Collector-to-base bias

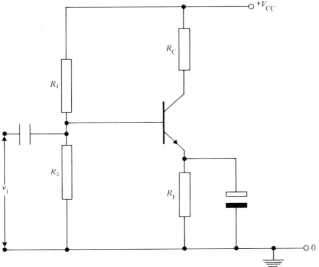

Fig. 4.12 A CE amplifier stage with self-biasing

most stable and the most frequently used biasing circuit. Again, the resistor R_E will provide ac negative feedback and a consequent reduction in gain. For this reason R_E is by-passed by a large capacitance which effectively short-circuits R_E to the ac signal but does not affect the dc stabilisation provided by R_E.

The dc equivalent circuit of the amplifier in fig. 4.12 is shown in fig. 4.13. This equivalent circuit is developed in two stages: fig. 4.13a shows the potential divider network replaced by V_{eq} in series with R_{BB} where $R_{BB} = R_1 R_2 /(R_1 + R_2)$, using Thévenin's theorem, and fig. 4.13b shows the transistor replaced by two constant current sources.

For the purpose of comparing the relative stabilities of various transistor circuits three stability factors are defined:

$$S = \frac{\partial I_C}{\partial I_{CBO}} \; ; \; M = \frac{\partial I_C}{\partial V_{BE}} \; \text{and} \; N = \frac{\partial I_C}{\partial h_{FE}}$$

Of these three, S is the most important and will be evaluated for the general circuit of fig. 4.12. From the equivalent circuit of fig. 4.13b certain relationships are apparent:

$$I_C = h_{FE} I_B + (h_{FE} + 1) I_{CBO} \quad (4.4)$$

$$V_{eq} - V_{BE} = R_{BB} I_B + R_E I_E \quad (4.5)$$

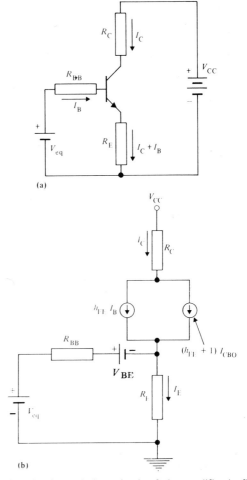

(a)

(b)

Fig. 4.13 The dc equivalent circuit of the amplifier in fig. 4.12

also, $I_E = I_C + I_B$ (4.6)

Differentiating equation (4.4) with respect to I_C, considering h_{FE} to be independent of I_C, gives

$$1 = \frac{1+h_{FE}}{S} + h_{FE} \frac{\partial I_B}{\partial I_C} \quad (4.7)$$

Re-writing equation (4.5), making use of equation (4.6),

$$V_{eq} = R_{BB} I_B + V_{BE} + (I_B + I_C) R_E \quad (4.8)$$

Differentiating equation (4.8) with respect to I_C and assuming that V_{BE} is constant, we have

$$\frac{\partial I_B}{\partial I_C} = \frac{-R_E}{R_E + R_{BB}} \quad (4.9)$$

Combining equations (4.7) and (4.9) gives

$$S = \frac{\partial I_C}{\partial I_{CBO}} = \frac{1 + R_{BB}/R_E}{1 + R_{BB}/[R_E(1+h_{FE})]} \quad (4.10)$$

Hence S varies between 1 for small values of R_{BB}/R_E and $(1 + h_{FE})$ as $R_{BB}/R_E \to \infty$. For high stability it is required that $R_E \gg R_1 R_2/(R_1 + R_2)$. The permissible value of S depends on both the transistor material and the application. Low leakage silicon transistor circuits tolerate a much higher value of S than comparable germanium transistor circuits.

4.7 Realistic approximations in circuit analysis

It is very valuable to be able to make realistic approximations which allow the operating point for a transistor circuit to be calculated quickly and allow the setting-up of a correct bias condition in the initial stages of circuit design. Device parameters are usually not known with sufficient accuracy to justify complex detailed analysis. The following simplifications will be applied:

(i) $I_E = I_C$.
(ii) $V_{BE} = 0.2V$ for germanium and 0.6V for silicon.
(iii) I_{CBO} will be neglected except where specifically required.
(iv) Any impedance in the emitter circuit is multiplied by $(1 + h_{FE})$ when viewed from the base terminal.
(v) Any impedance in the base circuit is divided by $(1 + h_{FE})$ when viewed from the emitter terminal.

Points (iv) and (v) greatly simplify circuit analysis and apply to ac circuits as well as to dc presuming that the appropriate value of h_{fe} is used. To justify this transformation of impedance from the emitter circuit to the base circuit and vice-versa, consider equation (4.5) relating to the equivalent base circuit of fig. 4.13b, namely,

$$V_{eq} - V_{BE} = R_{BB} I_B + R_E I_E$$

Writing

$$I_E = (1 + h_{FE}) I_B$$

$$V_{eq} - V_{BE} = I_B [R_{BB} + R_E (1 + h_{FE})] \quad (4.11)$$

Hence,

$$I_B = \frac{V_{eq} - V_{BE}}{R_{BB} + (1 + h_{FE}) R_E} \quad (4.12)$$

Equation (4.12) indicates that the emitter resistor R_E is effectively much larger when viewed from the base, its value being $(1 + h_{FE}) R_E$. By similar reasoning, a resistor in the base circuit is divided by $(1 + h_{FE})$ when viewed from the emitter terminal.

EXAMPLE 4.7a *Consider the amplifier stage of fig. 4.14a which utilises an n-p-n silicon transistor in CE connection. Calculate values for the resistors such that $I_C = 3mA$, $V_{CE} = 8V$, $V_E = 6V$ and $S = 10$.*

The Thévenin equivalent circuit is shown in fig. 4.14b in which the impedance transformation has also been made.

Because $I_E = I_C = 3mA$ and $V_E = 6V$,

$$R_E = 6/(3 \times 10^{-3}) = 2k\Omega.$$

Hence, the input resistance

$R_i = (1 + h_{FE}) R_E \simeq h_{FE} R_E = 200\text{k}\Omega$ as $h_{FE} = 100$.

If $V_{CE} = 8\text{V}$ and $V_E = 6\text{V}$, the voltage drop across R_C is $20 - 14 = 6\text{V}$ as the supply voltage $V_{CC} = 20$. Hence,

$R_C = 6/(3 \times 10^{-3}) = 2\text{k}\Omega$

From equation (4.10)

$$S = \frac{R_E + R_{BB}}{R_E + R_{BB}/(1 + h_{FE})} = \frac{2000 + R_{BB}}{2000 + R_{BB}/101} = 10$$

Hence,

$R_{BB} = 20\text{k}\Omega$

Referring to fig. 4.14b,

$V_B = V_E + 0.6\text{V} = 6.6\text{V}$

Hence,

$$V_{eq} = \frac{6 \times 20 \times 10^3}{200 \times 10^3} + 6.6 = 7.2\text{V}.$$

Now,

$$V_{eq} = 7.2\text{V} = \frac{V_{CC} R_2}{R_1 + R_2} = \frac{20 R_2}{R_1 + R_2}$$

But,

$$\frac{R_2}{R_1 + R_2} = \frac{7.2}{20} = \frac{R_{BB}}{R_1} = \frac{20 \times 10^3}{R_1}$$

Hence,

$$R_1 = \frac{20 \times 20 \times 10^3}{7.2} = 56\text{k}\Omega$$

Substituting, gives

$R_2 = 32\text{k}\Omega$.

EXAMPLE 4.7b *An n-p-n silicon junction transistor for which h_{FE} is 200 is used in a CE amplifier stage with a potential divider network (R_1 and R_2) and*

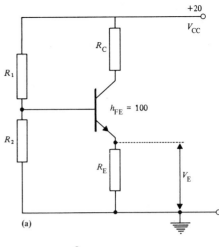

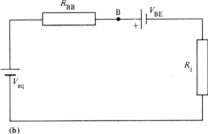

$V_{eq} = \dfrac{V_{CC} R_2}{R_1 + R_2}$; $R_{BB} = \dfrac{R_1 R_2}{R_1 + R_2}$

$V_{BE} = 0.6\text{V}$; $R_i = R_E (1 + h_{FE})$

Fig. 4.14 CE amplifier for example 4.7a

emitter resistor R_E to establish the dc bias conditions (fig. 4.15). Assuming that I_{BB}, the current flowing in the potential divider network, is greater than $10\,I_B$ where I_B is the base current, calculate the values of the direct currents I_E and I_B and the steady voltages V_B, V_C and V_{CE}.

$V_B = 20 \times 10/57 = 3.5\text{V}$.
For silicon, the junction voltage $V_{EB} = 0.6\text{V}$.
Hence,
$V_E = 3.5 - 0.6 = 2.9\text{V}$.
The emitter current I_E is therefore given by
$I_E = (2.9/10^3)\text{ A} = 2.9\text{mA}$.
As $I_E = I_C$, the steady voltage at the collector is
$V_C = 20 - 2.7 \times 2.9 = 12.2\text{V}$
Hence,
$V_{CE} = 12.2 - 2.9 = 9.3\text{V}$.
As $h_{FE} = 200$
$I_B = (2.9/200)\text{mA} = 14.5\mu\text{A}$.
Note that $I_{BB} = 20\text{A}/(57 \times 10^3) = 350\mu\text{A}$; the assumption that I_B is negligible compared with I_{BB} is therefore valid.

80 Electronics

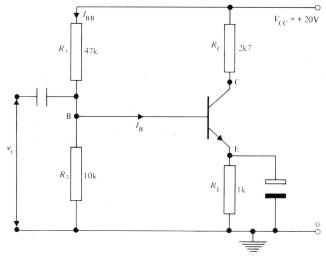

Fig. 4.15 CE amplifier stage for example 4.7b

EXAMPLE 4.7c *A silicon p-n-p transistor with $h_{FE} = 150$ is used in the CE amplifier stage of fig. 4.16. Determine the bias currents and voltages.*

This is an example of the transformation of impedance.

$$I_C = I_E = \frac{V_{CC} - V_{EB}}{R_E + R_B/(1 + h_{FE})}$$

(Note, the resistance R_B in the base circuit is divided by $(1 + h_{FE})$ when viewed from the emitter terminal).

$$I_E = \frac{12 - 0.6}{10^6/150 + 2.2 \times 10^3} = \frac{11.4}{8.9} = 1.3 \text{mA}$$

$V_E = 1.3 \times 2.2 = 2.9\text{V},$

$V_B = 2.9 + 0.6 = 3.5\text{V},$

$V_C = 12 - 2.2 \times 1.3 = 9.1\text{V}$

4.8 An equivalent circuit for a bipolar transistor: a small signal low-frequency model

To simplify the calculations required to examine the ac behaviour of a bipolar transistor, an equivalent circuit is used. In this equivalent circuit, no

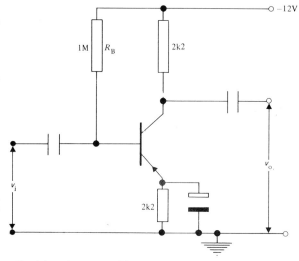

Fig. 4.16 The CE amplifier stage of example 4.7c

information concerning the dc bias level is included: it is assumed that the device is already biased in the active region, i.e. with the emitter-base (EB) junction forward-biased and the collector-base (CB) junction reverse-biased.

The transistor – an active non-linear device – may be replaced by resistive elements and a constant current source provided that calculations are restricted to deal with small variations about the operating point so that the characteristics may be assumed to be linear over this limited region. For this reason, this analysis does not include the operation of power transistors in the output stage of an amplifier.

Capacitance is rarely included in an equivalent circuit unless it is necessary to assess the effect it may have on behaviour over a specific frequency region.

Fig. 4.17 shows an electrical model of a bipolar transistor in CE connection which is valid at frequencies below 100kHz. Following conventional practice, lower case letters and subscripts are used for small signal parameters.

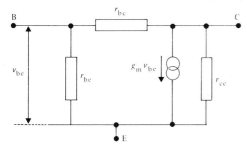

Fig. 4.17 Electrical model of a bipolar transistor in CE connection

Consider the features included within fig. 4.17 as well as those intentionally omitted:

r_{bc} is the resistance of the reverse-biased BC junction. Because r_{bc} exceeds $10^8 \Omega$, it will be neglected henceforth.

C_{bc}, the capacitance of the CB junction, is small and its effect is negligible at frequencies below 100kHz, so it is not included.

C_{be}, the sum of the diffusion and depletion region capacitances of the EB junction, where the diffusion component accounts for the stored charge in the base region, is small and has a negligible effect at frequencies below 100kHz so it is not included.

r_{be}, the input resistance, and r_{ce}, the output resistance may be related by consideration of the characteristic curves of the bipolar transistor in CE connection (fig. 4.5).

r_{be} is the dynamic resistance of the EB junction. When viewed from the emitter terminal, this dynamic resistance has a value of $26\text{mV}/I_E$. Transforming this into the base circuit:

$$r_{be} \equiv h_{ie} = (1 + h_{fe}) 26\text{mV}/I_E.$$

A small ac signal applied between base and emitter sees r_{be} (or better, h_{ie}) as the input resistance of the transistor.

r_{ce} is the equivalent resistance between the emitter and the collector terminals. An approximate value is obtained from the CE characteristics (fig. 4.5) where

$$r_{ce} = 1/h_{oe} = \left(\frac{\Delta V_{CE}}{\Delta I_C}\right)_{I_B}$$

The constant current $g_m v_{be}$ produced by the generator (the transistor) is defined by the equation

$$i_c = g_m v_{be} \qquad (4.13)$$

where g_m is usually expressed in mA/V.

Hence, we are able to draw (making use of Thévenin's theorem) the very useful equivalent circuit of fig. 4.18. As

$$g_m = i_c/v_{be} = h_{fe} i_b/v_{be} = h_{fe}/r_{be}$$

it follows that,

$$g_m = \frac{h_{fe} I_E}{(1 + h_{fe}) 26\text{mV}} \qquad (4.14)$$

and because h_{fe} is large compared with unity, we can write

$$g_m = 38 I_E \text{ mA V}^{-1} \qquad (4.15)$$

where I_E is the direct emitter current expressed in mA and 38 has the units V^{-1} (section 2.1).

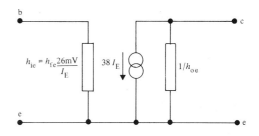

Fig. 4.18 An equivalent circuit for a bipolar transistor in CE connection

The output resistance, $1/h_{oe}$, for a small signal silicon planar transistor is of the order of $100\text{k}\Omega$ at $I_E = 1\text{mA}$; although it will be retained in the equivalent circuit, its effect will be shown to be negligible in most cases.

82 Electronics

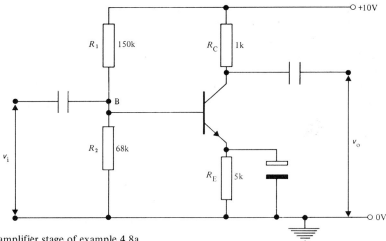

Fig. 4.19 The CE amplifier stage of example 4.8a

EXAMPLE 4.8a *An n-p-n silicon transistor with h_{fe} = 200 and $h_{oe} = 10^{-5}$ siemen is used in a CE amplifier stage (fig. 4.19). Draw an equivalent circuit for the amplifier stage, and calculate its input resistance and the voltage gain.*

Examining first the dc conditions in the amplifier stage of fig. 4.19. Assuming that $I_B \ll I_{BB}$, where I_{BB} is the current through the potential divider resistors R_1 and R_2,

$V_B = (10 \times 68/218)$ V = 3.1V.

The steady emitter voltage V_E is given by

$V_E = 3.1 - 0.6 = 2.5$V.

The steady emitter current I_E is therefore given by

$I_E = 2.5/(5 \times 10^3) = 0.5$ mA.

Hence,

$g_m = 38 I_E = 19$ mA V^{-1} as shown in the equivalent circuit of fig. 4.20.

The resistor R_C in this case of magnitude 1kΩ, is in parallel with the output resistance of 100kΩ (i.e. $1/h_{oe}$). The 100kΩ can therefore be ignored and the voltage gain

$$A_v = \frac{\text{output voltage}}{\text{input voltage}} = \frac{g_m \cdot v_i R_C}{v_i} = g_m R_C \quad (4.16)$$

$= 19 \times 10^{-3} \times 10^3 = 19.$

Note that the voltage gain is, in fact, numerically 38 times the dc voltage drop across R_C.

The input resistance of the transistor

$h_{ie} = h_{fe} \; 26\text{mV}/I_E = \dfrac{200 \times 26}{0.5} = 10.4\text{k}\Omega$

The equivalent circuit (fig. 4.20) shows the influence of R_1 and R_2 on the input resistance R_i of the stage, because

$$\frac{1}{R_i} = \frac{1}{68} + \frac{1}{150} + \frac{1}{h_{ie}} \quad (4.17)$$

$$= \frac{1}{68} + \frac{1}{150} + \frac{1}{10.4}$$

Therefore, $R_i = 8.5$ kΩ.

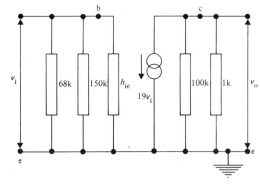

Fig. 4.20 The equivalent circuit of the amplifier stage in fig. 4.19

Example 4.8a illustrates the very convenient form of the equivalent circuit of fig. 4.18. A significant feature is that the term h_{fe} does not appear in the calculation of the voltage gain of the stage. Hence, *any* small signal transistor used in the amplifier circuit of fig. 4.19 would give a voltage gain of 19 *provided that the direct emitter current*, I_E *were the same*. Thus, if I_E is constant, the current gain, h_{fe}, of the transistor affects the input resistance of the stage but does not affect the voltage gain.

Before examining further equivalent circuits (the behaviour of a transistor can be simulated in a variety of ways, so resulting in a number of equivalent circuits) the validity of the assumptions made to arrive at the equivalent circuit of fig. 4.18 needs examination experimentally. Three relationships are of particular interest:

(i) $h_{ie} = (h_{fe}\ 26mV/I_E)\Omega$ *(4.18)*

(ii) $A_v = 38\ I_C R_C$ *(4.19)*

(iii) A_v is independent of h_{fe} provided that I_E is constant.

The circuit of fig. 4.21 has been used to investigate the small signal behaviour of an *n-p-n* silicon transistor in CE connection. The high quality (1%) 10kΩ resistor in the base lead enables the alternating base current i_b to be determined because

$$i_b = (v_i - v_2)/10^4$$

The direct emitter current I_E is selected by means of the variable resistor R_E in the emitter lead and measured by means of the milliammeter in the collector lead ($I_E \triangleq I_C$). The milliammeter is decoupled by the 100μF capacitor so its resistance does not add to R_C which is a high quality resistor of value 1.0kΩ to within 1%. With a sinusoidal input signal of frequency 1kHz both the 100μF capacitor and the 250μF capacitor connected across R_E may be considered as effective short-circuits for ac. For convenience, the input signal is arranged so that v_1 = 5.0mV and v_1, v_2 and v_3 are measured by means of an ac millivoltmeter. This experiment is repeated for each value of I_E selected. These observations allow the small signal behaviour of the transistor to be determined because

$$i_b = (v_1 - v_2)/10^4\ ;\ i_c = v_3/10^3,\ h_{fe} = i_c/i_b\ \text{and}$$

$$h_{ie} = v_2/i_b \quad (4.20)$$

and, finally, the voltage gain A_v of the stage is given by

$$A_v = v_3/v_1.$$

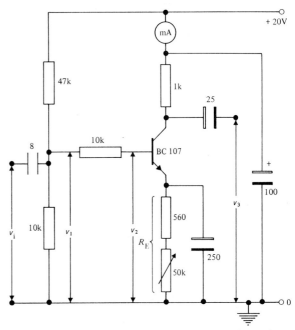

Fig. 4.21 Investigation of the small signal behaviour of an *n-p-n* silicon transistor in CE connection

Typical values obtained with a BC 107 transistor (an *n-p-n* silicon type) are recorded in table 4.1. With an emitter current I_E = 0.4mA, the BC 107 has a current gain h_{fe} of 200 and the amplifier provided a voltage gain of 15. When the BC 107 was replaced by a BC 109, another *n-p-n* silicon transistor with a current gain h_{fe} of 640, at the same emitter current (I_E = 0.4mA), the amplifier provided a voltage gain of 14. However, the input resistance (h_{ie}) of the BC 109 had a value of 46kΩ as compared with 14kΩ for the BC 107. Hence, the behaviour is exactly as indicated by the equivalent circuit of fig. 4.18 and stated in equation 4.19.

The experimental values recorded in table 4.1 were used to verify the two important relationships (4.19) and (4.18), namely,

$$A_v = 38\ I_C R_C$$

and

$$h_{ie} = h_{fe}\ 26mV/I_E$$

Table 4.1

Typical results for the BC 107 in the circuit of fig. 4.21
(signal frequency = 1.0 kHz)

I_E (mA)	v_1 (mV)	v_2 (mV)	v_3 (mV)	$i_b = \dfrac{v_1 - v_2}{10^4}$ µA	$i_c = v_3/10^3$ µA	$h_{fe} = i_c/i_b$	$h_{ie} = v_2/i_b$ kΩ	$A_v = v_3/v_2$
0.2	5.0	3.6	26	0.14	26	185	26	7.2
0.4	5.0	2.9	42	0.21	42	200	14	15
0.6	5.0	2.5	55	0.25	55	220	10	22
0.8	5.0	2.2	64	0.28	64	230	7.9	29
1.0	5.0	2.0	70	0.30	70	235	6.7	35
2.0	5.0	1.4	93	0.36	93	260	3.9	66
3.0	5.0	1.1	104	0.39	104	270	2.6	95
BC 109 in place of the BC 107								
0.4	5.0	4.1	58	0.09	58	640	46	14

The first of these relates the voltage gain A_v of the amplifier to the dc voltage drop across R_C, remembering that $I_C = I_E$. The second relates the input resistance h_{ie} to the current gain h_{fe} of the transistor for a constant value of I_E. The graph A_v against $I_C R_C$ (fig. 4.22a) is a straight line passing through the origin and of slope $37 V^{-1}$, whereas the graph of h_{ie} against h_{fe}/I_E (fig. 4.22b) is a straight line with a very small intercept on the h_{ie} axis and a slope of 27mV. Experiment therefore shows that the simple theory adequately represents the behaviour of the bipolar transistor over this operating region.

4.9 The hybrid or *h* parameters

The second equivalent circuit to be considered makes use of the hybrid or *h* parameters of the bipolar transistor. Again, the analysis is limited to frequencies below 100kHz.

These *h* parameters have already been mentioned with reference to the transistor characteristic curves and used in the previous transistor model. It is convenient to represent the behaviour of a transistor with these *h* parameters because it is easy to measure them, they lead to simple circuit analysis and are quoted by manufacturers for each type of transistor marketed.

Consider the CE connection of fig. 4.23. Let i_B, i_C, v_B and v_C represent the instantaneous voltages and currents.

Writing
$$v_B = f_1(i_B, v_C)$$
because v_B is some function f_1 of i_B and v_C, also
$$i_C = f_2(i_B, v_C)$$
where i_C is some function f_2 of i_B and v_C, from Taylor's expansion,

$$\Delta v_B = \left(\frac{\partial v_B}{\partial i_B}\right)_{V_C} \Delta i_B + \left(\frac{\partial v_B}{\partial v_C}\right)_{I_B} \Delta v_C$$

(4.21)

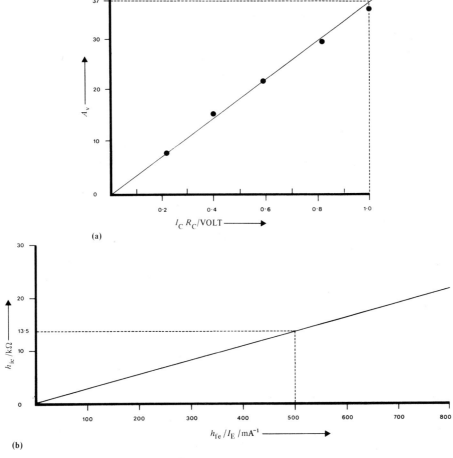

Fig. 4.22 (a) Graph of A_v against $I_C R_C$; slope = $37V^{-1}$ (b) Graph of h_{ie} against h_{fe}/I_E; slope = 27mV

and

$$\Delta i_C = \left(\frac{\partial i_C}{\partial i_B}\right)_{V_C} \Delta i_B + \left(\frac{\partial i_C}{\partial v_C}\right)_{I_B} \Delta v_C$$

(4.22)

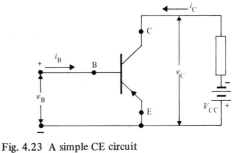

Fig. 4.23 A simple CE circuit

These equations (4.21) and (4.22) can be written in the forms

$$v_b = h_{ie} i_b + h_{re} v_c \qquad (4.23)$$

and

$$i_c = h_{fe} i_b + h_{oe} v_c \qquad (4.24)$$

where

$$h_{ie} = \left(\frac{\partial v_B}{\partial i_B}\right)_{V_C} = \text{input resistance (ohms)}$$

and

$$h_{re} = \left(\frac{\partial v_B}{\partial v_C}\right)_{I_B}$$ = reverse open-circuit voltage amplification (dimensionless)

$$h_{fe} = \left(\frac{\partial i_C}{\partial i_B}\right)_{V_C}$$ = forward current gain (dimensionless)

and

$$h_{oe} = \left(\frac{\partial i_C}{\partial v_C}\right)_{I_B}$$ = output conductance (siemen).

Remember that the lower case letters refer to small signal currents and voltages. Subscript i refers to the input while e designates CE connection. The parameters are termed 'hybrid' because they are not all alike dimensionally.

Note the limitations expressed by equations (4.23) and (4.24): thus, h_{ie} is the input resistance to ac when V_C is constant, i.e. when the output is short-circuited to ac. Similar limitations apply to the other parameters.

Equations (4.23) and (4.24) lead immediately to the hybrid model or equivalent circuit for the bipolar transistor. Fig. 4.24 shows the three transistor configurations, their hybrid models and the corresponding parametric equations. These circuits and equations are valid for either the n-p-n or p-n-p transistors and are independent of the type of load or the method of providing the dc bias.

Each equivalent circuit represents diagrammatically the behaviour of the transistor, expressed mathematically by the parametric equations. This is seen to be the case if Kirchhoff's current and voltage laws are applied to the input and output of each equivalent circuit.

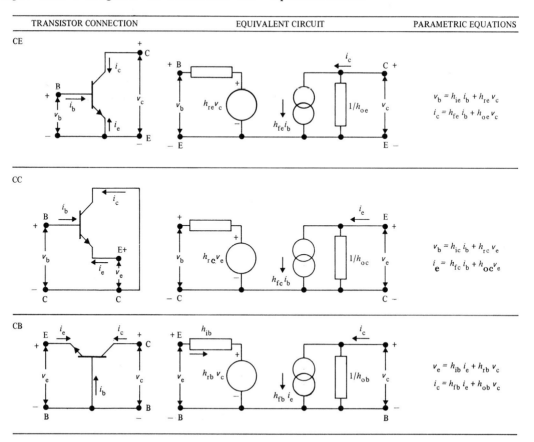

Fig. 4.24 Bipolar transistor configurations and their hybrid equivalent circuits and parametric equations

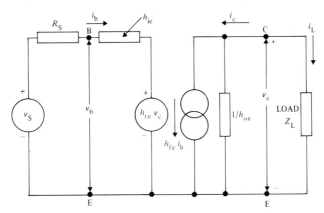

Fig. 4.25 Hybrid equivalent circuit of the basic CE amplifier

4.10 The basic CE amplifier

Assuming that a transistor in CE connection is correctly biased and that small sinusoidal voltage signals are applied from a source of voltage v_S of internal resistance R_S, the behaviour of the amplifier can be represented and analysed by means of the equivalent circuit of fig. 4.25. The quantities of interest are the current gain A_i, the input impedance Z_i, the voltage gain A_v and the output impedance Z_o.

Current gain, A_i

$$A_i = i_L / i_b = -i_c / i_b \qquad (4.25)$$

But,

$$i_c = h_{fe} i_b + h_{oe} v_c$$

and

$$v_c = -i_c Z_L$$

Hence,

$$i_c = h_{fe} i_b - h_{oe} i_c Z_L$$

or

$$i_c (1 + h_{oe} Z_L) = h_{fe} i_b$$

Therefore,

$$A_i = -h_{fe} / (1 + h_{oe} Z_L) \qquad (4.26)$$

Input impedance, Z_i

$$Z_i = v_b / i_b \qquad (4.27)$$

But,

$$v_b = h_{ie} i_b + h_{re} v_c$$

and

$$v_c = -i_c Z_L = A_i i_b Z_L$$

Hence,

$$v_b = h_{ie} i_b + h_{re} A_i i_b Z_L$$

Therefore,

$$Z_i = h_{ie} + h_{re} A_i Z_L \qquad (4.28)$$

It is an important feature that the input impedance is a function of the load impedance.

Voltage gain, A_v

$$A_v = v_c / v_b \qquad (4.29)$$

As,

$$V_c = A_i i_b Z_L$$

$$A_v = A_i Z_L / Z_i \qquad (4.30)$$

Output admittance, $Y_o = 1/Z_o$

From the right-hand network (fig. 4.25)

$$Y_o = h_{oe} - h_{fe} i_b / v_c$$

The input network gives

$$i_b = h_{re} v_c / (h_{ie} + R_S)$$

Hence,

$$Y_o = h_{oe} - h_{fe} h_{re} / (h_{ie} + R_S) \quad (4.31)$$

Note that the output admittance is a function of the source resistance. When the transistor is driven from a high impedance source, R_S tends to infinity and $Y_o = h_{oe}$.

The effect of the source internal resistance R_S on the voltage gain. Because the source provides the alternating base current i_b, the source resistance can influence the voltage gain. Fig. 4.26, representing the input circuit, shows that some voltage is dropped across R_S and so the true voltage gain A_{vs} is lower than that calculated by means of equation (4.30).

$$A_{vs} = v_c / v_s = \frac{v_c v_b}{v_b v_s} \quad (4.32)$$

But,

$$v_b = v_s Z_i / (Z_i + R_S)$$

Hence,

$$A_{vs} = A_v Z_i / (Z_i + R_S)$$
$$= A_i Z_L / (Z_i + R_S) \quad (4.33)$$

If $R_S \to 0$

$$A_{vs} = A_v$$

Similarly, the current gain A_i obtained when a source resistance R_S is present may be shown to be

$$A_{is} = A_i R_S / (Z_i + R_S) \quad (4.34)$$

If $R_S \to \infty$,

$$A_{is} = A_i.$$

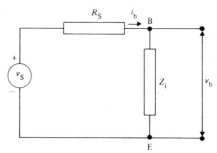

Fig. 4.26 Input circuit of the CE amplifier

The circuit analysis carried out for the CE connection can be readily applied to other configurations by using the equivalent circuits shown in fig. 4.24. Of special importance is the fact that the expressions for A_v, A_{vs} and A_{is} do not contain hybrid parameters. The relationships derived from these quantities are consequently valid whatever equivalent circuit is used, and even at high frequencies where the h parameters become frequency-dependent.

4.11 Realistic approximations to the CE amplifier model

The detailed expressions derived in section 4.10 for the current gain, voltage gain, input and output impedances are valuable aids to the understanding of transistor behaviour. In practical cases, nevertheless, they are rarely used in these forms. The hybrid parameters of a device are only known with sufficient accuracy to make such detailed calculations worthwhile when they have been measured under the conditions in which the device will operate. Manufacturers quote a range of values within which fall the values of the parameters for their devices. As an example, the current gain h_{fe} at $I_C = 1\text{mA}$ of a particular silicon *n-p-n* transistor may be quoted to be within the range 250 to 450.

Clearly a number of approximations may be made which enable the values of A_i, A_v, Z_i and Z_o to be obtained quickly without sacrificing accuracy. The hybrid equivalent circuit of a CE amplifier (fig. 4.25) will be simplified in the following ways:

(a) The load on a CE amplifier is usually resistive and of magnitude between perhaps 1kΩ and 8kΩ. The resistive load R_L could be a resistance R_C in series in the collector lead or it could be R_C in parallel with some coupling components. The load R_L is in parallel with the output resistance of the transistor $(1/h_{oe})$. As $(1/h_{oe})$ has a value of the order of $10^5 \Omega$, whenever $R_L h_{oe} \leqslant 0.1$, h_{oe} will be neglected.

(b) If h_{oe} is omitted, $i_c = h_{fe} i_b$ and the voltage generator is of magnitude

$$h_{re} v_c = h_{re} i_c R_L = h_{re} h_{fe} R_L i_b$$

Typical values would be $h_{re} = 10^{-4}$ and $h_{fe} = 200$. Hence, if R_L is not very large, the voltage generator is negligible compared with $h_{ie} i_b$. The simplified equivalent circuit for the CE amplifier therefore takes the form of fig. 4.27.

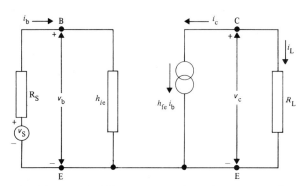

Fig. 4.27 The simplified equivalent circuit for a CE amplifier

From the equivalent circuit of fig. 4.27, the quantities of interest are determined.

Current gain, A_i

$$A_i = i_L / i_b = -i_c / i_b = -h_{fe} i_b / i_b = -h_{fe}.$$

Thus,

$$A_i = -h_{fe}.$$

Input impedance, Z_i

$$Z_i = v_b / i_b = h_{ie} i_b / i_b = h_{ie}$$

Thus,

$$Z_i = h_{ie}.$$

Voltage gain A_v

$$A_v = v_c / v_b = -i_c R_L / (i_b h_{ie}) = -h_{fe} R_L / h_{ie}$$

Thus,

$$A_v = -h_{fe} R_L / h_{ie}.$$

Output impedance, Z_o. Although the simplified circuit of fig. 4.27 has an infinite output resistance, the actual value depends on the source internal resistance R_S. Thus, Z_o normally has a value of approximately 60kΩ. As the load impedance Z_L is in parallel with this output impedance, Z_o has an effective value of Z_L.

The relationships derived for the CE amplifier are summarised in table 4.2.

EXAMPLE 4.11a *A CE amplifier stage is based on an n-p-n silicon transistor with the following parameters: $h_{ie} = 4k\Omega$, $h_{re} = 10^{-4}$, $h_{fe} = 200$ and $h_{oe} = 2.0 \times 10^{-5}$ siemen. The load resistor has a value of 1.0kΩ. If the source resistance is negligible, calculate values for (a) the current gain, (b) the input resistance (c) the voltage gain, (d) the output resistance and (e) the power gain achieved. What errors would be involved if the simplified form of the equivalent circuit were used?*

(a) Current gain $A_i = -h_{fe}/(1 + h_{oe} R_L)$
$= -200/(1 + 2 \times 10^{-5} \times 10^3) = -196$

(b) Input resistance $R_i = h_{ie} + h_{re} Z_L A_i$

$= 4000 - 10^{-4} \times 10^3 \times 196 = 3980 \Omega$

(c) Voltage gain $A_v = A_i R_L / R_i = -196 \times 10^3 / (3.98 \times 10^3)$
$= -49$

(d) Output impedance $Z_o = 1/Y_o$

$Y_o = h_{oe} - h_{fe} h_{re} / h_{ie} = 2 \times 10^{-5} - 200 \times 10^{-4} / 4000$
$= 1.5 \times 10^{-5}$

Table 4.2

CE amplifier relationships utilising h parameters

Quantity	Relationship from the detailed equivalent circuit	Approximate form
A_i	$-h_{fe}/(1 + h_{oe} Z_L)$	$-h_{fe}$
Z_i	$h_{ie} + h_{re} Z_L A_i$	h_{ie}
A_v	$A_i Z_L / Z_i$	$-h_{fe} Z_L / h_{ie}$
Y_o	$h_{oe} - \dfrac{h_{fe} h_{re}}{h_{ie} + R_S}$	$1/Z_L$

so $Z_o = 1/(1.5 \times 10^{-5}) = 66.7\text{k}\Omega$

This output impedance is in parallel with $R_L = 1\text{k}\Omega$, so the effective $Z_o = R_o = 1\text{k}\Omega$

(e) Power gain $= A_v A_i = 49 \times 196 = 9.6 \times 10^3$

With a simplified model,

$A_i = -h_{fe} = -200$

$R_i = h_{ie} = 4000\Omega$

$A_v = h_{fe} R_L / h_{ie} = -200 \times 10^3/4000 = -50$

Output resistance $= R_L = 1\text{k}\Omega$

In this particular example, the error involved in using the simplified CE model never exceeds 2%.

EXAMPLE 4.11b *A silicon n-p-n transistor with $h_{fe} = 150$ is used in the CE amplifier circuit of fig. 4.28. The output circuit includes a coupling capacitor C_1. Derive the dc conditions in this amplifier, draw an equivalent circuit as regards ac and determine values for A_v and A_{vs}.*

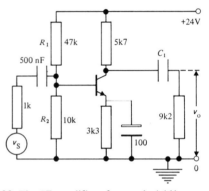

Fig. 4.28 The CE amplifier of example 4.11b

With regard to the dc bias conditions, assume $I_{BB} \gg I_B$.

$V_B = 24 \times 10/(10 + 47) = 4.2\text{V}$

$V_E = V_B - V_{BE} = 4.2 - 0.6 = 3.6\text{V}$

$I_C = I_E = 3.6/3.3 = 1.1\text{mA}$

$V_C = 24 - 5.7 \times 1.1 = 17.7\text{V}$

$I_{BB} = 24/(57 \times 10^3) = 420\mu\text{A}$

$I_B = 420/150 = 2.8\mu\text{A}$

Hence, the assumption that $I_{BB} \gg I_B$ is valid.

Concerning the ac behaviour,

$$h_{ie} = \frac{(h_{fe}+1)\,26\text{mV}}{1.1} = 3.6\text{k}\Omega$$

The effective load resistance R_L is given by assuming that the impedance of the coupling capacitor C_1 is negligible at the frequencies considered. Then,

$$\frac{1}{R_L} = \frac{1}{5.7} + \frac{1}{9.2}$$

Hence,

$$R_L = 3.5\text{k}\Omega$$

Ignoring the source resistance R_S, the voltage gain is

$$A_v = -h_{fe}\,R_L/h_{ie} = \frac{-150 \times 3.5}{3.6} = -146.$$

The equivalent circuit for ac is given in fig. 4.29.

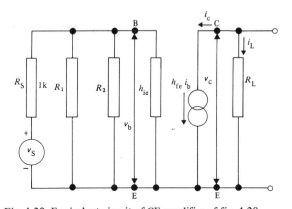

Fig. 4.29 Equivalent circuit of CE amplifier of fig. 4.28

The resistors R_1, R_2 and h_{ie} can be combined and written R_i where

$$\frac{1}{R_i} = \frac{1}{47} + \frac{1}{10} + \frac{1}{3.6}$$

Hence,

$$R_i = 2.5\text{k}\Omega$$

From equation (4.33)

$$A_{vs} = A_v\,R_i/(R_i + R_S)$$

Hence,

$$A_{vs} = -\frac{146 \times 2.5}{2.5+1} = -146 \times 2.5/3.5$$

$$= -104.$$

EXAMPLE 4.11c *An n-p-n silicon transistor used in a CE amplifier has the following parameters at the operating point: $h_{ie} = 2200\,\Omega$, $h_{re} = 2 \times 10^{-4}$, $h_{fe} = 290$ and $h_{oe} = 30 \times 10^{-6}$ siemen. If the load resistance $R_L = 3\text{k}\Omega$ and the source internal resistance $R_s = 500\,\Omega$, determine the current gain, the voltage gain and the input and output impedances. Compare the results obtained with those which are based on the simple CE model*

Current gain $A_i = -h_{fe}/(1 + h_{oe}R_L)$

$$= -290/(1 + 30 \times 10^{-6} \times 3 \times 10^3)$$

$$= -266$$

Input impedance $Z_i = h_{ie} + h_{re}R_L A_i$

$$= 2200 - 2 \times 10^{-4} \times 3 \times 10^3 \times 266$$

$$= 2040\,\Omega$$

Output impedance $Z_o = \dfrac{1}{h_{oe} - \dfrac{h_{fe}\,h_{re}}{h_{ie}+R_S}}$

$$= \frac{1}{30 \times 10^{-6} - \dfrac{290 \times 2 \times 10^{-4}}{2200+500}}$$

$$= 118\text{k}\Omega$$

This is the output resistance of the transistor and is in parallel with the load resistor of value $3\text{k}\Omega$.

Voltage gain $A_v = A_i R_L/Z_i$

$$= -266 \times 3 \times 10^3 / 2040 = 391$$

Insufficient information is available to calculate A_{vs} for this amplifier. Z_i represents the input resistance to the transistor but the effect of the biasing network, which will really determine the input resistance to the stage, is unknown.

Making use of the simplified CE model

$A_i = -h_{fe} = -290$

$Z_i = h_{ie} = 2200 \, \Omega$

$A_v = -h_{fe} R_L / h_{ie} = -290 \times 3 \times 10^3 / 2200$

$= -395$

The effective output resistance of the stage is the load resistance $R_L = 3 \, \text{k}\Omega$.

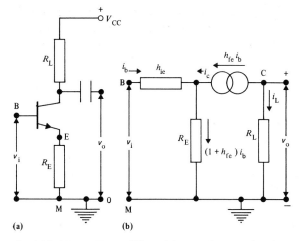

Fig. 4.30 (a) A CE amplifier with an unby-passed emitter resistor; (b) the equivalent circuit

4.12 A CE amplifier in which the emitter resistor is not by-passed

In the CE amplifiers already discussed the resistance R_E in the emitter lead, which contributed to the dc bias condition and also provided thermal stability, was always by-passed by a capacitor of large capacitance offering a negligible impedance at the operating frequency. Removing this capacitor reduces greatly the voltage gain of the stage, because negative feedback (see Appendix B) is introduced. In spite of this, an emitter resistor R_E without a by-pass capacitor, is often used. A number of beneficial effects result from such a circuit which more than compensate for the voltage gain reduction produced by ac negative feedback. Fig. 4.30a shows the circuit of a CE amplifier of this form and fig. 4.30b is the simplified equivalent circuit.

In this simplified equivalent CE circuit (fig. 4.30 b), it is again assumed that $R_L h_{oe} \leq 0.1$ so that h_{oe} can be neglected, and also that the voltage generator $h_{re} v_c$ is small compared with $h_{ie} i_b$ and is neglected.

from fig. 4.30b

Current gain $\quad A_i = i_L / i_b = -i_c / i_b = -h_{fe} i_b / i_b$

$\qquad\qquad\qquad = -h_{fe} \qquad\qquad (4.35)$

Hence, the current gain is unaffected by the presence of R_E.

Input resistance $R_i = v_i / v_b = h_{ie} + (1 + h_{fe}) R_E \quad (4.36)$

In this case, the input resistance is increased by the term $(1 + h_{fe}) R_E$.

With $R_E = 1 \, \text{k}\Omega$ and $h_{fe} = 250$,

$R_i = (h_{ie} + 251) \, \text{k}\Omega$

This input resistance is very large compared with h_{ie} which would have a value of a few $\text{k}\Omega$. Thus, the emitter resistance increases greatly the input resistance.

Voltage gain $A_v = A_i R_L / R_i = -h_{fe} R_L /$

$\qquad\qquad [h_{ie} + (1 + h_{fe}) R_E] \qquad (4.37)$

The presence of R_E greatly reduces the voltage gain. As $(1 + h_{fe}) R_E \gg h_{ie}$ and also $h_{fe} \gg 1$, we can write equation (4.37)

$A_v = -h_{fe} R_L / h_{fe} R_E = -R_L / R_E \qquad (4.38)$

Equation (4.38) is particularly important because – subject to the approximations made – we have an amplifier of which the voltage gain is independent of the parameters of the transistor. If R_L and R_E were stable resistors, the same voltage gain would be achieved even if a transistor with a low current gain h_{fe} were substituted for one with a high current gain.

This voltage gain stability is invaluable: the magnitude of the voltage gain is readily sacrificed because it can be made up simply by adding a further stage.

The output resistance of the stage is R_L, as in the conventional case with a by-pass capacitor across R_E.

EXAMPLE 4.12a *The circuit of fig. 4.31 was set up to investigate the effect of using a resistor R_E in the emitter lead with no by-pass capacitor. The direct emitter current I_E was 2.8mA. A sinusoidal signal generator provided the input signal at a frequency of 1kHz and, by means of an ac millivoltmeter, the voltages v_1, v_2 and v_3 were determined to be respectively 5.0mV, 4.8mV and 4.8mV. Calculate from these measurements the values of the base current, the collector current, the current gain h_{fe}, the voltage gain, and the input resistance of the transistor.*

The alternating base current $i_b = (v_1 - v_2)/10^4 = 0.02\mu A$.
The alternating collector current $i_c = v_3/10^3 = 4.8\mu A$.
The current gain $h_{fe} = i_c/i_b = 4.8/0.02 = 240$.
The input resistance of the transistor = $v_2/i_b = 4.8 \times 10^{-3}/(0.02 \times 10^{-6}) = 240k\Omega$.
The voltage gain $A_v = v_3/v_2 = 4.8/4.8 = 1$.

Using these calculated values we can examine how well the values deduced by means of the simplified model agree with those determined experimentally.

Theoretically, the input resistance of the transistor
$$= h_{ie} + (1 + h_{fe})R_E$$
$$\simeq (1 + h_{fe})R_E = 241k\Omega$$

This agrees well with the experimentally determined value of 240kΩ.
The voltage gain is given theoretically by
$$A_v = R_L/R_E = -1$$
and the experimentally determined value for A_v is 1 and, indeed, the signal is shown to be inverted by the use of a cathode ray oscillograph.

One last feature is noteworthy. The input resistance of the transistor is 240kΩ but the input resistance of the amplifier stage is much lower and determined, in fact, by the potential divider network R_1 and R_2. The input resistance R_i of the stage is given by

$$\frac{1}{R_i} = \frac{1}{47} + \frac{1}{10} + \frac{1}{240}$$

so, $R_i \simeq 8k\Omega$

EXAMPLE 4.12b. *A four-stage transistor amplifier is designed so that each stage produces a voltage gain of 10. If the gain is not stabilised by negative feedback, the spread of the parameter values of the transistors could cause a 10% reduction in the gain of each stage. Calculate the overall voltage gain of this amplifier in the worst case.*

Intended gain = 10^4 = 80dB
Voltage gain in worst case = 9^4 = 6.6 x 10^3 = 76dB
The gain in this case is only 66% of the intended value.

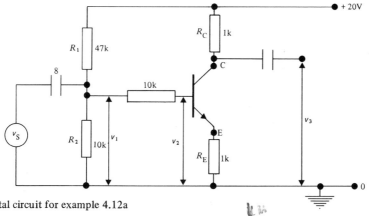

Fig. 4.31 Experimental circuit for example 4.12a

4.13 The paraphase amplifier

In the CE amplifier of fig. 4.31, R_E is unby-passed and $R_C = R_E = 1\text{k}\Omega$. If the output signal from the emitter E is examined simultaneously with that from the collector C (by means of a double-beam oscilloscope) one observes (a) that the two signals are equal in amplitude (this is not surprising because $i_e = i_c$ and $R_E = R_C$) and (b) that the two signals are 180° out-of-phase.

Hence, from a single input signal voltage v_2, the amplifier will provide two outputs of equal amplitude: one from the collector which is 180° out-of-phase with v_2 and one from the emitter which is in-phase with v_2.

A CE amplifier operating in this way (with $R_E = R_C$) is called a 'phase-splitter' or a paraphase amplifier. It is of particular value at the output stage of an amplifier when two power transistors of the same type are used to drive a loudspeaker. Each transistor provides one-half of the input waveform.

4.14 The CC amplifier or emitter-follower

The circuit of a common-collector (CC) amplifier with the load resistor R_L connected between the emitter and earth is shown in fig. 4.32. This configuration is called an emitter-follower and is similar in operation and behaviour to the source-follower discussed in section 3.12. As the name suggests, the emitter will follow any voltage signal applied at the input. The special features of this circuit are:

(i) a voltage gain less than but close to unity;
(ii) an output signal in-phase with the input;
(iii) a high input resistance: several hundred kilohm is typical;
(iv) a low output resistance: typically of the order of 30Ω.

From Fig. 4.33,

Current gain $A_i = i_L/i_b = -i_e/i_b = (1 + h_{fe})i_b/i_b$

i.e. $A_i = 1 + h_{fe}$ \hfill (4.39)

Input resistance $R_i = v_b/i_b = h_{ie} = (1 + h_{fe})R_L$ \hfill (4.40)

Though h_{ie} may be a few kilohm, R_i is very large. Thus, if $h_{fe} = 250$ and $R_L = 1\text{k}\Omega$, $R_i \simeq 250\text{k}\Omega$.

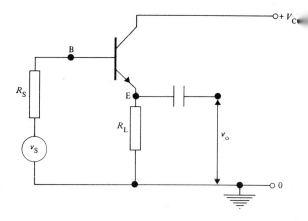

Fig. 4.32 A CC amplifier or emitter-follower (The base-biasing network is not shown)

Fig. 4.33 shows the simplified hybrid model for the CC hybrid amplifier.

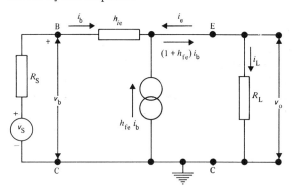

Fig. 4.33 Simple equivalent circuit for a CC amplifier

Voltage gain $A_v = i_L R_L/v_b = -i_e R_L/v_b$

$= (1 + h_{fe})i_b R_L/v_b$

$= \dfrac{(1 + h_{fe})R_L}{R_i} = A_i R_L/R_i$ \hfill (4.41)

This value is obviously close to unity for we could write

$A_v = \dfrac{(1 + h_{fe})R_L}{h_{ie} + (1 + h_{fe})R_L}$

or

$$1 - A_v = h_{ie} / R_i \quad (4.42)$$

Output impedance Z_o. The open-circuit voltage in fig. 4.33 is v_s and the short-circuit output current is $(1 + h_{fe}) i_b$ or $(1 + h_{fe}) v_s/(h_{ie} + R_S)$ Hence, the output admittance of the transistor is

$$Y_o = 1/Z_o = (1 + h_{fe}) / (h_{ie} + R_S) \quad (4.43)$$

so that

$$Z_o = \frac{h_{ie} + R_S}{1 + h_{fe}}$$

R_S is usually large. If $R_S = 50\text{k}\Omega$ and $h_{fe} = 250$, $Z_o = 200\Omega$ and the output resistance of the stage is Z_o in parallel with R_L.

4.15 A Darlington pair

When a very high input resistance transistor circuit is required, a Darlington pair (fig. 4.34) is often used.

The input resistance of the transistor T_2 constitutes the emitter load for the transistor T_1. Although two transistors may be selected and connected in this way, most manufacturers provide this composite pair in a single package with the conventional three terminals.

A Darlington pair in an emitter-follower circuit is shown in fig. 4.35.

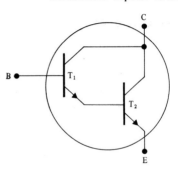

Fig. 4.34 A Darlington pair

Summarising the properties of the Darlington pair:

(a) The overall current gain is approximately equal to the product of the individual current gains ($h_{fe} = h_{fe1} \times h_{fe2}$). A value of $h_{fe} = 10^4$ is common for such a composite pair.

(b) The input resistance of a Darlington pair in emitter-follower connection (fig. 4.35) is given approximately by the relationship

$$R_i = h_{fe} / h_{oe}$$

where the transistors are of the same type, each with a current gain h_{fe}. A value of $R_i > 5\text{M}\Omega$ is therefore readily achieved.

Because of its high input resistance a Darlington pair is often used in each arm of a difference

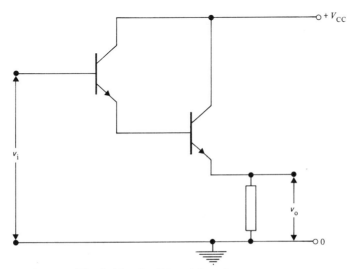

Fig. 4.35 A Darlington emitter-follower (The dc bias circuit is not shown)

amplifier to form the first stage of an operational amplifier (Chapter 5). A power Darlington pair is available capable of producing a collector current of 5A with an overall current gain $h_{fe} = 1000$. Such a device is often used as the series control element in a voltage stabilising network or in the output stage of an audio amplifier.

4.16 High input-resistance amplifiers

The emitter-follower configuration provides an amplifier with an input resistance of perhaps $500\mathrm{k}\Omega$ (section 4.14); using a Darlington pair (section 4.15) this can be increased to several megohm. The problem still has to be faced, however, that the dc biasing network, usually in the form of a potential divider R_1 and R_2, lowers the input resistance of the stage. The bootstrap method (described with a jgfet in common-source connection, section 3.14b) is shown in the emitter-follower stage of fig. 4.36 based on a bipolar transistor and can provide an input resistance of several megohm. With a bootstrapped Darlington pair this could be increased to several hundred megohm. This topic is not examined in more detail here because in most cases a field-effect transistor in source-follower connection is normally used as the input stage when a very high input resistance is required. This may be followed by an emitter-follower (based on a bipolar transistor) to provide a low output resistance, appreciable current gain and an overall voltage gain of, perhaps, 0.9.

4.17 Some useful circuits

(i) *A constant current source* is designed to have a very high (ideally infinite) output resistance. For a junction transistor in common-base configuration, the output resistance is very high (section 4.3); this feature is used in the circuit of fig. 4.37. The desirable consequence of a high output resistance is that any change in the load resistance does not affect the current which therefore remains constant at its selected value. Variations in temperature and in the input voltage can also affect the current flowing and the influence of these variables must be minimised. In the current source shown in fig. 4.37, a constant current of up to 10mA can be provided; the temperature effects are reduced by

(a) the use of a silicon transistor
(b) the two 5.6 V Zener diodes in series. The temperature coefficient $\frac{\partial V_Z}{\partial T}$ of these diodes is approximately zero.
(c) the current is limited to 10mA so that negligible heating occurs at the reverse-biased junction. The transistor should also be mounted on a small heat sink.

The output characteristics of a silicon transistor in CB connection (fig. 4.3) show that the collector current I_C is virtually constant and independent of the collector-base voltage (V_{CB}) provided that the emitter current I_E is constant. To achieve this, and yet allow the constant value of I_E to be selected, the

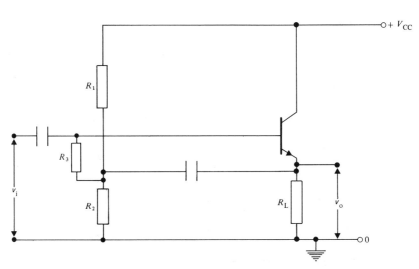

Fig. 4.36 The bootstrap principle used on an emitter-follower

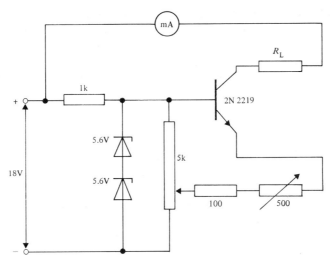

Fig. 4.37 A constant current source

emitter-base voltage (V_{EB}) is obtained from the Zener stabilised supply. The voltage V_{CB} with the load resistance $R_L = 0$ is the voltage across the 1kΩ resistor, i.e. approximately $18 - 11.2 = 6.8V$. Hence the load current should remain constant until the voltage drop across the load approaches 6.8V. The resistors in the emitter lead not only provide a fine adjustment for selecting I_E but also provide dc negative feedback.

Examples of the use of such a constant current source of this type are: to generate a linear voltage ramp, in general conductivity measurements, to provide a direct-reading ohmmeter from voltage measurements across R_L, to provide the current through a Hall probe.

Its performance should be compared with that of a jgfet constant current source (section 3.6) and that of an operational amplifier, especially with a current-booster amplifier (section 5.7i).

(ii) *A stabilised voltage supply.* A Zener diode provides a reference voltage when it is operated continuously in a reverse breakdown condition with the current limited to avoid over-heating the junction. That portion of the characteristic beyond the 'knee' is the most useful and the reference voltage, nominally V_Z, is most stable if the diode current is fairly constant so that the operating point makes very small excursions along the characteristic which, unfortunately, deviate from the ideal vertical line. The gradient of the Zener characteristic ($\partial V_Z / \partial I$) in the region of the operating point is called the 'slope resistance' of the diode and should ideally be zero.

The simple Zener voltage stabiliser is discussed in section 2.6. If the load current is greater than can be provided by this circuit, or if the load resistance is likely to change considerably, a series-control transistor operating in emitter-follower connection provides a much improved voltage stabiliser.

A voltage stabilising circuit is designed to produce a constant output voltage V_o in spite of a change (ΔV_i) in the input voltage, a change (ΔI_L) in the load current and a change (ΔT) in the ambient temperature.

The performance of any voltage stabiliser can be expressed by the following equation:

$$\Delta V_o = S_V \Delta V_i + R_o \Delta I_L + S_T \Delta T \qquad (4.44)$$

where the coefficients S_V, R_o and S_T are defined as

$$S_V = \left(\frac{\Delta V_o}{\Delta V_i}\right)_{I_L, T} \quad : \text{the stabilisation factor}$$

$$R_o = \left(\frac{\Delta V_o}{\Delta I_L}\right)_{V_i, T} \quad : \text{the output resistance}$$

$$S_T = \left(\frac{\Delta V_o}{\Delta T}\right)_{I_L, V_i} \quad : \text{the temperature coefficient}$$

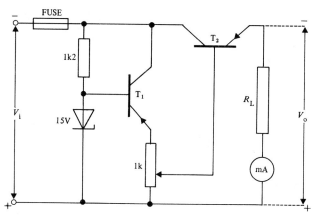

Fig. 4.38 A voltage stabiliser utilising a series control transistor

In a well-designed circuit, S_V, R_o and S_T are all as small as possible.

An input voltage change ΔV_i may be due to a change in the ac line voltage or may be due to ripple because of inadequate filtering. The ability of the circuit to inhibit these variations from affecting the output voltage V_o is governed by S_V while the output resistance R_o determines how constant the output voltage remains when the current demanded from the circuit varies.

A series control transistor used in emitter-follower connection (fig. 4.38) decreases the output resistance of the circuit and improves the stabilisation.

In fig. 4.38, both transistors T_1 and T_2 are used in emitter-follower connection. In this way, the current through the Zener diode is kept fairly constant while the Zener voltage is reproduced across the 1 kΩ potentiometer, so allowing the output voltage to be selected.

A voltage stabiliser which makes use of the emitter-follower principle can be much improved if a difference amplifier is incorporated. In the circuit of fig. 4.39, a fraction b of the output voltage is fed to one input of the difference amplifier, where $b = R_2/(R_1 + R_2)$. The reference voltage V_R is fed to the other input. The difference amplifier of voltage gain A, where A is large, will attempt to reduce the difference between bV_o and V_R to zero.

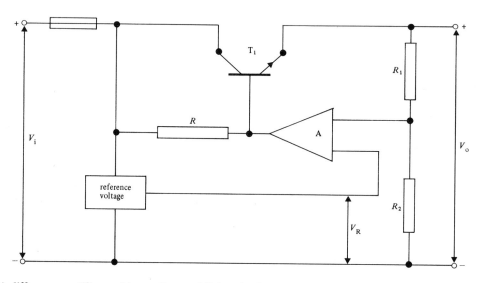

Fig. 4.39 A difference amplifier used in a voltage stabilising circuit

If this is achieved,

$$V_R = bV_o = R_2 V_o / (R_1 + R_2)$$

i.e.

$$V_o = V_R (1 + R_1/R_2)$$

It is now apparent that the output voltage may be selected with a minimum value $V_{o\ min} = V_R$.

A practical form of this voltage stabilising circuit is shown in fig. 4.40. The two circuits (fig. 4.38 and fig. 4.40) form ideal experimental units for laboratory use and clearly illustrate the principles employed. However, a number of features should be examined to improve further the performance of these units:

(a) A fuse is a comparatively slow-acting device. Transistors are often damaged before the fuse is blown. Electronic means of current limiting is needed so that the circuits are not damaged if the output terminals are short-circuited.

(b) The output transistor may be replaced by a power Darlington.

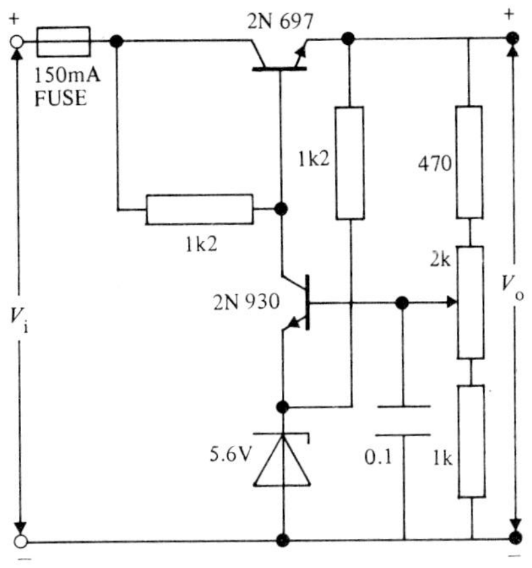

Fig. 4.40 Voltage stabiliser incorporating a difference amplifier

(c) The single transistor in fig. 4.40, which is serving as a difference amplifier, can be replaced by a high-gain differential operational amplifier (*see* Exercise 5.8).

Stabilised voltage supplies are available in integrated circuit or in modular form. A single package, suitably mounted, can often provide an output current limited at, say, 2A with the output voltage selected by means of an external resistor.

Exercises 4

1. You are provided with two small-signal *p-n-p* junction transistors, one germanium and the other silicon. What tests would you make to identify the germanium device?

2. Sketch the input and the output characteristics of a junction transistor in (a) common-base and (b) common-emitter connection. Why is the CE connection favoured as a voltage amplifier?

3. For a transistor in CB connection, the current gain is less than unity. Explain how such a configuration can still be used as a voltage amplifier.

4. Define the leakage currents denoted by I_{CBO} and I_{CEO} and derive the relationship between them.

Draw the circuit and outline the operation of a transistor tester capable of measuring I_{CEO} and h_{FE} for small-signal transistors.

5. In describing the operation of a junction transistor, it is stated that the reverse-biased CB junction provides an accelerating field for the current carriers. Is this statement correct? If so, explain what appears to be a contradiction.

6. Germanium devices have larger leakage currents and are thermally less stable than the silicon equivalents. What factors may nevertheless encourage the use of a germanium device in preference to one of silicon?

7. What is meant by 'thermal runaway'? Explain why this is a hazard with germanium junction transistors, less so with silicon and never with a jgfet.

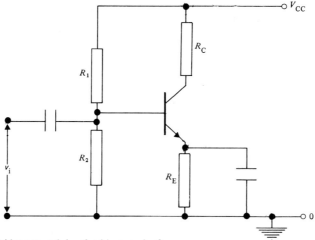

Fig. 4.41 The potential divider network involved in exercise 9

8. Discuss the choice of the operating point of a transistor used in a CE amplifier, and the various methods of establishing the correct dc bias. Comment on the thermal stability of each biasing circuit.

9. Fig. 4.41 shows a potential divider network used to establish the dc bias on a transistor in CE connection.
Draw an equivalent circuit to represent the dc behaviour of the circuit and derive an expression for the stability factor S where $S = \partial I_C / \partial I_{CBO}$.

10. Construct a table showing the CB, CE and CC connections for a transistor. By means of the hybrid parameters, give an equivalent circuit for each and the relevant parametric equations.

11. Outline the special features of

    (a) a Darlington pair,
    (b) an emitter-follower,
    (c) a bootstrap circuit, and
    (d) a paraphase amplifier.

12. By replacing a transistor in CE connection with its hybrid parameter model, derive expressions for (a) the current gain, A_i, (b) the input impedance, Z_i, (c) the voltage gain, A_v and (d) the output impedance, Z_o.

13. Under what circumstances can the hybrid equivalent circuit for a transistor in CE connection be simplified?
Draw the simple equivalent circuit and compare the new values of A_i, Z_i, A_v and Z_o with those obtained by means of the exact model.

14. By means of a simple transistor model show that, if the direct emitter current I_E is constant, a transistor with a high current gain, h_{fe}, used in a CE amplifier stage will provide the same voltage gain as a transistor with a low value of h_{fe}.
What feature changes when a transistor with a high current gain replaces one with a low current gain in a CE amplifier stage?

15. In a voltage stabilising circuit utilising a series control transistor, the output voltage can be selected. Under what conditions is the power dissipated in the series control transistor a maximum?
Suggest a method of reducing the risk of damage to this output transistor.

16. In a CE amplifier stage, how can you arrange that the voltage gain is independent of the transistor parameters?

17. Fig. 4.42 shows an n-p-n silicon transistor used in a CE amplifier stage. If the quiescent direct emitter current I_E is 0.45 mA, make estimates of the following:
    (a) The steady voltages V_E, V_C, V_B and the direct currents I_B and I_{BB}.
    (b) The voltage gain of this amplifier.
    (c) The voltage gain when the 250 μF capacitor is removed.

If $h_{oe} R_C \leqslant 0.1$, draw an equivalent circuit for the amplifier and calculate the input resistance of the stage shown in fig. 4.42.

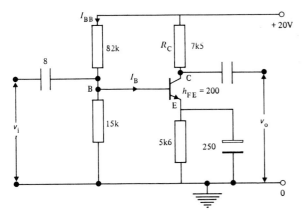

Fig. 4.42 The CE amplifier of exercise 13

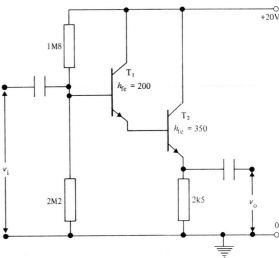

Fig. 4.43 Amplifier circuit of exercise 20

18. In an equivalent circuit, representing the small signal behaviour of a transistor, several components included in the original circuit diagram are omitted. Explain why this is done and state under what conditions these omissions are valid. Include in your discussion the dc power supply, the biasing network, by-pass capacitors and junction capacitance.

19. (i) By examining the alternating components of the base and collector currents, explain the inverting nature of a CE amplifier.
    (ii) The output signal from a CE stage is an amplified replica of the low power input signal. What is the source of this additional power?
    (iii) Why is it that the resistor in the collector lead of a CE amplifier is denoted by R_C and only on occasions is referred to as the load on the amplifier?

20. An amplifier circuit (fig. 4.43) is based on the use of two n-p-n silicon transistors.
    (a) What type of amplifier is it?
    (b) Calculate the dc bias conditions, i.e. the voltages V_{B1}, V_{E1}, V_{E2} and the current I_{E2}.
    Indicate for this amplifier, without attempting detailed calculations, the magnitudes of (c) its voltage gain and (d) its input resistance.
    For what purpose would such an amplifier be used?

21. An n-p-n silicon transistor is used in the circuit of the CE amplifier stage of fig. 4.44. Calculate the dc bias voltages and currents.
    Establish that the following statements (i) and (ii) merely form alternative methods of performing these calculations.

    (i) any resistance in the emitter circuit is multiplied by $(h_{FE} + 1)$ when viewed from the base terminal;
    (ii) any resistance in the base circuit is divided by $(h_{FE} + 1)$ when viewed from the emitter terminal.

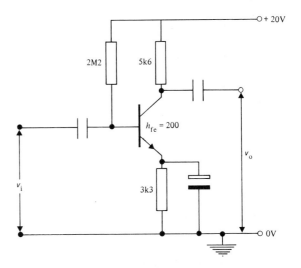

Fig. 4.44 The CE amplifier stage of exercise 21

5 · Operational Amplifiers

5.1 Introduction

The term 'operational amplifier' was introduced originally to denote a class of stable, high gain amplifiers designed to simulate the mathematical 'operations' of multiplication, division, differentiation and integration required for the development of analogue computers. The operation to be performed is chosen merely by connecting suitable inputs and feedbacks to the amplifier. In the analogue computer, voltage is used as the analogue of a numerical quantity. By using operational amplifiers to amplify these voltages (which is analogous to multiplication by a constant), to provide the differential with respect to time and to sum voltages, the resulting output voltage from the analogue computer can be made to represent the solution of a differential equation.

Compact, high performance, directly-coupled (dc) amplifiers capable of very stable operation over a range of frequencies, including zero (i.e. direct current) have become readily available at low cost. Their use is no longer confined to analogue computers: they are so versatile that they are being used in electronic measuring instruments and control circuits. It is often more economical to design electronic circuitry around these ready-made amplifiers than around discrete components. Not only does this avoid the detailed analysis necessary for good amplifier design, the resulting system will be more compact, more readily assembled and inherently more reliable.

It is not necessary to know the intricate internal circuit details of operational amplifiers in order to use them effectively. Nevertheless, the facilities provided by these units and the terms used to specify their performance have to be understood in principle. The amplifier is the most important basic circuit in electronics: by its use one can create most of the important function generators. In this chapter, we shall study the operational amplifier as a unit performing in a near perfect manner over the limited region within which we confine its operation. With suitable feedback components connected to form a sub-system, its role will be examined in sinusoidal and non-sinusoidal oscillators, voltage sweep circuits, constant voltage and constant current sources. It must be emphasised that design based on these sub-systems involves more than merely interconnecting packaged modules from manufacturers. How well the operational amplifier performs and the quality and reliability of the resulting system is often only limited by the ingenuity of the designer.

5.2 Structure and operating notes

At present operational amplifiers are available in two main forms:—

(a) *Modular*: discrete components are selected, mounted on small printed-circuit boards and the whole unit is encapsulated. The dimensions of these modules are approximately 25 x 25 x 12mm. They are termed 'hybrid' when they contain a mixture of discrete components and small integrated circuits.

(b) *Integrated circuit* (IC): the complete amplifier is constructed on a single silicon slice (monolithic integrated circuit) in the manner described in Appendix A. The amplifier may be mounted on a TO 116 encapsulation (a 14-pin plastic dual in-line package) or in a TO 99 metal can (an 8-pin, reduced height TO 5).

An amplifier is represented diagrammatically by a triangular arrow-head pointing in the direction of data flow, i.e. input to output. A differential operational amplifier with a seven-pin base is shown in fig. 5.1. The function of the terminal connections will be examined although the numbers and positions shown in fig. 5.1 have no special significance.

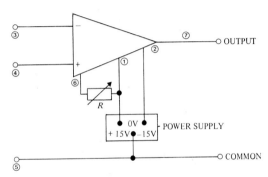

Fig. 5.1 Terminals on a differential operational amplifier

Terminals 1 and 2 are the power supply terminals. In general, the amplifier must be supplied with two stabilised dc voltages, e.g. +15V and −15V, with respect to the common line. This enables the unit to amplify a dc input signal which may be positive or negative with respect to the common reference level. These supply voltage levels are not critical although the manufacturer specifies the voltages at which the reported performance figures were recorded. Most amplifiers will function satisfactorily with reduced gain at voltages down to +9 and −9. Batteries of adequate capacity may be used or modular mains-operated power supplies are available providing +15V−0−−15V at three output terminals. It is essential that the 0V terminal of the power supply be connected to the common line. It is also advisable to provide a radio-frequency by-pass by connecting a 0.1 μF capacitor between each power supply terminal and earth. The quiescent current from the power supply may be typically 3mA whereas the maximum current drawn may be 12mA; operational amplifiers demanding much less power than this are also available.

Terminal 3 is the inverting input terminal. A voltage applied between this terminal and the common line appears in an amplified and inverted form between the output terminal and the common line. For a dc input signal, inversion means a change of sign; for an ac input signal, inversion means a change of phase by 180° i.e. the output signal is in anti-phase with the input signal.

In circuit diagrams, the inverting terminal is identified by a negative sign. For an amplifier to be termed 'operational' it must invert the input signal.

Terminal 4 is the non-inverting input terminal, identified in circuit diagrams by a positive sign. Any input signal applied between this terminal and the common line appears at the output in an amplified form and in-phase with the input signal.

If this input terminal 4 is connected to the common line, the amplifier is left with one input terminal and is said to be 'single-ended'. Difference amplifiers with two inputs are much more versatile so they will be examined more fully and the single-ended version will be treated as a special case.

Terminal 5 is the common line terminal. It forms the reference level for both the input and the output voltage. It may or may not be connected to earth but the zero point of the power supply must be connected to it. There may or may not be a pin labelled 'common' on the amplifier base. Even if a pin is provided, it is rarely connected internally to the amplifier.

Terminal 6 is sometimes called the 'balance terminal' because it allows an external variable resistor R (perhaps of maximum value 1kΩ) to be connected as shown in fig. 5.1 and used to zero the amplifier input off-set voltage.

Terminal 7 is the output terminal where the output voltage V_o is developed relative to the common line.

An eighth pin is provided if frequency-compensating components have to be connected externally. The manufacturers state the values required when the amplifier is operating at a particular gain. When a different gain is selected the frequency-compensating components must be changed if the amplifier is to operate in a stable manner.

It will be assumed henceforth that the correct components have been connected or that frequency compensation has been incorporated in the operational amplifier itself as is very often the case.

5.3 The ideal differential operational amplifier

An ideal differential or difference amplifier is sensitive to a difference of potential between its two input terminals but will not respond to any voltage applied to both the input terminals simultaneously. The ability of the amplifier to ignore or reject this 'common mode signal' is very valuable and will be examined in more detail later. An ideal operational amplifier is furthermore assumed to have the following characteristics:—

(a) *Infinite gain.* When the gain of an amplifier is very high, the performance of a sub-system which makes use of the amplifier depends only on the input and feedback networks. Typically, the dc open loop voltage gain, A_{VOL}, (the voltage gain without feedback) may be 10^5 or 10^6 or more in practice.

(b) *Infinite input impedance.* With this condition satisfied, the input current to the amplifier is zero. By the use of field-effect transistors and good design, a differential input impedance (impedance between the two input terminals) may be as high as $10^{12}\Omega$ in practice. Furthermore, the common mode input impedance (the result of the parallel combination of each input terminal to the common line) may also be $10^{12}\Omega$.

(c) *Infinite band width.* The bandwidth of an amplifier is the frequency range over which the gain is virtually constant. If this extends from zero to infinity, the amplifier will respond equally well to dc or ac signals of any frequency. Also, no distortion or phase change with frequency would occur. In actual amplifiers, the gain is reduced progressively at higher frequencies to ensure stability. For some operational amplifiers with internal frequency compensation, amplification without distortion can be obtained for signals at frequencies up to 100kHz and the frequency at which the voltage gain is unity (i.e. 0dB) may be 3MHz.

(d) *Zero output impedance.* With this condition satisfied, the output would be unaffected by alteration of the load.

(e) *Zero voltage offset.* This implies that, when the voltage between the input terminals is zero, the output voltage will be zero.

5.4 Negative feedback

Operational amplifiers make use of external components arranged to provide negative feedback (*see* Appendix B) except in those occasional cases when they are required to perform a switching action. The feedback path links the output to the inverting input terminal. Whereas an amplifier operating without feedback is said to be 'open-loop', one operating with feedback is said to be in 'closed-loop': the symbol A_{VCL} denotes the closed-loop voltage gain.

In a simple example of negative feedback (fig. 5.2) the feedback resistor R_2 links the output with the point X — the input terminal of the inverting amplifier.

An amplifier is said to be fully operational if it is stable (i.e. it will not oscillate) when the output is connected directly to the inverting input terminal. This implies that stability will be maintained when a capacitor is used as the feedback component even though this effectively short-circuits the two terminals to high frequency signals.

Assuming the amplifier in fig. 5.2 to be ideal, any current arriving at point X must flow through the feedback resistor R_2. If there were several alternative signal paths connected to X instead of only one via the resistor R_1, the sum of the several currents arriving at X would flow through the feedback path. For this reason, the inverting input terminal (point X) is often referred to as the amplifier *summing junction*.

If the input voltage V_1* tends to drive the potential at the point X positive, the right-hand end of resistor R_2 assumes a negative potential because of the inverting action of the amplifier. A current I_f will flow in R_2 in a direction to reduce the positive excursion at the point X to a very low value. With the amplifier having a voltage gain of the order of 10^6, an

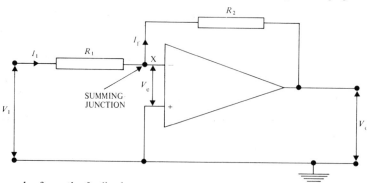

Fig. 5.2 A simple example of negative feedback

excursion of a few microvolts at X is sufficient to cause the output voltage to swing through its entire range — perhaps 10V on either side of zero. The self-adjusting nature of the negative feedback system has the effect of holding the potential at X to a very low value, i.e. $V_e \to 0$, no matter what values are chosen for R_1 and R_2. Because the point X is thus always maintained at a negligible potential with respect to the other input terminal, it is called a *virtual earth*. Henceforth, the point X at the input to the inverting amplifier will be referred to as the *summing junction* or the virtual earth, whereas the input terminal will refer to a point at the left-hand side of the resistor R_1.

5.5 The inverting and non-inverting configurations: the closed-loop voltage gain equations

The two basic ways in which negative feedback may be applied to a differential amplifier are shown in fig. 5.3 where (a) is the inverting configuration and (b) is the non-inverting configuration. In each case the output is connected via the resistor R_2 to the summing junction of the amplifier.

If the voltage difference between the input terminals (fig. 5.3a) $V_e = 0$,

$$I_i = V_i/R_1 \text{ and } I_f = -V_o/R_2.$$

Because the current flow to the input of the amplifier is zero, $I_i = I_f$
Therefore $V_i/R_1 = -V_o/R_2$
and the closed loop gain

$$A_{VCL} = V_o/V_i = -R_2/R_1 \qquad (5.1)$$

where the negative sign indicates the inverting nature of the amplifier.

In the case of the circuit of fig. 5.3b, if no current flows into the amplifier, the voltage at point X is V_X given by

$$V_X = V_o R_1/(R_1 + R_2)$$

---

*V_1 is to be used for the input voltage when the inverting terminal (−) of the operational amplifier is involved and V_2 when the non-inverting terminal (+) is involved. Often, but not necessarily, the non-inverting terminal is connected to the common-line so that V_1 can be written as V_i, the input voltage, which will be the signal voltage as produced by the transducer or other device which it is desired to amplify.

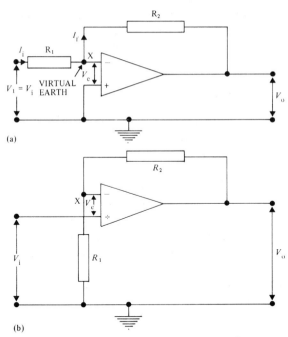

(a)

(b)

Fig 5.3. The two basic negative feedback circuits: **(a)** the inverting configuration; **(b)** the non-inverting configuration

As $V_e = 0$,
$V_i = V_X = V_o R_1/(R_1 + R_2)$
and the closed loop gain $A_{VCL} = V_o/V_i = (R_1 + R_2)/R_1$

Therefore $A_{VCL} = 1 + R_2/R_1 \qquad (5.2)$

5.6 Negative feedback and linearity

Equations (5.1) and (5.2) show that the closed loop voltage gains depend only on the input and feedback components: in the cases presented on the values of the resistors R_1 and R_2. Hence negative feedback enables the linearity of a good amplifier to be improved beyond recognition. A manufacturer will usually quote the minimum dc open loop gain of an operational amplifier at a given temperature, say 20°C. The voltage gain A_{VOL} will vary with temperature, with supply voltage and, most important of all, it will vary with the magnitude of the input signal. However, we can now see that this non-linear behaviour on open-loop is not as serious as it might appear because, with negative feedback, the closed-loop gain, A_{VCL} (which is that required in practice) is dependent only on R_1 and R_2 in the case of the circuits of fig. 5.3. By using precision resistors with negligible temperature coefficients for R_1 and R_2, the performance of the amplifier can be made extremely linear and, moreover, this linearity is further improved as the negative feedback is increased, i.e. as A_{VCL} is reduced.

Much design literature has been published under the general heading of 'linear microcircuits' because of the extensive use of operational amplifiers as linear amplifiers in integrated circuit (microcircuit) form.

It must be recalled that equations (5.1) and (5.2) are valid only if the open loop gain, A_{VOL}, is infinitely large. As large values of A_{VOL} are a feature of available operational amplifiers, the errors introduced by using these equations are usually negligibly small. These errors, resulting from finite loop gains, are examined in detail in section 5.11.

Apart from inversion, the chief difference between the circuit configurations of fig. 5.3 is the input resistance presented to the source of the input signal. In the inverting configuration (fig. 5.3a), feedback prevents the voltage at X from changing whatever the input signal; because the point X is virtually at earth potential, and the input current I_i is determined by the resistor R_1 which is therefore the input resistance. Further evidence that R_1 is the input resistance is provided by the use of the Miller theorem (Appendix C). In the non-inverting configuration, no current flows into the amplifier and the input resistance is theoretically infinite.

Replacing the resistances R_1 and R_2 by impedances Z_1 and Z_2 (fig. 5.4a) is considered from the equivalent circuit of this amplifier (fig. 5.4b) in which the double-headed arrow XE indicates the significance of the virtual earth. This arrow links the point X to earth even though no current could flow in this fictitious connection. Applying a low-frequency sinusoidal signal voltage v_i the same current i_i must flow through Z_1 and Z_2 in fig. 5.4b. Therefore,

$$i_i = v_i / Z_1$$

and

$$A_{VCL} = -i_i Z_2 / i_i Z_1 = -Z_2 / Z_1 \qquad (5.3)$$

Under these conditions we speak of the *transfer function* of the amplifier rather than the gain. In general, the transfer function is a complex number.

5.7 Some linear dc applications of operational amplifiers

The applications of operational amplifiers are considered by suggesting a number of introductory experiments which enable the characteristics of these amplifiers to be examined. The experimental techniques explored should be adopted subsequently with more complex circuits.

For introductory work, a suitable general purpose differential operational amplifier* has a dual fet input, a differential input impedance of $10^{12} \Omega$ and a common-mode input impedance of $10^{12} \Omega$ and is provided with internal frequency compensation, short-circuit protection and a dc open loop gain of about 10^5 (100dB).

Power to the amplifier is provided by a stabilised $+15V - 0 - -15V$ supply with the zero point connected to the common line (even though this is not shown) and a $0.1\mu F$ capacitor should be connected between each power supply terminal and earth.

For measuring input and output voltages, a good quality digital voltmeter (DVM) is recommended though it should be stressed that the practice of accepting three figures from such an instrument is dubious if resistors with a tolerance of only 5% are

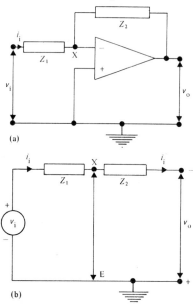

Fig 5.4. Impedances Z_1 and Z_2 with an inverting amplifier: (a) the feedback amplifier circuit; (b) the equivalent circuit

---

*An example is the modular E 78 manufactured by Computing Techniques Ltd. This operational amplifier has a dual fet input and a differential input impedance of $10^{12}\Omega$.

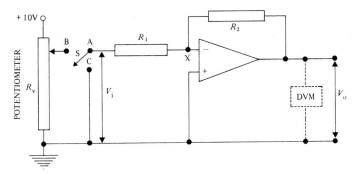

Fig. 5.5 Experiment with a unity gain sign inverter

used. Before any measurements are made, the residual off-set voltage should be neutralised by means of the external variable resistor (R connected between terminals 6 and 1 in fig. 5.1; for the E 78, the maximum value of R should be 1kΩ).

(a) *A unity gain sign inverter circuit* (fig. 5.5) has resistors R_1 and R_2 each set at 100kΩ and the input (point A) initially connected to the common line (the two-way switch S is thrown so as to link A with C). With a DVM connected between the output terminal and the common line, the variable resistor R between terminals 1 and 6 (see fig. 5.1) is adjusted to ensure that the output voltage (indicated by the DVM) is zero. This zero setting should also be checked with R_2 = 100kΩ and R_1 = 10kΩ and also at intervals throughout the experiments because the zero off-set voltage is somewhat temperature sensitive. With $R_1 = R_2$ = 100kΩ, the input voltage V_i is set at various values within the range 0 to ±10V by connecting A to B on throwing the switch S and making use of the potentiometer R_v.

At each setting, the input voltage V_i and the corresponding output voltage V_o are both measured by a DVM. As $R_2 = R_1$ and $A_{VCL} = -R_2/R_1 = -1$, the voltage gain is unity and the signal is inverted.

(b) *Gain control in the inverting configuration.* Referring again to the circuit of fig. 5.5, the resistor R_2 is set at 1MΩ whereas R_1 is set at 10kΩ. Hence $A_{VCL} = -10^6/10^4 = -100$.

The potentiometer R_v is now modified to provide small dc voltages V_i of less than 200mV. Again V_i and V_o are measured at a number of settings of V_i. The input voltage may be positive or negative with respect to the common line:

each should be examined. A graph of the measured output voltage V_{OM} against the calculated value of the output voltage $V_{OC} = -100 V_i$ (fig. 5.6) shows that the behaviour is linear and that the output voltage saturates; for this amplifier at a value of about ±14V. If linear behaviour is to be obtained, the input signal should always be maintained at a level below that which will cause saturation of the output.

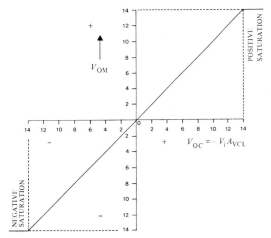

Fig. 5.6 Typical amplifier transfer characteristic (For the E78 operational amplifier R_1 = 10kΩ and R_2 = 1MΩ)

In some switching applications it is very convenient to have the output switch from negative saturation to positive saturation for a small input voltage change (section 5.4) Many operational amplifiers incorporate internal protection against input overload conditions. Where such protection is not provided, silicon diodes may be connected externally to the amplifier input terminals as shown in fig. 5.7. The amplifier performance is not thereby impaired because feedback never allows the voltage between the input terminals to exceed a few millivolts.

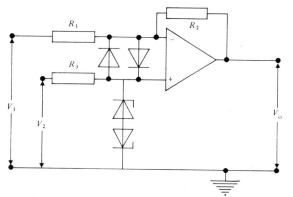

Fig. 5.7 Overload protection using silicon diodes

For a given amplifier there is a maximum common-mode voltage, ie. a maximum allowable voltage at both inputs of a differential amplifier with respect to earth. Where excessive common-mode voltage could occur a pair of Zener diodes can provide protection if connected as shown in fig. 5.7. This type of circuit also serves to increase the reliability of the system.

(c) *Summation.* A number of voltages may be applied simultaneously to the summing junction (fig. 5.8). If voltages V_1, V_2 and V_3 are each applied via an input resistor R_1, the current at the summing junction is $(V_1 + V_2 + V_3)/R_1$.
If the feedback resistor is of the same value, the feedback current

$$-V_o/R_1 = (V_1 + V_2 + V_3)/R_1$$

Therefore $V_o = -(V_1 + V_2 + V_3)$.

From equation (5.1) it is seen that any one of the input voltages can be multiplied by a constant so that the output voltage could represent z, where

$$5x + 4y = -z$$

This behaviour can be readily demonstrated using the circuit of fig. 5.9.

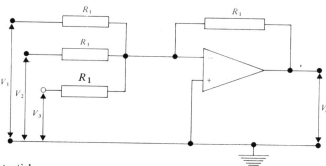

Fig. 5.8 Summing input potentials

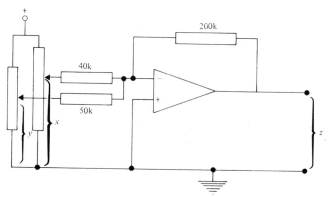

Fig. 5.9 Circuit for demonstrating multiplication and summation

(d) *Output current and output resistance.* The circuits described so far have indicated only voltage measurements at the output so the current demanded from the amplifier has been minimal. The output resistance of an operational amplifier in the open loop state may be of the order of tens of ohms. With negative feedback applied via resistor R_2, the output resistance is decreased by a factor A_{VOL}/A_{VCL}. Hence an output resistance as small as $10^{-2}\,\Omega$ under operating conditions is not unusual. Performance figures of operational amplifiers are generally quoted with a specified load resistance, say $2k\Omega$. Many of these amplifiers have the output current limited to, perhaps, 5mA although on occasions 10mA is available. A current of 10mA should be regarded as the maximum available from an operational amplifier. If additional power is required a booster amplifier (*see* section 5.7i) can be added.

The experiments described previously should be repeated with the operational amplifier providing a current of 5mA in a load resistor connected between the output terminal and the common line. No detectable change in performance will be observed, indicating a very low value of the amplifier's output resistance.

(e) *Attenuated feedback.* It is often convenient to connect a potential divider network across the output and to feedback only a fraction of the output voltage (fig. 5.10).

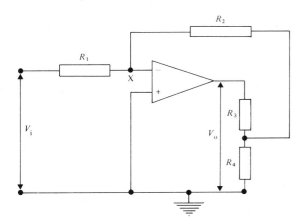

Fig. 5.10 Attenuated feedback

Equating the currents at the summing junction X:

$$V_i/R_1 = -V_o R_4/R_2(R_3 + R_4)$$

The closed loop gain

$$A_{VCL} = V_o/V_i = -(R_2/R_1)[(R_3 + R_4)/R_4]$$

A specific example illustrates the value of this circuit configuration. Suppose a value of A_{VCL} of 10^3 is required with an input resistance of not less than $10k\Omega$. This would demand that $R_1 = 10k\Omega$ and $R_2 = 10^3 \times 10^4 = 10M\Omega$ in a normal inverting configuration. But a resistance of $10M\Omega$ with a tolerance of better than 5% is difficult to obtain: without a low tolerance, the gain is not accurately known. With the circuit of fig. 5.10 where $R_1 = 10k\Omega$, $R_2 = 200k\Omega$, $R_3 = 9.8k\Omega$ and $R_4 = 200\Omega$, the conditions are readily obtained and the gain is predetermined without difficulty to within 1%.

(f) *The non-inverting configuration.* The circuit of fig. 5.11 is relevant (cf. fig. 5.3b).

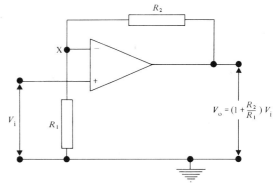

Fig. 5.11 Gain in the non-inverting configuration

Applying dc input voltages between 0 and ±5V the gain relationship (equation 5.2) can be verified. With $R_1 = R_2 = 100k\Omega$, the non-inverting nature of the amplifier can be examined. With $R_1 = 1k\Omega$ and $R_2 = 1M\Omega$, $A_{VCL} = 1000$ (the unity in equation (5.2) being negligible). With small dc input signals, a transfer characteristic similar in form to that of fig. 5.6 is obtainable. Note that with $R_2 = 0$ in equation (5.2) the gain is unity. Such an arrangement – termed a *unity gain voltage follower* – is shown in fig. 5.12. Its function is similar to that of an emitter follower in that it serves as a buffer amplifier or impedance changer.

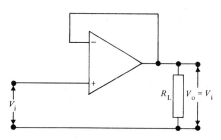

Fig. 5.12 A unity gain voltage follower

(g) *The difference amplifier.* In the circuit of fig. 5.13, a dc voltage of magnitude V_1 is applied to the inverting amplifier and one of magnitude V_2 to the non-inverting amplifier.

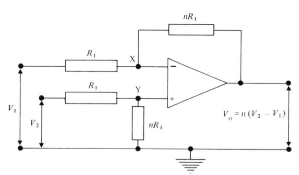

Fig. 5.13 The difference amplifier

The voltage at the point Y

$$= V_2 n R_1 / [R_1 (1 + n)] = V_X, \text{ the voltage at point X} \quad (5.4)$$

Also

$$\frac{V_1 - V_X}{R_1} = \frac{V_X - V_o}{nR_1} \quad (5.5)$$

Therefore, $V_X (1 + n) = n V_1 + V_o$

Substituting for V_X from equation (5.3) gives

$$\frac{V_2 n (1 + n)}{1 + n} = n V_1 + V_o$$

Therefore $V_o = n (V_2 - V_1)$ \quad (5.6)

With $n = 1$, the circuit (fig. 5.13) acts as a unity gain, differential input, single-ended output amplifier. Operational amplifiers are often used in this form, but frequently with $n > 1$, to amplify the signals from transducers such as strain-gauge bridges, thermocouples or Hall probes. For a transducer which has a high output resistance, a voltage follower is used as a buffer amplifier for each signal; the circuit of fig. 5.12 is attached to each input. This arrangement, using three low-cost integrated circuit operational amplifiers, provides a high performance dc amplifier.

(h) *A constant current source.* The operational amplifier is used in a non-inverting configuration with a standard cell (SC) to provide the input signal (fig. 5.14). The total loading $(R + R_L)$ would be maintained at a value exceeding 1500Ω. The current provided by the amplifier will be of such a value that a voltage equal to the emf E of the standard cell is established across the adjustable resistor R.
For R it is advisable to use a high quality decade resistance box which contains resistors having a negligible temperature coefficient.
A constant current within the range 5μA to 10mA can be selected and measured by means of

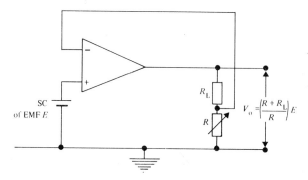

Fig. 5.14 A circuit to provide a constant current of less than or equal to 10mA

a DVM across the accurately known resistor R. An alternative method of control is to vary the input voltage obtained, for example, from a precision Zener stabilised supply source.

When a constant current in excess of 10mA is required, the booster amplifier circuit of fig. 5.15 can be used in a modified form.

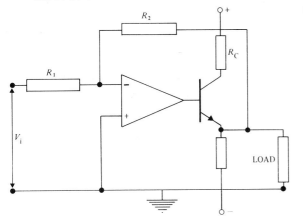

Fig. 5.15 A current booster utilising an emitter follower in the feedback loop

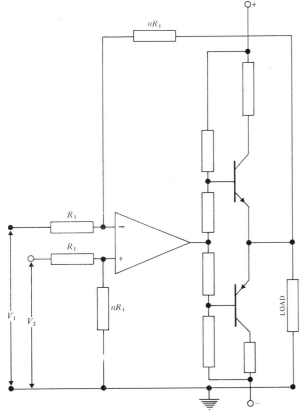

Fig. 5.16 A complementary pair used as a current booster

The circuit of fig. 5.14 may also be used to provide a constant voltage, the output voltage being given by $(1 + R_L/R)E$. The output voltage V_o would be limited to about 12V and the current available would be small, so arranged that the total current drawn from the amplifier did not exceed 10mA.

(i) *Current booster.* In general, the current available from an operational amplifier is limited to 10mA. This can be increased by making use of the emitter-follower principle if the emitter-follower is included in the feedback loop. In fig. 5.15, the power for the current booster is obtained from a separate dc supply. By the use of a pair of complementary transistors (fig. 5.16) positive and negative output currents can be obtained. Current booster modules are available commercially to extend the power role of the operational amplifier.

5.8 AC operation

Although the stages of an operational amplifier are directly coupled so that it can be used to amplify steady or zero frequency signals, this design does not affect its low frequency ac performance. The gain equations (5.1) and (5.2) are valid for low-frequency sinusoidal signals provided that the resistors R_1 and R_2 are non-reactive.

In the circuit of fig. 5.17, the input is from a signal generator providing a sinusoidally alternating emf. This circuit can be used to examine the ac behaviour of the operational amplifier in the inverting configuration.

Using a closed loop gain A_{VCL} of 1, 10 and 100 with $R_1 = 10\text{k}\Omega$ and a sinusoidal alternating input signal of 50mV rms of which the frequency can be

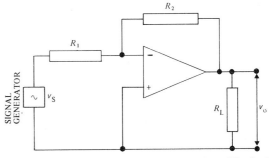

Fig. 5.17 Alternating input to an operational amplifier in the inverting configuration

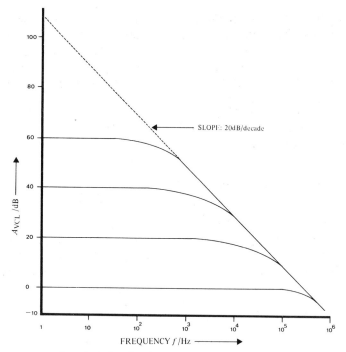

Fig. 5.18 Typical frequency-response characteristics of a closed-loop operational amplifier

varied and is known, the gain equation, the transfer characteristic and the frequency response of the amplifier can all be examined. The load resistor R_L should exceed 2kΩ. An ac millivoltmeter is ideal for measuring the input and output alternating voltages and a double-beam oscillograph is useful for comparing the waveforms and the phase of these voltages. From the transfer characteristic, it will be seen that the amplifier is again operating in a linear fashion provided that the input signal amplitude is small enough to prevent saturation at the output. The frequency response curve (fig. 5.18) is examined in more detail in section 5.12, but it is apparent that the frequency range over which the voltage gain is substantially constant (i.e. the bandwidth of the amplifier) increases as the negative feedback is increased.

If the alternating input signal to the operational amplifier is superimposed on a dc level, a blocking capacitor C_1 (fig. 5.19) may be incorporated. To determine the value of the capacitance of C_1 required, consider that the closed loop gain A_{VCL} of the amplifier is given by equation (5.3):

$$A_{VCL} = -Z_2/Z_1 = -R_2 \bigg/ \sqrt{R_1^2 + 1/\omega^2 C_1^2}$$

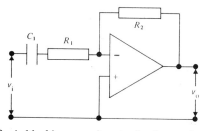

Fig. 5.19 A blocking capacitor in the input circuit of an ac amplifier

Defining the gain in dB by $10 \log_{10} (v_o/v_1)^2$, (where the fact that the input and output impedances are not equal is ignored) the gain will be 3dB down when

$$R_2^2 \big/ (R_1^2 + 1/\omega^2 C_1^2) = R_2^2/2R_1^2$$

i.e. when $R_1 = 1/\omega C_1$.

Hence there will be a low frequency fall in gain of 3dB at a frequency f_1 given by

$$f_1 = 1/2\pi C_1 R_1.$$

The upper frequency limit will be determined by the frequency compensation built into the operational amplifier. It is important to realise that the low frequency performance has been ruined solely by the capacitor C_1. It is assumed that at high frequencies the impedance of C_1 is negligible. If C_1 is an electrolytic capacitor, this is not necessarily the case and the published performance characteristics of the capacitor at high frequencies must be examined.

5.9 Some wave-shaping and non-linear applications

In many applications the operational amplifier is used to change the nature or form of a signal. A number of these circuits called 'signal conditioners' or 'function generators' will be examined.

(a) *The integrator.* The input voltage signal v (a function of time) is applied via the resistor R whereas the feedback component is a capacitor C (fig. 5.20a).

In the equivalent circuit (fig. 5.20b) of this operational integrator, the double-headed arrow XE indicates the virtual earth at X. Because the current v/R flowing through the input resistor must pass through the feedback capacitor, we have

$$i = v/R \text{ and } v_o = -\frac{1}{C}\int_0^t i\,dt$$

Therefore
$$v_o = -\frac{1}{RC}\int_0^t v\,dt \qquad (5.7)$$

The output voltage v_o is therefore proportional to the time integral of the input voltage. If the input voltage were constant, say $v = V$, the output would be a linear ramp voltage because then

$$v_o = -Vt/RC.$$

This type of voltage sweep circuit, called a *Miller integrator*, is used extensively in time-base circuits and in digital voltmeters. The time constant $T = CR$ is called the 'characteristic time' of the integrator. Writing

$$v_o/t = -V/T \qquad (5.8)$$

the magnitude of CR determines the rate of change of output in volts/second for a 1 volt input.

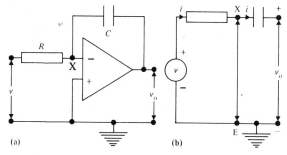

Fig. 5.20 (a) The operational integrator; (b) the equivalent circuit

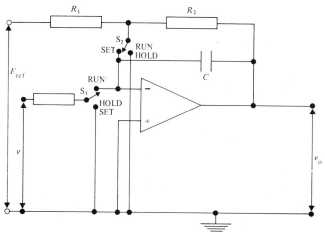

Fig. 5.21 The principle of run, set and hold states of an integrator

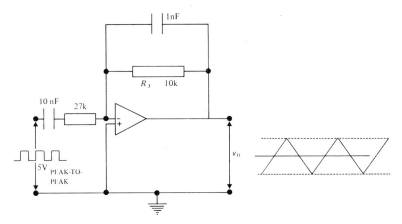

Fig. 5.22 A triangular waveform generator [the component values given are for operation at a basic frequency (repetition rate) of 100kHz]

The basic operational integrator (fig. 5.20) is usually fitted with circuits to establish the initial conditions at the output and to re-set the integrator at some chosen point on the ramp. These facilities are shown in principle in fig. 5.21.

With the two switches S_1 and S_2 both in the 'set' position, the initial output of the amplifier can be chosen because

$$v_{o_{t=o}} = -(R_2/R_1)E_{ref} \qquad (5.9)$$

With S_1 and S_2 both in the 'run' position, the output of the amplifier is given by

$$v_o = v_{o_{t=o}} - \frac{1}{CR}\int_o^t v\,dt \qquad (5.10)$$

With S_1 and S_2 both in the 'hold' position, the output should remain constant at the value recorded when the switching occurred.

(b) *Linear triangular waveform generator.* An immediate application of the integrator (fig. 5.20) is that of converting a square wave input into a triangular wave. Fig. 5.22 shows the circuit of an integrator operating in this way with the resistor R_3 providing dc feedback to maintain the output symmetrical about the common line or earth.

(c) *The basic differentiator.* By interchanging the positions of the resistor and the capacitor in the simple integrator of fig. 5.20, a circuit is obtained for which the output voltage is proportional to the time derivative of the input voltage. Such an operational differentiator together with its equivalent circuit are shown in fig. 5.23.

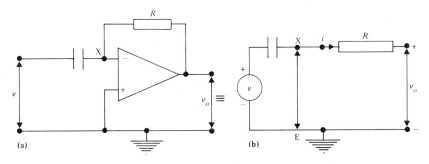

Fig. 5.23 (a) The basic operational differentiator and (b) the equivalent circuit

The input voltage v at any instant of time t is given by

$v = Q/C$ where Q is the charge on the capacitor C.

But, $\quad i = \dfrac{dQ}{dt} = C\dfrac{dv}{dt}$

and $\quad i = -v_o/R$

Therefore $-v_o/R = C\dfrac{dv}{dt}$

i.e. $v_o = -RC\dfrac{dv}{dt}$ \hfill (5.11)

Unfortunately the operational differentiator in this simple form is inherently unstable and very susceptible to random noise. To minimise these difficulties, the circuit is modified by the addition of R_1 and C_1 (fig. 5.24). These components are added to ensure the necessary frequency response stability and also to limit the gain at high frequencies (much higher than the operating frequency) so as to reduce considerably the output noise.

Comparison between the practical differentiating circuit (fig. 5.24) and the circuit of fig. 5.22 shows immediately that any difference in behaviour must be determined by the component magnitudes and the operating frequency.

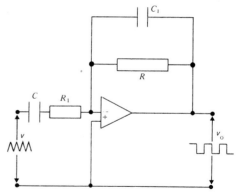

Fig. 5.24 A practical differentiator

In the circuit of fig. 5.24, if $C = 1\mu F$, $R_1 = 510\Omega$, $C_1 = 10nF$ and $R = 100k\Omega$, it will act as a differentiator at low frequencies, an integrator at high frequencies and as a proportional amplifier at a limited frequency range (fig. 5.25).

(d) *The astable multivibrator.* The circuit of an astable or free-running multivibrator (fig. 5.26) has two quasi-stable states: in one state the output is at the positive saturation level; in the other, at the negative saturation level. Note that positive feedback has been introduced in an operational amplifier so oscillation would be expected.

Immediately the voltage supply to the circuit (fig. 5.26) is switched-on the amplifier output

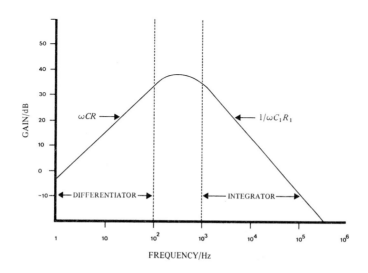

Fig. 5.25 The frequency response of the practical differentiator of fig. 5.24 ($C = 1\mu F$; $R_1 = 510\Omega$; $C_1 = 10nF$; $R = 100k\Omega$)

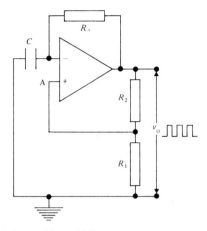

Fig. 5.26 An astable multivibrator

will saturate. Suppose it reaches positive saturation at the voltage $V^+_{o\ sat}$. A fraction β of this output [where $\beta = R_1/(R_1 + R_2)$] is fed-back to the non-inverting input terminal labelled A. The voltage at A is therefore $\beta V^+_{o\ sat}$. The capacitor C will now begin to charge through the resistor R and when the voltage across it reaches $\beta V^+_{o\ sat}$ (or some minute voltage in excess of this value) the output will swing to the negative saturation voltage level $V^-_{o\ sat}$. The transition is so fast that the voltage at X is virtually unchanged but the voltage at A is now $\beta V^-_{o\ sat}$. The capacitor C will now start to discharge through R until the operational amplifier switches again when the voltage at X is $\beta V^-_{o\ sat}$. If the saturation voltage levels are symmetrical about the common line the time period T of the multivibrator will be given by

$$T = 2RC \log_e [1 + (2R_1/R_2)] \qquad (5.12)$$

The voltage waveforms at the various points are shown in fig. 5.27.

EXAMPLE 5.9 *Derive an expression for the time period of oscillation of the multivibrator shown in fig. 5.26. Assume that the positive and negative saturation levels are of equal magnitude. Calculate the frequency of the multivibrator if $C = 0.1 \mu F$, $R = 16k\Omega$, $R_1 = 20k\Omega$ and $R_2 = 80k\Omega$.*

The time period will be determined by the time taken for the capacitor C to charge from, say, $\beta V^-_{o\ sat}$ to

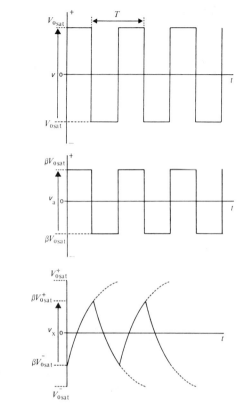

Fig. 5.27 Voltage waveforms associated with an astable multivibrator

$\beta V^+_{o\ sat}$ through the resistor R. This time is $T/2$, where T is the period (fig. 5.27).

The expression for the time t for a capacitor C to charge from V_1 to V_2 from a source of constant voltage V_{ref} is given by

$$t = CR \log_e \frac{V_{ref} - V_1}{V_{ref} - V_2} \qquad (5.13)$$

where R is the resistance in series with C.

Hence,

$$\frac{T}{2} = CR \log_e \frac{V^+_{o\ sat} - \beta V^-_{o\ sat}}{V^+_{o\ sat} - \beta V^+_{o\ sat}} \qquad (5.14)$$

But $V^+_{o\ sat} = -V^-_{o\ sat}$

Therefore, $T = 2CR \log_e \left[\dfrac{1+\beta}{1-\beta} \right]$

Therefore, $T = 2CR \log_e [1 + (2R_1/R_2)]$

because $\beta = R_1/(R_1 + R_2)$.

Substituting $C = 10^{-7}$ F, $R = 16 \times 10^3 \Omega$ and R_1/R_2
$= 0.25$.

Therefore, $T = 2 \times 10^{-7} \times 16 \times 10^3 \log_e 1.5$
$= 1.29 \times 10^{-3}$ s

Therefore, the frequency $f = 10^3/1.29 = 770$Hz.

(e) *A squaring circuit.* In the multivibrator circuit of fig. 5.26, the output from the operational amplifier was switched automatically by the *RC* network from the positive saturation level to the negative voltage saturation level. This behaviour immediately suggests that an operational amplifier could serve to transform any input waveform into a rectangular waveform (fig. 5.28) of the same frequency. The amplitude of the input signal should be limited and ideally the waveform should be symmetrical about the common line voltage. However, if the input signal is superimposed on some dc level, the dc voltage level at the non-inverting terminal could be varied about the common line voltage by means of a potential divider. In this way, a symmetrical output could be obtained.

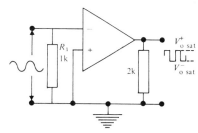

Fig. 5.28 A squaring circuit

The input resistance is virtually R_1 and may be chosen to match the source impedance. Provided that the input amplitude is sufficient to cause the switching, the amplitude of the output signal is constant and independent of the input signal. Any fraction of the output signal may be selected by means of a potential divider across the output in place of the 2kΩ load resistor shown in fig. 5.28

(f) *A crystal-controlled oscillator.* Positive feedback (Appendix B) is provided via the crystal (section 6.6) to the non-inverting input (fig. 5.29). Selection of the dc operating point is by choice of the values of the resistors R_1 and R_2 whereas the capacitor C_1 decouples R_1 at the oscillation frequency in order to remove negative feedback. The reactance X_C of C_1 should be chosen so that $X_C < R_1/500$. The positive feedback voltage is developed across the resistor R_3 which should have a value of approximately $R_1 R_2/(R_1 + R_2)$.

(g) *A logarithmic amplifier.* An amplifier which provides an output voltage proportional to the logarithm of the input voltage can be obtained by the use of a diode as the feedback element in an operational amplifier (fig. 5.30).

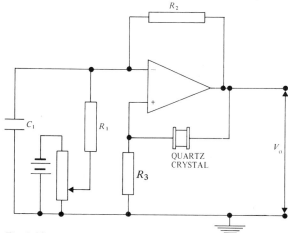

Fig. 5.29 A crystal controlled oscillator

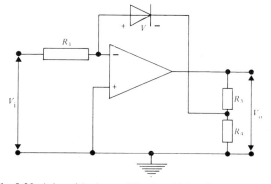

Fig. 5.30 A logarithmic amplifier in which a diode is used as the feedback element

The current-voltage relationship for a *p-n* junction (section 2.1) is

$$I = I_s [\exp(eV/kT) - 1] \qquad (5.15)$$

where I_s is the reverse saturation current, V is the voltage across the junction, k is the Boltzmann constant, and T is the absolute temperature. At 300K, $kT/e \simeq 26$mV; hence for voltages in excess of about 100mV, equation (5.15) can be written in the simpler form:

$$I = I_s \exp(eV/kT)$$

Therefore $\log_e (I/I_s) = eV/kT$

and $V = (2.3\, kT/e) \log_{10} (I/I_s)$ (5.16)

Assuming that the operational amplifier in fig. 5.30 is ideal, the voltage across the diode is given by

$$V = -R_4 V_o/(R_3 + R_4) = 2.3\,(kT/e) \log_{10}(I/I_s) \qquad (5.17)$$

Note that $I = V_i/R_1$.

The load resistors R_3 and R_4 are included in the manner shown in fig. 5.30 to allow a convenient output signal to be obtained. The factor 2.3 (kT/e) has a value of approximately 60mV per current decade (equation 5.17). It is often convenient to arrange for the output voltage to increase by one volt per current decade and this can be achieved readily by the correct choice of R_3 and R_4.

It would appear that the straightforward circuit of fig. 5.30 could be used over, perhaps, seven decades of current. Unfortunately, more complex circuits are necessary to achieve high performance over such a wide range.

The validity of equation $I = I_s \exp(eV/kT)$ should be examined under constant temperature conditions for a number of diodes over the current range 10^{-2} to 10^{-8} A. Some diodes behave much better than others in this respect so that manufacturers often select and market diodes which are particularly suited for logarithmic amplifiers.

Equation (5.17) indicates that V_o is dependent on the temperature; I_s is also a function of temperature.

An *n-p-n* transistor is often used to replace the diode in the circuit of fig. 5.30. The collector is connected to the summing junction, the base is earthed and the emitter is connected to the junction between the resistors R_3 and R_4. The base-emitter junction is really being utilised because the voltage between the collector and base is effectively zero.

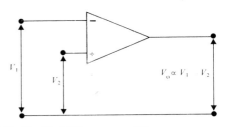

Fig. 5.31 The ideal difference amplifier

5.10 Common-mode rejection ratio

The output voltage V_o from an ideal differential operational amplifier is directly proportional to the input difference voltage $V_1 - V_2$ (fig. 5.31). Hence when the two input voltages V_1 and V_2 are equal in magnitude and phase, V_o should be zero. However, such perfect balance does not exist in real differential amplifiers. The output voltage from such an actual amplifier may be expressed as the sum of two terms, i.e.

$$V_o = A_D (V_1 - V_2) + M(V_1 + V_2) \qquad (5.18)$$

where the first term is proportional to the difference between the two input voltages whereas the second is directly proportional to their sum. A_D is called the *differential gain* whereas M is a constant. If the two inputs are equal (i.e. $V_1 = V_2$) the output from the real amplifier is

$$V_o = 2MV_1 \qquad (5.19)$$

The *common-mode gain* A_C is defined as the ratio of the output voltage to the common-mode input voltage. Hence with $V_1 = V_2$

$$A_C = V_o/V_1$$

Therefore, $A_C = 2MV_1/V_1 = 2M$.

Equation (5.18) can hence be written

$$V_o = A_D(V_1 - V_2) + A_C[(V_1 + V_2)/2] \quad (5.20)$$

The ratio of the differential gain A_D to the common-mode gain A_C is called the *common-mode rejection ratio* (CMRR). This term, often denoted by the letter C serves as a figure of merit for the difference amplifier.

Rewriting equation (5.20)

$$V_o = \underset{\underset{\text{difference term}}{\downarrow}}{A_D(V_1 - V_2)} + \underset{\underset{\text{sum or common-mode term}}{\downarrow}}{(A_D/C)[(V_1 + V_2)/2]} \quad (5.21)$$

The sum term may be regarded as an error term because it is desirable to have the output proportional to the voltage difference alone. This error term is small if the common-mode rejection ratio C is made very large. With good amplifier design, values of C exceeding 10^4 can be achieved.

EXAMPLE 5.10 *A differential operational amplifier has a differential gain of A_D. If the input signal voltages are initially $V_1 = +100\mu V$, $V_2 = -100\mu V$ and later are $V_1 = +1100\mu V$ and $V_2 = 900\mu V$, calculate the percentage difference in the output voltage obtained for the two sets of input signals, (a) when the common-mode rejection ratio C is 100 and (b) when $C = 10000$.*

(a) Initially: $V_1 = +100\mu V$, $V_2 = -100\mu V$ so $(V_1 - V_2) = 200\mu V$ and $(V_1 + V_2) = 0$. Substituting these values together with $C = 100$ into equation (5.21) gives

$$V_o = A_D 200\mu V + (A_D/100)0 = 200 A_D \mu V.$$

Subsequently: $V_1 = +1100\mu V$, $V_2 = +900\mu V$ so $(V_1 - V_2) = 200\mu V$, and $(V_1 + V_2) = 2000\mu V$. Substituting these values together with $C = 100$

into equation (5.21) gives

$$V_o = [A_D 200 + (A_D/100)(2000/2)]\mu V$$
$$= 200 A_D(1 + 0.05)\mu V.$$

Hence the difference = 5%.

(b) Using the same values as in (a) except that $C = 10000$ we have:

initially: $\quad V_o = 200 A_D \mu V$

subsequently: $V_o = A_D 200 + (A_D/10000) \times (2000/2) \mu V$

$$= 200 A_D(1 + 0.0005) \mu V.$$

Hence the difference is 0.05% which is negligible.

5.11 Finite gain error, loop gain and feedback factor

An ideal operational amplifier is taken to provide an infinite gain (section 5.3) so that negative feedback would maintain the voltage across the amplifier input terminals at an infinitesimally small value. In the practical case of an amplifier which provides a finite gain, a small voltage V_e will exist between the input terminals (fig. 5.32).

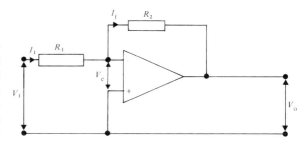

Fig. 5.32 Inverting amplifier with finite open loop gain

When the current flowing into the amplifier with finite gain is zero, in the inverting configuration (fig. 5.32):

$$(V_1 - V_e)/R_1 = (V_e - V_o)/R_2 \quad (5.22)$$

The input error voltage V_e may be defined by

$$V_e = -V_o/A \quad (5.23)$$

where A is written in place of A_{VOL}, the open-loop

voltage gain of the amplifier.

Substitution from equation (5.23) into equation (5.22) yields

$(V_1 + V_o/A)/R_1 = (-V_o/A - V_o)/R_2$

Therefore $V_o(1/AR_1 + 1/AR_2 + 1/R_2) = -V_1/R_1$

and the closed-loop voltage gain is

$$A_{VCL} = \frac{V_o}{V_1} = -\frac{R_2}{R_1}[1/\{1 + 1/A(1 + R_2/R_1)\}]$$
(5.24)

Writing $R_1/(R_1 + R_2) = \beta$, gives

$A_{VCL} = -(R_2/R_1)[1/(1 + 1/\beta A)]$ (5.25)

Usually $\beta A \gg 1$, hence

$A_{VCL} = -(R_2/R_1)(1 - 1/\beta A)$ (5.26)

Because $A \to \infty$, $A_{VCL} \to -R_2/R_1$, as derived in section 5.5 for the ideal operational amplifier.

In the non-inverting configuration (fig. 5.33):

$V_2 + V_e = V_o R_1/(R_1 + R_2)$ (5.27)

also $V_o = -A V_e$

Therefore, $V_2 - V_o/A = V_o R_1/(R_1 + R_2)$

Therefore, $V_2 = V_o[1/A + R_1/(R_1 + R_2)]$

and $A_{VCL} = \frac{V_o}{V_2} = [(R_1 + R_2)/R_1][1/(1 + 1/\beta A)]$

$= (1 + R_2/R_1)[1/(1 + 1/\beta A)]$

$\simeq (1 + R_2/R_1)(1 - 1/\beta A)$ (5.28)

Again, as $A \to \infty$, $A_{VCL} \to (1 + R_2/R_1)$ as derived in section 5.5.

The term

$1/(1 + 1/\beta A) \simeq 1 - 1/\beta A$

is sometimes called the 'finite gain error'. The quantity β is known as the *feedback fraction* and βA is the *loop gain*, which is clearly a very important

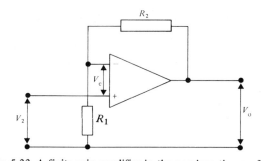

Fig. 5.33 A finite gain amplifier in the non-inverting configuration.

factor in determining the closed loop performance.

A large value of the loop gain provides a system behaviour dependent only on input and feedback components. Furthermore, four other important improvements are ensured. The first is that the closed loop stability is improved as βA is increased. This can be shown by considering the ratio $\Delta A_{VCL}/A_{VCL}$, where ΔA_{VCL} denotes a small change in A_{VCL}. It is easily seen that

$\Delta A_{VCL}/A_{VCL} = (\Delta A/A)/\beta A$

and, as ΔA_{VCL} needs to be small for good closed loop stability, clearly βA needs to be large.

The second improvement consequent upon increasing βA is reduction of the closed-loop output impedance, Z_{OCL}. This is given by

$Z_{OCL} = Z_{OOL}/\beta A$

where Z_{OOL} is the open-loop output impedance. Obviously, βA needs to be large for Z_{OCL} to be small.

The third is that the closed-loop distortion D_{CL} is reduced because

$D_{CL} = D_{OL}/\beta A$

where D_{OL} is the open-loop distortion.

Fourthly, the frequency response of the amplifier is improved.

EXAMPLE 5.11 *Derive a general expression relating the gain with feedback (A_{VCL}) of an amplifier to the open loop gain A. Distinguish between the cases of negative and positive feedback.*

Assuming that the feedback is negative and the loop gain is large, show that the performance is substantially independent of changes in that of the amplifier itself.

Use this analysis to arrive at an expression for the gain of an operational amplifier in the non-inverting configuration.

Consider the amplifier of open-loop gain A where the feedback network provides a fraction β of the output voltage V_o to the summing junction (fig. 5.34).

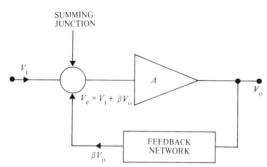

Fig. 5.34 Concerning example 5.11

Using the notation given in fig. 5.34,

$V_e = V_i + \beta V_o$

$V_o = A V_e = A(V_i + \beta V_o)$

Therefore, $V_o (1 - \beta A) = A V_i$

Therefore, $A_{VCL} = V_o / V_i = A/(1 - \beta A)$.

Note that no inversion has been assumed to take place in the amplifier.

If $|1 - \beta A| > 1$, the feedback is negative and $A_{VCL} < A$.
If $|1 - \beta A| < 1$, the feedback is positive and $A_{VCL} > A$.

When the feedback is large and the loop gain βA is large, assuming an inverting amplifier of gain $-A$ and that the feedback network is resistive such that

$A_{VCL} = A/(1 + \beta A)$

As $|\beta A| \gg 1$,

$A_{VCL} \simeq 1/\beta$

so A_{VCL} is dependent on the values of the resistances in the feedback network but is virtually independent of changes in A.

In the case of a non-inverting operational amplifier (see fig. 5.3b)

$$A_{VCL} = 1/\beta = \frac{R_1 + R_2}{R_1} = 1 + R_2/R_1$$

5.12 Frequency response

The operational amplifier with its directly coupled stages can amplify dc signals and performs well with low-frequency ac signals. Due to capacitance at the junctions of the active devices and between neighbouring conductors, the gain decreases as the frequency increases.

It is convenient to represent the high-frequency behaviour of the amplifier without feedback by an equivalent circuit (fig. 5.35).

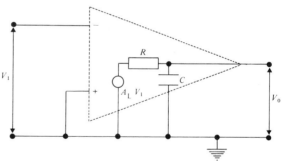

Fig. 5.35 The high-frequency equivalent circuit of an operational amplifier

The input voltage V_1 produces an output voltage $A_L V_1$ where A_L is the low-frequency open-loop gain of the amplifier. At low frequencies the reactance X_C of the capacitance C is large compared with R so the output voltage which appears across it is virtually $A_L V_1$. At higher frequencies the divider network provided by R and C causes the output voltage to be smaller.

The gain of the amplifier $A = V_o/V_1$.

But $V_o = \dfrac{A_L V_1 (-jX_C)}{R - jX_C}$

Therefore $A = A_L/(1 + j 2 \pi f R C)$

The *corner frequency* f_1 is defined as a specific frequency where

$$f_1 = 1/2\pi RC$$

At this frequency f_1,

$$A = A_L / [1 + j(f/f_1)]$$

Therefore, $A = \dfrac{A_L \exp(-j\phi)}{\sqrt{1 + (f/f_1)^2}}$ (5.29)

where $\phi = \tan^{-1}(f/f_1)$.

The magnitude of the voltage gain $|A|$ expressed in dB is usually plotted against frequency to yield the frequency-response curve (also called a Bode plot) of which an example is shown in fig. 5.36.

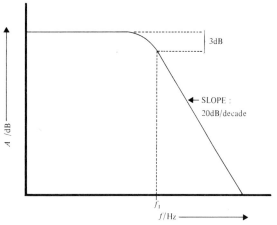

Fig. 5.36 The frequency response curve for an amplifier of the type shown in fig. 5.35

The gain against frequency behaviour (fig. 5.36) is given by

$$|A| = A_L / \sqrt{1 + (f/f_1)^2}$$

and is called a *first order frequency response*.

Writing the gain in dB, the gain equation takes the form

$$A = 20\log_{10} A_L - 20\log_{10}[1 + (f/f_1)^2]^{1/2} \quad (5.30)$$

Three regions are of particular interest:—

At low frequencies (i.e. $f \ll f_1$), $A = A_L$

At a frequency $f = f_1$, $A = A_L - 10\log_{10} 2$

Therefore, $A = A_L - 3$ where A and A_L are in dB. Hence, at the corner frequency f_1, the gain has dropped by 3dB from its low frequency value. The corner frequency is thus sometimes called the '3dB frequency'.

At much higher frequencies (i.e. $f \gg f_1$), from equation (5.30)

$$A = 20\log_{10} A_L - 20\log_{10}(f/f_1)$$

The rate at which the gain falls off at frequencies greater than f_1 is of special interest. Thus, it is required to find ΔA between f_A and f_B where $f_B > f_A \gg f_1$.

From equation (5.30) it is seen that

at $f = f_A$, $A_{f_A} = A_L - 20\log_{10}(f_A/f_1)$,

and

at $f = f_B$, $A_{f_B} = A_L - 20\log_{10}(f_B/f_1)$.

Therefore, $\Delta A = A_{f_A} - A_{f_B} = 20\log_{10}(f_A/f_B)$

$$= -20\log_{10}(f_B/f_A)$$

Suppose that $f_B = 10 f_A$, then

$$\Delta A = -20\text{dB},$$

i.e. the gain decreases (*rolls off*) at a rate of 20dB/decade.
Suppose that $f_B = 2f_A$, then

$$\Delta A = -20\log_{10} 2 = -6,$$

and the gain rolls off at −6dB/octave.

It is frequent practice to utilise two straight lines (fig. 5.37a) as an approximation to the frequency response curve. The origin of the term 'corner frequency' for f_1 is now apparent.

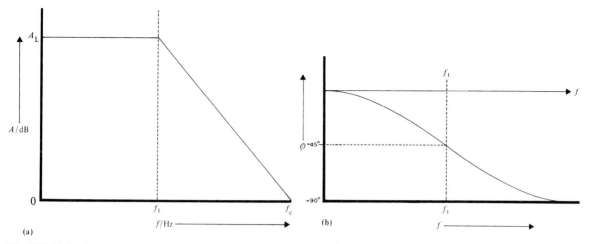

Fig. 5.37 **(a)** Straight line approximation to the frequency response curve; **(b)** phase angle against frequency

The response curve crosses the frequency axis (i.e. the amplifier gain has dropped to unity, or 0dB) at a frequency f_C called the *crossing* or *transition frequency*.

A graph of the phase angle ϕ against frequency is shown in fig. 5.37b. As $\phi = \tan^{-1}(f/f_1)$ it follows that $\phi = -45°$ at the corner frequency $f = f_1$. Because $\tan \phi \to \infty$ as $f \to \infty$, the maximum phase shift between the output and the input voltage is 90°. An amplifier of this type must be stable for all values of resistive feedback because the phase shift cannot exceed 90°.

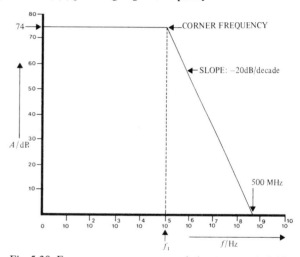

Fig. 5.38 Frequency-response curve relating to example 5.12

EXAMPLE 5.12 *An operational amplifier with a first-order frequency response has a low-frequency voltage gain of 5000 and a corner frequency $f_1 = 100kHz$. Sketch the frequency response curve of this amplifier and calculate its voltage gain at 1MHz.*

The low frequency voltage gain A_L expressed in dB

$= 20 \log_{10} 5000 = 20 \times 3.70 = 74$dB.

At the crossing frequency $f = f_C$, the amplifier will have a gain of unity, then,

$0 = 20 \log_{10} A_L - 20 \log_{10}(f_C/f_1)$

Therefore, $f_C = A_L f_1 = 5000 \times 100$kHz $= 500$MHz.

At $f = 1$MHz, i.e. one decade above the corner frequency, the gain will have fallen by 20dB.

Therefore gain at 1MHz = 54dB.

The frequency-response curve required is in fig. 5.38.

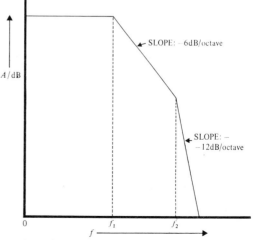

Fig. 5.39 A frequency response curve with two corner frequencies

We have assumed for simplicity that there is one corner frequency. For some amplifiers, the frequency response is more complex: the case where there are two corner frequencies, f_1 and f_2, is shown in fig. 5.39. The gain-frequency relationship could then be represented by

$$A = A_L/[(1 + jf/f_1)(1 + jf/f_2)]$$

where f_1 is the first corner frequency and f_2 is the second corner frequency. Between frequencies f_1 and f_2 the gain rolls off at $-6\text{dB}/\text{octave}$ whereas above f_2 the gain rolls off at $-12\text{dB}/\text{octave}$.

5.13 Bandwidth, gain bandwidth product and closed loop frequency response

The band of frequencies over which the gain is approximately constant is called the *bandwidth* of the amplifier. For those we have been discussing the gain is constant over the frequency range from 0 to f_1, so the bandwidth is f_1. A large bandwidth is obviously desirable: its magnitude is determined by amplifier design and construction.

Limitations due to bandwidth are particularly noticeable when attempting to amplify pulses. If the input signal is of rectangular waveform so containing the sum of odd harmonics, the first corner frequency of the amplifier may be above the first and third harmonic but below the fifth. In this case the amplifier will act as a filter whereby the fifth and higher harmonics are amplified less than the lower harmonics so that distortion is introduced.

Closed loop frequency response. To investigate the frequency response of an amplifier having negative feedback, assume a first-order frequency-response curve of form similar to that of fig. 5.37.

Denote the open loop gain of the amplifier by A, the open loop gain at low frequency by A_L and the closed loop gain (i.e. with feedback) by A_F (not A_{VCL}, as previously), we have

$$A = A_L/(1 + jf/f_1)$$
and
$$A_F = A/(1 + \beta A)$$

where β is the feedback factor.

Therefore, $A_F = \dfrac{A_L/(1 + jf/f_1)}{1 + \beta A_L/(1 + jf/f_1)} = \dfrac{A_L}{1 + \beta A_L + jf/f_1}$

$$(5.31)$$

Therefore, $A_F = \dfrac{A_L/(1 + \beta A_L)}{1 + jf/[f_1(1 + \beta A_L)]} = \dfrac{A_{FL}}{1 + jf/f_{1F}}$

where A_{FL} is the low-frequency closed loop gain. and f_{1F} is the closed-loop corner frequency.

With $f_{1F} = f_1(1 + \beta A_L)$
it is apparent that the closed-loop bandwidth is greater than the open-loop bandwidth. The frequency-response curves for A and A_F are shown in fig. 5.40.

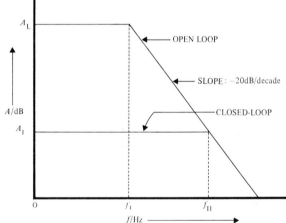

Fig. 5.40 Frequency response curves for open- and closed-loop amplifiers

The *gain-bandwidth product* is given by

$$A_{FL}f_{1F} = A_L f_1 (1 + \beta A_L) / (1 + \beta A_L) = A_L f_1$$

Hence the gain-bandwidth product of the closed-loop amplifier is equal to the gain-bandwidth product of the open-loop amplifier. This is only true for amplifiers having a first order frequency response, i.e. having a roll-off of $-20\text{dB}/\text{decade}$ or $-6\text{dB}/\text{octave}$.

EXAMPLE 5.13 *An operational amplifier has a low frequency voltage gain of 10^4 and a corner frequency $f_1 = 10\text{kHz}$. With a resistive feedback network it has an ideal closed loop gain of 10. Calculate the closed loop corner frequency and sketch the open- and closed-loop frequency response curves, assuming a first-order frequency response.*

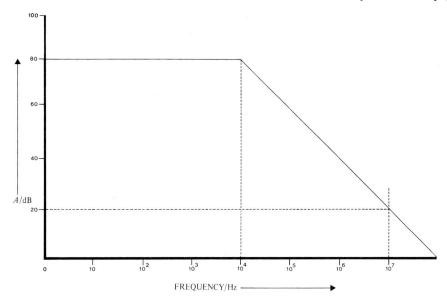

Fig. 5.41 Frequency response curve for the amplifier in example 5.13

$A_L = 10^4 = 20 \log_{10}(10^4) = 80\text{dB}$
The closed-loop gain = $10 = 20\text{dB} = 1/\beta$
The closed-loop corner frequency

$$f_{1F} = f_1(1 + \beta A_L) = 10^4(1 + 10^{-1} \times 10^4)$$
$$= 10^7 = 10\text{MHz}$$

An alternative approach. Because the gain-bandwidth products are equal

$$A_{FL}f_{1F} = A_L f_1$$

Therefore, $20 \log_{10} 10 + 20 \log_{10} f_{1F} = 20 \log_{10}(10^4) + 20 \log_{10}(10^4)$
Therefore, $20 + 20 \log_{10} f_{1F} = 160$
Therefore, $\log_{10} f_{1F} = 140/20 = 7$
Therefore, $f_{1F} = 10^7 \text{Hz} = 10\text{MHz}$.
The frequency-response curves required are shown in fig. 5.41.

5.14 Slew rate

The presence of internal capacitance limits the rate at which the output voltage can change. The maximum possible time rate of change of output voltage is called the *slew rate* and is usually specified in volts per microsecond. Hence the maximum frequency at which an amplifier can usefully operate is determined by both the slew rate and the bandwidth. However, with a sinusoidal input signal, whereas the bandwidth limits the gain at the higher frequencies, slew rate results in distortion of the waveform. With a non-sinusoidal signal, both the slew rate and the bandwidth can contribute to distortion of the waveform.

A voltage of sinusoidal waveform is represented by
$v = V_P \sin \omega t$
The time rate of change of v is given by

$$\frac{dv}{dt} = V_P \omega \cos \omega t$$

and this has a maximum value when $t = 0$, corresponding to the points where the sinusoidal waveform crosses the time axis.

$$\left(\frac{dv}{dt}\right)_{max} = 2\pi f V_P$$

For the amplifier to operate without distortion, the slew rate S must exceed $2\pi f V_P$, i.e.

$$S \geqslant 2\pi f V_P \qquad (5.32)$$

It follows, therefore, that slew rate distortion (i.e. distortion created by an inadequate slew rate) depends on both the frequency and the amplitude of the output voltage. Because the slew rate is constant for a given amplifier, the amplifier design can provide for either a large output voltage or a good high frequency performance, but not both.

Conventional operational amplifiers have a slew rate of 0.5 to 1.0V/μs; special purpose types are available with a slew rate of 200V/μs.

EXAMPLE 5.14. *An amplifier has a slew rate of 5V/μs. If the input is sinusoidal, what is the maximum frequency at which the amplifier can operate to produce an undistorted output signal of amplitude 2.0V?*

From equation (5.32.)

$$f = 5/2\pi V_p = 5/(2\pi \times 2 \times 10^{-6}) \simeq 4 \times 10^5 \text{ Hz}.$$

5.15 Further examples

(i) *In a laboratory experiment, a student is using a transducer which provides an output of 2.0mV and has an internal resistance of 500Ω. He decides to use an operational amplifier in the inverting mode to provide a gain of 50. The amplifier is known to have the following characteristics:* –

Open loop gain A = 5000
Input resistance R_i = 100kΩ
Output resistance R_o = 100Ω

Using the gain equation $A_{VCL} = -R_2/R_1$ for an ideal amplifier he sets up the circuit of fig. 5.42 and measures the output voltage with a digital voltmeter having an input resistance of 10MΩ. Instead of the expected output of 100mV he records a very much lower output voltage and yet can find nothing wrong with his circuit.
Draw the equivalent circuit of his amplifier, calculate the output voltage he measures, explain why his

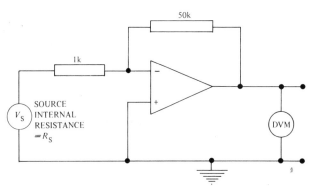

Fig. 5.42 Circuit of amplifier as set up by student (example 5.15(i))

design is useless and suggest a circuit which will behave in the way he intended.

The equivalent circuit to the amplifier arrangement of fig. 5.42 is shown in fig. 5.43.

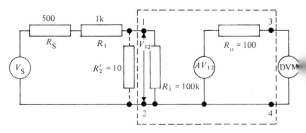

Fig. 5.43 Circuit equivalent to amplifier of fig. 5.42

The Miller resistor R'_2 across the input terminals of the amplifier (*see* Appendix C) is $R_2/(1-A)$

$$= \frac{50 \times 10^3}{1 + 5000} \simeq 10\Omega$$

(note that the open loop voltage gain A is negative). Hence the input resistance R_i (= 100kΩ) of the amplifier can be neglected.
The chain of resistors R_S, R_1 and R'_2 provides a voltage divider such that

$$V_{12} = \frac{V_S \times 10}{500 + 1000 + 10} = \frac{2 \times 10^{-3} \times 10}{1.5 \times 10^3}$$
$$= 1.33 \times 10^{-5} \text{V}$$

Hence the output voltage V_o as recorded by the digital voltmeter $= AV_{12} = -5000 \times 1.33 \times 10^{-5}$V, so that

$$V_o = -66.6 \text{mV}.$$

The student assumed that A_{VCL}
$= -R_2/R_1$

$$= -50 \times 10^3/10^3 = -50,$$

which would give

$$V_o = -100\text{mV}.$$

If the internal resistance of the transducer were 1000Ω the output would be −50mV.
These calculations indicate that the amplifier set up in this form is useless because its output varies with

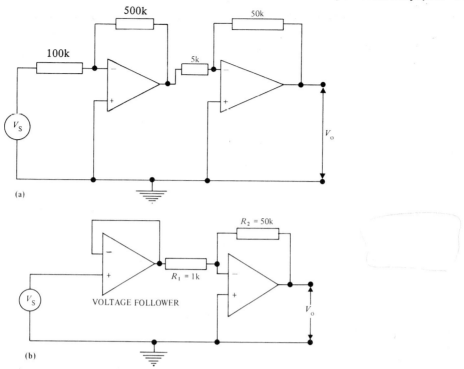

Fig. 5.44 (a) The first solution (NB signal inversion occurs at each amplifier); (b) The second solution: a voltage follower is used as a buffer amplifier (NB the output signal is inverted)

the resistance of the transducer. The basic fault is that the input resistance of the amplifier $(R_1 + R'_2)$ is only 510Ω and therefore of the same order of size as that of the transducer. The source resistance should be very small compared with the input resistance of the amplifier.

A first solution is to use two operational amplifiers where the first has R_i (the effective input resistance) = $100k\Omega$ so that the gain of this amplifier would be

$$A_{VCL} = -R_2/R_1 = -\frac{5 \times 10^5}{10^5} \times 2 = -5$$

This could be followed by a second amplifier with a gain of 10 (fig. 5.44a).

An alternative solution is to use the first operational amplifier as a unity gain voltage follower. This has a very high input resistance (perhaps $10M\Omega$) and a low output resistance. The second amplifier with a gain of 50 would then provide the required output (fig. 5.44b).

Using the equivalent circuit (fig. 5.43) to calculate V_{12} in the case of fig. 5.44a gives

$$V_{12} = \frac{2 \times 10^{-3} \times 100}{10^5 + 500 + 100} \simeq 2\mu V$$

$$V_o = -5 \times 10^3 \times 2 \times 10^{-6} V = -10mV$$

This value is equal to that given by the ideal amplifier equation (section 5.5)

$$V_o = -(R_2/R_1) \times 2mV = -\left(\frac{5 \times 10^5}{10^5}\right) \times 2 = -10mV.$$

The amplifier is behaving correctly because $R_1 \gg R_S$. The following stage will also behave as an ideal amplifier providing a further voltage gain of 10 to give the expected output of 100mV. In the same way the amplifier after the voltage follower (fig. 5.44b) would behave in an ideal manner.

(ii) *The operational amplifier circuit (fig. 5.45) comprises the following components:—*

Operational amplifier: open loop gain $A = 10^4$; input resistance $R_i = 100k\Omega$

Feedback components: $R_1 = 10k\Omega$; $R_2 = 100k\Omega$.

The input voltage $V_i = 0.5 V$.

Use the Miller theorem (Appendix C) to establish the equivalent circuit, the input resistance R_i of the stage and the output voltage V_o

The Miller theorem enables the feedback resistor R_2 to be replaced by a fictitious resistor R_2' of magnitude $R_2/(1 - A)$ across the input terminals and another fictitious resistor of magnitude $R_2/(1 - 1/A)$ across the output terminals, where A is the voltage gain of the amplifier. The equivalent circuit is hence that of fig. 5.46.

The resistor R_2' is said to be fictitious because it could never be produced experimentally: it is merely a mathematical device to aid the analysis. In the inverting amplifier of fig. 5.45, A is negative so that

$$R_2' = 10^5/(1 + 10^4) \simeq 10\Omega$$

The Miller resistor of value 10Ω is in parallel with the input resistance R_i of $100k\Omega$; R_i can therefore be neglected. The input circuit is thus a simple voltage divider comprising R_1 and R_2'. It is apparent that the input resistance to the amplifier stage is R_1 as previously established (section 5.6), which, in this case, is $10k\Omega$. The voltage across R_2' is hence

$$V_{12} = \frac{0.5 \times 10}{10^4 + 10} \simeq 500\,\mu V$$

The output voltage $AV_{12} = 10^4 \times 5 \times 10^{-4} = 5V$.

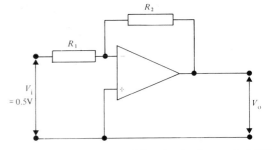

Fig. 5.45 The operational amplifier circuit of example 5.15(ii)

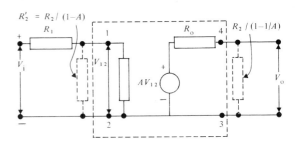

Fig. 5.46 Equivalent circuit of the amplifier of fig. 5.45

As, from equation (5.1)

$$A_{VCL} = R_2/R_1$$

Therefore, $A_{VCL} = 10^5/10^4 = 10$,

so that $V_o = 10 \times 0.5 = 5V$.

Exercises 5

1. What characteristics is the ideal operational amplifier assumed to have?
   Assuming that the amplifier is ideal and that the input and feedback components are resistors, derive an expression for the closed loop voltage gain in (a) the inverting and (b) the non-inverting configuration.

2. Define and explain the significance of the following terms: *virtual earth, summing junction, common mode voltage, open loop and closed loop, monolithic integrated circuit*.

3. An operational amplifier is to be used in the inverting configuration with an input resistance of $10k\Omega$ and a closed loop voltage gain of 20.

   (a) Assuming that the amplifier is ideal, calculate the value of the input and feedback resistor.
   (b) If the resistors calculated in case (a) are used with an operational amplifier having a finite open loop gain of 60dB, determine the voltage gain.

4. Define the term *common mode rejection ratio* (CMRR). A differential operational amplifier has an open loop voltage gain, A, of 4000 and a CMRC, C, of 1000. If the dc voltage between the inverting input terminal and the common line is 6mV and that between the non-inverting terminal and common is 4mV, calculate the output voltage.

5. A differential operational amplifier has an open loop voltage gain of 10^3 and a CMRR of 10^4.

Initially, the common-mode signal voltage is zero and a voltage difference of 1.0mV exists between the input terminals. Calculate the output voltage. If the common-mode voltage is increased while the difference voltage is maintained at 1.0mV, what value of the common-mode voltage will cause the output to increase by 1%?

6. The operational amplifier shown in fig. 5.47 has an open-loop voltage gain, A, of 10^4 and an input resistance of 50kΩ. If the source resistance is negligible and the source voltage is 1.0V, use the Miller theorem to derive the input resistance of the stage and to calculate the output voltage. Draw the equivalent circuit of this amplifier.

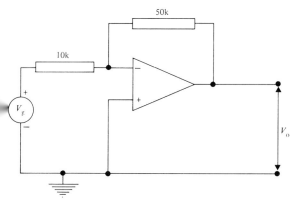

Fig. 5.47 Concerning exercise 6

7. Fig. 5.48 is the circuit diagram of a laboratory instrument designed around a 741 operational amplifier. Examine this circuit carefully and answer the following:—
   (a) For what purpose was the circuit designed?
   (b) In what configuration is this amplifier operating, and why was this configuration chosen?
   (c) What is the closed-loop voltage gain (A_{VCL}) of this amplifier?
   (d) What is the function of the potentiometer P?
   (e) What is the function of diodes D_1 and D_2?
   (f) Label the resistors and switch positions which are not marked already on the diagram.
   (g) What advantage does this instrument have over a moving-coil meter not utilising an operational amplifier?
   (h) Suggest how this circuit could be modified to enable it to function with an ac input.

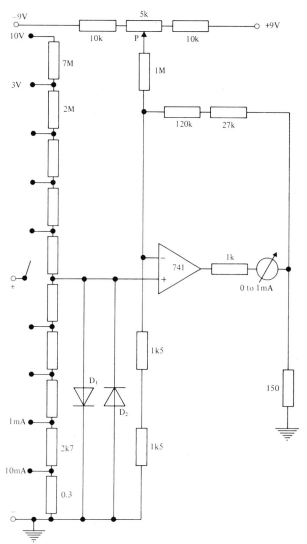

Fig. 5.48 Concerning exercise 7

8. Examine the circuit of fig. 5.49 and answer the following:—
   (a) For what purpose was this circuit designed?
   (b) Discuss the role of the operational amplifier in this circuit.
   (c) The operational amplifier has only one voltage supply; why is this satisfactory in this circuit?
   (d) Explain why the Darlington pair (T_1 and T_2) has been used in this circuit.
   (e) Explain the function of the bipolar transistor T_3.

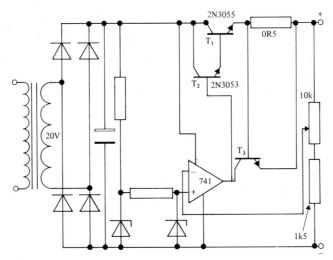

Fig. 5.49 Concerning exercise 8

9. Draw the circuit and analyse the behaviour of an operational amplifier used as a logarithmic amplifier.

10. An equivalent circuit for an operational amplifier is shown in fig. 5.50. Use this equivalent circuit to derive the expression

$$A = A_L \exp(-j\phi)\big/\sqrt{1 + (f/f_1)^2}$$

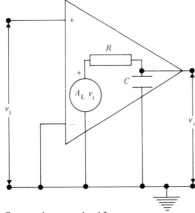

Fig. 5.50 Concerning exercise 10

where $\phi = \tan^{-1}(f/f_1)$, A is the open-loop gain at a frequency f and A_L is the low frequency open-loop gain. Define the corner frequency f_1 in terms of R and C and also in terms of the decrease in gain with frequency.

11. An operational amplifier has a low frequency open-loop gain $A_L = 3000$ and a corner frequency $f_1 = 1\text{MHz}$.
    (a) Calculate the transition frequency, i.e. the frequency at which the gain has dropped to unity.
    (b) Sketch the frequency response curve.
    (c) Calculate the gain, expressed in dB, at a frequency f = 100MHz.

12. The open-loop frequency response of an operational amplifier is designed to show a first-order characteristic such that

$$A = A_L/(1 + jf/f_1)$$

where A_L is the open-loop gain at low frequency, A is the open-loop gain at frequency f and f_1 is the corner frequency. Discuss, with analysis, the significance of this characteristic, and show that A rolls-off at -20dB/decade.

13. Establish that for an amplifier having a first order frequency response, the gain-bandwidth product of the closed-loop amplifier is constant and equal to the open loop gain-bandwidth product.
    An amplifier has a low frequency open-loop gain of 10^4 and a corner frequency f_1 of 1kHz. With a resistive feedback, the amplifier has an ideal closed loop gain of 100. Calculate the closed-loop corner frequency and sketch the frequency response curve.

14. Sketch the frequency response curve of an operational amplifier which has a low frequency open-loop gain A_L of 5000, a first corner frequency $f_1 = 1$MHz and a second order frequency $f_2 = 10$MHz. Write the equation which expresses the open-loop gain A in terms of A_L, f_1 and f_2.

15. Discuss the use of logarithmic units to express power and voltage ratios in amplifiers.

16. What is meant by the 'slew rate' of an amplifier? An amplifier has a slew rate of 100V/μs. If the input signal is sinusoidal, what is the maximum frequency at which the amplifier can operate to produce an undistorted output signal of amplitude 5.0V?

17. Draw the circuit diagram for and outline the operation of
    (a) a unity gain inverter, (b) a unity gain voltage follower, (c) a current-booster to increase the current available from an operational amplifier.

18. Draw the circuit diagram for and analyse the behaviour of (a) an operational differentiator and (b) an operational integrator.

6 · Waveform Generators

6.1 Waveform information

A voltage waveform displays voltage as a function of time: usually the voltage axis is vertical and the time axis is horizontal. To observe such a waveform an oscilloscope is used if the voltage change with time is rapid whereas for slower variations, an ultraviolet (uv) recorder or a pen recorder is convenient. In an analogue system, the voltage waveform contains the information of value: for example, the voltage at any instant may be the analogue of the angular position of a rotating shaft, the temperature of a surface or any other piece of information which has been converted into an electrical signal by means of a transducer. Amplifiers, storage systems, display and measuring systems of many designs have been developed for such transducers, where a prime requirement is that such systems do not introduce distortion or noise (i.e. the addition of fictitious information) to the original signal.

A simple example in sound is the pure note which may be converted into a sinusoidal voltage waveform by means of a microphone, stored or recorded on disc, tape or film sound track to be reproduced as required (at a chosen power level) with exactly the waveform of the original signal. The more complex voltage waveforms containing features characteristic of the human voice or an orchestra are handled in the same way by means of appropriate analogue circuits.

If the waveform information representing sound is to be transmitted in radio-communication, it is superimposed on a carrier wave of radio frequency: amplitude, frequency or phase modulation is practised.

In about 1960 it was recognised that information could be stored, transmitted, processed or displayed more efficiently, faster and at lower cost if it were in digital form. Semiconductor technology has provided the required compact, reliable digital integrated circuits at low cost so that very rapid processing of information in digital form is now routine.

The signals from the transducers are generally in analogue form. These signals are amplified, converted into digital form by an analogue-to-digital (A to D) converter, rapidly processed within a minicomputer and subsequently converted back to analogue form (by means of a D to A converter) to perform some control function.

6.2 Definitions and some common waveforms

The term *voltage pulse* may be defined as a waveform which is non-sinusoidal and which usually has sharp leading and trailing edges. The waveform may or may not be periodically recurrent: it is normally produced by a switching circuit, i.e. an active device is switched from a non-conducting to a fully conducting or saturated state.

Some common waveforms each of pulse height E volts are shown in fig. 6.1. The perfect rectangular waveform (fig. 6.1a) is used to define a number of terms relating to waveforms of this type. Thus, the amplitude of the pulse is referred to as the *pulse height* or *peak value*; the first edge (increase to the peak value) is called the *leading edge* and the edge formed when the voltage drops is termed the *trailing edge*. The duration of the pulse is the *pulse width* and denoted by t_p. The time interval between successive leading edges is the *pulse repetition time* (prt): a term normally used when the pulses occur at regular intervals. A series of successive pulses is termed a *pulse train*. The number of pulses per second in a regular train of pulses is the *pulse repetition rate* (prr) or *pulse repetition frequency* (prf).

A square wave is a special case of a rectangular waveform and occurs when $t_p = t_2$ (fig. 6.1a).

The rectangular waveform (fig. 6.1a) never occurs in practice: instead a waveform similar to that shown in fig. 6.2 is involved. The features of such a rectangular waveform are specified by three para-

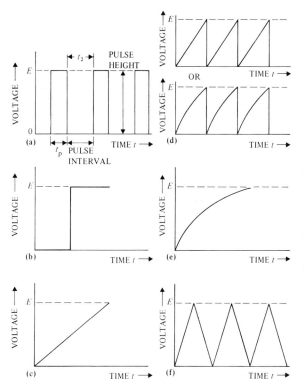

Fig. 6.1 Some common waveforms: (a) rectangular; (b) step or impulse; (c) linear voltage ramp; (d) sawtooth; (e) exponential; (f) triangular

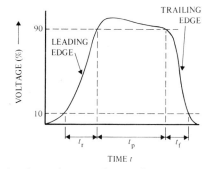

Fig. 6.2 An observed rectangular waveform

meters: the rise time (t_r), the pulse duration (t_p) and the decay time (t_f).

The rise time (t_r) is the time required for the voltage to rise from 10 to 90% of the maximum value.

The pulse duration (t_p) is the time for which the voltage is in excess of 90% of its maximum value.

The decay time (t_f), sometimes called the 'fall time', is the time taken for the trailing edge to fall from 90 to 10% of the maximum value.

6.3 The Barkhauszen criterion for oscillation

An oscillator is an amplifier with positive (or regenerative) feedback. Fig. 6.3 shows an amplifier of open-loop gain A and an inverting feedback network not connected to form a closed loop. With an input signal voltage v_i applied directly to the amplifier an output voltage v_o appears, where $v_o = Av_i$. The output from the feedback network is $-\beta A v_i$ (Appendix B). If $-\beta A v_i = v_i$, with the feedback loop connected, the amplifier would continue to provide an output signal even if the external input signal source were removed. Consider the special case in which the circuit operates in a linear fashion and the amplifier or the feedback network contains reactive components. Under such circumstances, only a sinusoidal oscillation will preserve its waveform.

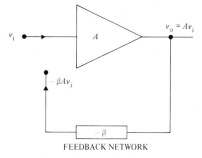

Fig. 6.3 An amplifier with a feedback loop not closed

Oscillation may be sustained if $-A\beta = 1$. This condition of unity loop gain is known as the *Barkhauszen criterion*. The condition implies that $|A\beta| = 1$ and that the phase of $-A\beta$ is zero or an integral multiple of 2π.

Consider a practical case: if $|\beta A| < 1$ and the external source of input voltage is removed, when the loop is closed, oscillations will begin but will die away rapidly. If $|\beta A| > 1$, when the external source is removed, the amplitude of the oscillation will continue to increase until non-linearity in the amplifier limits the amplitude. A steady state can only exist if the loop gain is unity. Hence, in every practical oscillator, the loop gain is initially slightly larger than unity and the amplitude of the oscillations is limited by the onset of non-linearity, or when $|A\beta| = 1$. The frequency of the oscillations will be the frequency for which the total phase shift is zero or $2n\pi$, where n is an integer. This last statement infers that every oscillator is a phase-shift oscillator, although the term 'phase-shift oscillator' is normally reserved for an oscillator containing one active amplifying device and a passive phase-shifting network (section 6.4).

The basic feedback equation (Appendix B) is

$$A_F = A/(1 + \beta A) \qquad (6.1)$$

where A_F is the gain with feedback, A is the open-loop gain and β is the feedback factor.

This equation (6.1) suggests that an increase in gain can be obtained by using positive feedback. However, this method of positive feedback to provide additional gain is never employed in amplifiers because the resulting amplifier would be inherently unstable. (The benefits of negative feedback in an amplifier are considerable and more than compensate for the gain loss.). Positive feedback is only used in the design of oscillators when an unstable state is a requirement. Equation (6.1) indicates that, as $\beta A \to -1$, $A_F \to \infty$, an obvious indication of instability.

Positive feedback amplifiers fall into two main classes: (a) sinusoidal oscillators and (b) square-wave generators and trigger circuits. Some sinusoidal oscillators will be discussed first.

6.4 A phase-shift sinusoidal oscillator based on a jgfet

The circuit (fig. 6.4) contains one active device – a jgfet – in a CS amplifier configuration and a feedback network consisting of three cascaded RC sections. The amplifier itself is inverting so that any voltage appearing at the drain electrode is 180° out-of-phase with the initiating voltage signal at the gate. Consider a capacitance C in series with a resistance R. The alternating current through this combination leads in phase with the alternating voltage across it by an angle α given by

$$\tan \alpha = 1/\omega CR$$
$$\text{or} \quad \alpha = \tan^{-1}(1/\omega CR) \qquad (6.2)$$

At a particular frequency f, it is clearly possible to choose the values of C and R so that $\alpha = 60°$*. The use of three such sections in cascade can therefore be

---

*A convenient value for explanation; in fact, the three RC sections do not have to be identical and rarely are. Indeed, for other RC networks, the frequency will adjust itself so that the part of the signal providing a phase shift of 180° will be further amplified in preference to any other.

used to provide an overall phase shift of 180° and at this frequency, the feedback signal will be in phase with the original gate signal so that positive or regenerative feedback will occur. Providing that the amplification of the jgfet is sufficiently large, the circuit will oscillate at this frequency. Analysis shows that the frequency of oscillation of the circuit shown in fig. 6.4 is given by

$$f = 1/(2\pi RC\sqrt{6}) \qquad (6.3)$$

At this frequency, $\beta = 1/29$; to satisfy the Barkhauszen criterion ($|\beta A| = 1$), A must have a value of at least 29.

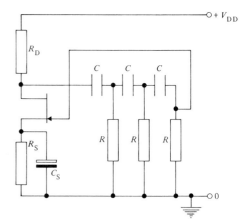

Fig. 6.4 A phase-shift oscillator based on a jgfet

6.5 A phase-shift sinusoidal oscillator based on a bipolar transistor

Fig. 6.5 is the circuit of a phase-shift oscillator which makes use of a bipolar transistor as the active device. The final resistance in the phase-shift network is provided by the input resistance of the transistor. Collector-to-base biasing is used for simplicity in spite of the fact that this introduces some ac negative feedback.

A good quality sinusoidal voltage waveform is produced by a phase-shift oscillator and the circuit will oscillate immediately it is connected, provided that the quiescent point (operating point) is chosen near the central linear region of the characteristics, say at $V_C \simeq V_{CC}/2$. Waveform distortion or no oscillation results from an incorrect choice of operating point. To provide a variable-frequency phase-shift oscillator, the three capacitors are usually varied simultaneously. Such a variation keeps constant the input impedance of the phase-shifting network and also the magnitude of β and $A\beta$, so that the amplitude of the oscillations remains unchanged as the frequency is changed.

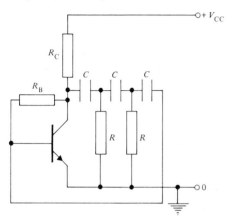

Fig. 6.5 A phase-shift oscillator based on a bipolar transistor

EXAMPLE 6.5a. *In the phase-shift oscillator of fig. 6.5 the transistor is a BC 109 with $h_{fe} = 420$, $V_{CC} = 4.5V$, $R_B = 2.2M\Omega$, $R_C = 5.6k\Omega$, $R = 10k\Omega$ and $C = 0.01\mu F$. The dc voltage V_{CE} is measured to be 2.3V. The circuit provides a good sinusoidal waveform of frequency 640Hz. Determine the input resistance of the transistor. Compare the value so found with an alternative method of evaluating this input resistance.*

From equation (6.2) for one RC section

$$\alpha = \tan^{-1}\left(\frac{1}{2\pi \times 640 \times 10^4 \times 10^{-8}}\right) = 68°\ 5'$$

As there are two similar RC sections, the total phase-shift is $136°\ 10'$. The remaining phase-shift is $(180° - 136°\ 10') = 43°\ 50'$ is that required for oscillation and is provided by the combination of C with R_i, where R_i is the input resistance of the transistor. Hence,

$$R_i = \frac{10^8}{2\pi \times 640 \tan 43°\ 50'} = 2.6 \times 10^4\ \Omega$$

The input resistance of the transistor can also be estimated if the direct emitter current I_E is known.

This is given by

$$I_E = \frac{V_{CC} - V_{CE}}{5.6 \times 10^3} = \frac{4.5 - 2.3}{5.6 \times 10^3} = 0.4\text{mA}.$$

As R_B is large, its effect on the input resistance can be ignored.
The input resistance $R_i = h_{ie}$ is given by

$$R_i = \frac{h_{fe}\ 26\text{mV}}{I_E} = \frac{420 \times 26}{0.4} = 2.7 \times 10^4\ \Omega$$

Note that the input resistance of the bipolar transistor calculated from the direct emitter current agrees within 5% of that deduced from the measured oscillator frequency.

EXAMPLE 6.5b. *Repeat the calculations for example 6.5a for the case where a BC 107 transistor is used with $h_{fe} = 280$, $V_{CC} = 7V$, $R_C = 5.6k\Omega$ $R_B = 2.2M\Omega$, $R = 10k\Omega$ and $C = 0.01\mu F$ where the steady voltage $V_{CE} = 4.4V$ and the circuit oscillates at a frequency of 760Hz. Again, compare the value of the input resistance of the transistor calculated from the oscillator performance with that calculated from the direct emitter current.*

Proceeding as in example 6.5a, the phase-shift across the first two RC sections is $2 \times 64°\ 30' = 129°$.

Hence,

$$R_i = \frac{1}{2\pi fC \tan \alpha} = \frac{10^8}{2\pi \times 760 \times \tan 51°}$$

$$= 1.7 \times 10^4\ \Omega$$

The direct emitter current $I_E = \frac{7 - 4.4}{5.6 \times 10^3} = 0.46\text{mA}.$

Again, ignoring the effect of R_B on the input resistance,

$$R_i = h_{ie} = \frac{h_{fe}\ 26\text{mV}}{I_E} = \frac{280 \times 26}{0.46} = 1.6 \times 10^4\ \Omega.$$

On comparing the performances of the two oscillators involved in these examples 6.5a and 6.5b, in the second case, a transistor with a lower current gain (h_{fe} = 280) is used in place of one with a current gain of 420, but the direct emitter current is maintained approximately the same. The input resistance of the amplifier stage decreased from $2.7 \times 10^4 \, \Omega$ in the first case to $1.6 \times 10^4 \, \Omega$ in the second and, as a result, the frequency of the oscillator increased from 640Hz to 760Hz to satisfy the condition of zero phase-shift around the closed loop.

6.6 A crystal-controlled sinusoidal oscillator

When an electric field E is applied to a solid, the strain component X created in the direction of E can be expressed by the equation:

$$X = g_1 E + g_2 E^2 + g_3 E^3 + \cdots \quad (6.4)$$

where g_1, g_2, g_3, g_n are called the strain coordinates.

In general, only the first two terms of this series are of significance. The production of strain dependent on the even powers of E is called *electrostriction* and is a property of all solids. The first term, on the other hand, is a property of relatively few crystalline solids, all of which lack lattice symmetry: the phenomenon involved is called *piezoelectricity*.

The creation of an electric field by strain consequent upon the fact that a mechanically strained crystal becomes electrically polarised is called the *direct piezoelectric effect*; the production of strain in a crystal on the application of an electric field is the *converse piezoelectric effect*.

Quartz, tourmaline and Rochelle salt are naturally-occurring piezoelectric crystals. Many synthetic piezoelectric materials have been produced, one of which, ammonium dihydrogen phosphate, is crystalline and frequently used in crystal microphones. Synthetic piezoelectric ceramics are used in such diverse fields as the generation of ultrasonic waves, pick-up heads in record players and to provide ignition in a butane gas lighter.

Quartz is the crystalline material of particular importance in oscillators. Quartz crystals occurring naturally are in the form of hexagonal pyramids (fig. 6.6a). The Z axis, and any axis parallel to this, is the optic axis of the crystal and is electrically useless. Fig. 6.6b shows a section through the pyramid with the three electric axes (XOX) and the three mechanical axes (YOY) marked. A common crystal cut is the X-cut shown in fig. 6.6c. If conducting electrodes are deposited on to the faces of this slice which are perpendicular to the $X_1 O X_1$ axis, the strain X created by an electric field E is given by

$$X = 2.15 \times 10^{-12} E \quad (6.5)$$

where E is the electric field strength in V/m.

The crystal is mounted and operated at its resonant frequency determined by its thickness, elastic properties and mounting.

EXAMPLE 6.6a. *The density of quartz is 2.654×10^3 kg m^{-3} and the Young modulus for quartz in the $X_1 O X_1$ direction is 8.0×10^{10} Nm^{-2}. Calculate the thickness required of an X-cut crystal if it is to have a fundamental frequency of vibration of 1.0MHz presuming that it is mounted with both its metal coated faces free to vibrate.*

The velocity v of compression waves in quartz is given by

$$v = \sqrt{E/\rho}$$

where E is the appropriate Young modulus and ρ is the density. Hence,

$$v = \sqrt{8 \times 10^{10}/(2.65 \times 10^3)} = 5.5 \times 10^3 \text{ ms}^{-1}.$$

The wavelength λ of such a wave for a frequency f of 1MHz is given by

$$\lambda = v/f = 5.5 \times 10^3/10^6 = 5.5 \times 10^{-3} \text{ m}.$$

If the plate is supported with each face free to vibrate, the node being the mid-point, in the fundamental mode, the crystal thickness x will be $\lambda/2$.

Hence, the required thickness of the crystal slice is

$$x = \frac{5.5 \times 10^{-3}}{2} = 2.75 \text{mm}.$$

The frequency range from about 4kHz to 10MHz can readily be covered by means of quartz crystal slices vibrating in their fundamental modes. For frequencies above 10MHz, the plates required would be very thin and fragile. However, quartz slices are prepared and mounted so as to accentuate the harmonic modes and enable frequencies up to several 100MHz to be obtained.

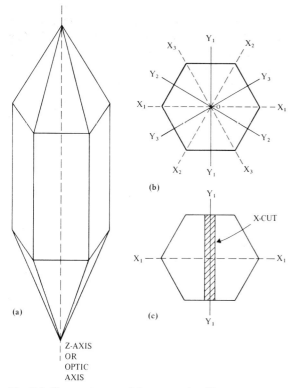

Fig. 6.6 The quartz crystal, its axes and an X-cut

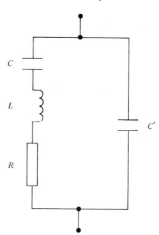

Fig. 6.7 The equivalent electrical circuit of a quartz crystal

Quartz possesses ideal properties for providing frequency standards. It is extremely hard and stable and final specimens can be ground and polished to a precision demanding optical interference methods of measurement. The resonant frequency of a given crystal is a function of temperature: it may have a positive or a negative temperature coefficient depending on the crystal cut. Cuts are available which have a negligible temperature coefficient over a specific temperature range. For high quality work, the crystal is aged, mounted in a vacuum and operated in a constant temperature enclosure to provide a frequency stable to 1 part in 10^7. The conducting electrodes on the crystal faces are usually of sputtered gold. Fine frequency adjustments can be made by depositing further gold whilst the frequency is being monitored. In this way, a frequency tolerance of a few parts per million can be achieved.

The behaviour of a quartz crystal can be represented by an electrical equivalent circuit (fig. 6.7). The inductance L, the capacitance C and the resistance R are the analogues respectively of the mass, the compliance and the viscous damping of the mechanical system. The capacitance C' represents the actual electrical capacitance between the plated faces of the the crystal.

The inherent Q of a crystal slice is almost infinite but even with very careful mounting *in vacuo* and with sputtered gold electrodes, the Q value is reduced to perhaps 100,000. For normal mountings, Q values of between 10,000 and 20,000 are common. It should be recalled that with a conventional circuit employing inductance and capacitance, a Q value exceeding 200 is difficult to achieve.

EXAMPLE 6.6b. *A quartz crystal with a resonant frequency f of 105kHz is represented by the equivalent circuit shown in fig. 6.7. If L = 115H, C = 0.02pF, R = 5kΩ and C' = 2.5pF, calculate the Q value for the crystal.*

The Q value of the circuit is $2\pi fL/R$.
Hence,

$$Q = \frac{2\pi \times 105 \times 10^3 \times 115}{5 \times 10^3} = 15{,}000.$$

This example 6.6b has been chosen to give realistic values to the components constituting the equivalent circuit and to indicate the very high Q value obtainable.

A circuit which illustrates admirably the features of a crystal-controlled oscillator (fig. 6.8) is based on a 465kHz crystal (which is readily available), is easy to construct, performs well and can be battery operated. The p-n-p silicon transistor is operated in CE connection with the dc operating point

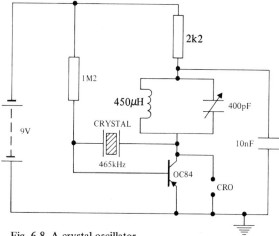

Fig. 6.8 A crystal oscillator

determined by the 1.2MΩ resistor in a fixed bias connection. Adjustment of the variable capacitor in the parallel *LC* network in the collector lead affects the amplitude of the oscillation but not the frequency. When the resonant frequency of the *LC* network is equal to the natural frequency of the quartz crystal, a maximum amplitude is obtained.

The crystal is connected between the collector and the base terminals of the transistor whereas an *LC* oscillating circuit together with a series resistor of 2.2kΩ is connected between the collector and the voltage supply of 9V. As the ohmic resistance of the inductance *L* is very small compared with 2.2kΩ, it is the series resistor which is primarily responsible for the voltage drop between the 9V supply and the steady potential on the collector.

As soon as the circuit is connected, a pd and so an electric field is established across the crystal maintaining it in a strained state and also the capacitor *C* of the *LC* oscillating circuit is charged up. This *LC* circuit will begin to oscillate so introducing a variation in the collector potential which causes small changes in the dimensions of the crystal due to converse piezoelectric effect. As the collector potential alternates so the crystal becomes strained in opposite senses, so the direct piezoelectric effect will operate to affect the transistor base potential. Hence, once the crystal is oscillating its behaviour will govern the frequency at which the circuit oscillates. The mechanical oscillations of the crystal would (without any means of sustaining them) die out because of damping. The transistor amplifier serves to maintain these mechanical oscillations via the converse piezo-electric effect. The power required from the transistor to maintain these oscillations, is small because the damping of the crystal is small.

With this circuit, the waveform can be observed with a cathode ray oscilloscope (CRO connected as shown in fig. 6.8). Frequency measurements are made either with a digital frequency meter or a heterodyne frequency meter: the frequency stability is so great that it could not be appreciated from observations by means of a CRO alone.

Another crystal oscillator in which use is made of an operational amplifier to provide the positive feedback is described in section 5.9.

6.7 A Wien bridge type of sinusoidal oscillator

A Wien capacitance bridge is an ac bridge used to measure capacitance in terms of resistance and frequency. The frequency-dependent balance condition is used in a design of oscillator (the Wien bridge oscillator) of which the principles of operation are discussed with reference to fig. 6.9.

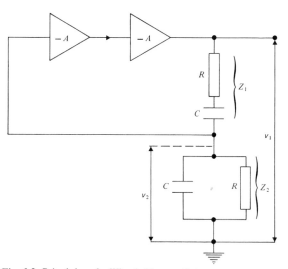

Fig. 6.9 Principles of a Wien bridge oscillator

If a non-inverting amplifier, normally provided by coupling two inverting stages, is used in conjunction with a Wien bridge, the bridge will determine the frequency at which the phase shift around the loop is zero. The amplifier must provide sufficient gain to compensate for the attenuation in the bridge network, shown below to be three.

In fig. 6.9

$$\frac{v_1}{v_2} = 1 + Z_1/Z_2 = 1 + \frac{(R + 1/j\omega C)(1 + j\omega CR)}{R}$$

$$= 3 + j(\omega CR - 1/\omega CR) \qquad (6.6)$$

The voltages v_1 and v_2 are in-phase when the imaginary part is zero, provided that the ratio v_1/v_2 is positive.

Then $\omega CR = 1/\omega CR$
or $\omega = 1/CR$
So,

$$f = \frac{\omega}{2\pi} = \frac{1}{2\pi CR}$$

The attenuation produced by the network is given by the ratio of the amplitudes of the in-phase components and is shown to be

$$v_2/v_1 = 1/3.$$

The Wien bridge oscillator is widely used at audio and low radio frequencies where large inductors and capacitors would be needed for the resonant feedback type of oscillator. Only two components need be altered to change frequency without affecting the attenuation.

A Wien bridge oscillator circuit based on a transistor amplifier of very low gain (fig. 6.10) utilises two stages where each transistor stage introduces a phase-shift of 180°, giving an overall phase-shift by the coupled stages of 360°. The Wien bridge is across the output terminals: the voltage across one section (impedance Z_2) of this bridge provides the feedback signal. With the component values and the p-n-p transistors (ACY 21) denoted in fig. 6.10, this oscillator provides a good sinusoidal voltage waveform observable with a CRO connected across the resistor R_8. The frequency of oscillation f is given by

$$f = 1/2\pi RC = 1/(2\pi \times 10^{-8} \times 4.7 \times 10^3) \simeq 3.3 \text{kHz}.$$

Several further features of this oscillator (fig. 6.10) are of interest. The resistors R_1 and R_{10} provide a potential divider network which determines the quiescent point of transistor T_1. The steady potential

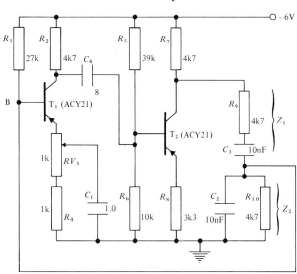

Fig. 6.10 A Wien bridge oscillator based on a two-stage bipolar transistor amplifier

at the point B, the base of T_1, is

$$V_B = -6 \times 4.7/31.7 = -0.9\text{V}.$$

If one assumes a voltage drop of 0.2V across the base-emitter junction (the ACY 21 is a germanium transistor) the steady voltage at the emitter of T_1 is -0.7V, the emitter current $I_E = 0.35$mA and the collector voltage $V_C = -4.4$V. Hence the dc operating (quiescent) point is well placed for linear operation. To reduce the gain of the first stage, ac negative feedback can be introduced by means of the voltage across the variable resistor RV_3. Also, with no by-pass capacitor across the resistor R_8 in the emitter lead of T_2, the voltage gain A_{v2}, of this second stage is low; it is given by

$$A_{v2} \simeq R_7/R_8 = 4.7/3.3 = 1.4$$

6.8 A Wien bridge sinusoidal oscillator based on an operational amplifier

The operational amplifier is used in a non-inverting configuration (fig. 6.11) and, assuming an infinite open-loop gain, the closed-loop gain is

$$A_{VCL} = 1 + R_2/R_1.$$

If $A_{VCL} = 3$, i.e. $R_2 = 2R_1$, oscillations are just maintained. Negative feedback is applied to the amplifier via resistors R_2 and R_1 to reduce the loop

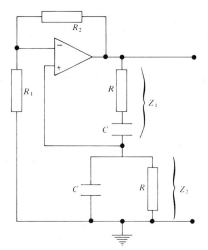

Fig. 6.11 A Wien bridge oscillator based on an operational amplifier

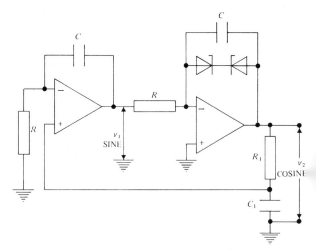

Fig. 6.12 A quadrature oscillator based on two operational amplifiers

gain to unity and so ensure a sinusoidal waveform. To obtain oscillations of constant amplitude, a non-linear resistor is used for R_1 so that the loop gain is dependent on the amplitude of the oscillations. A component is required of which the resistance increases as the current through it increases. A jgfet, operating as a voltage-controlled resistor, can also be used to provide this amplitude stability.

6.9 A quadrature oscillator to provide two signals 90° out-of-phase

Two operational amplifiers can be connected in cascade (fig. 6.12) to produce two sinusoidal voltage signals in quadrature (i.e. one 90° out-of-phase with the other). The first amplifier is connected as a non-inverting integrator and provides a sine-wave output whereas the second amplifier is connected as an inverting integrator and provides a cosine-wave output.

The operation of the feedback loop is governed by the differential equations

$$RC \frac{dv_1}{dt} = v_2 \text{ and } RC \frac{dv_2}{dt} = -v_1 \qquad (6.7)$$

The solutions of these equations can be written

$$v_1 = a \sin \omega t \text{ and } v_2 = a \cos \omega t \qquad (6.8)$$

where $\omega = 1/RC$.

As $\omega = 2\pi f$, the frequency f is given by

$$f = 1/2\pi RC \qquad (6.9)$$

The resistor R_1 is slightly larger in value than R to ensure sufficient positive feedback for oscillation. The Zener diodes in the feedback loop of the second operational amplifier limit the amplitude of the output signal but are found to create negligible waveform distortion.

6.10 The transistor switch

In earlier work with transistor amplifiers the need was stressed to operate the transistor within its linear or active region. With the CE amplifier, the operating point of the transistor was chosen to be near the centre of the active region so that the transistor was never driven into a saturated or a cut-off state. Such an amplifier is said to operate under Class A conditions to distinguish it from those in which the collector current flows for only a fraction of the period of the input waveform.

In the following sections, concerned with square-wave generators and trigger circuits, the switching action of the transistor is most important. Fig. 6.13 shows a junction transistor as an ideal switch: effectively open or 'OFF' (fig. 6.13a) when the emitter-base junction is reverse-biased and effectively closed or 'ON' (fig. 6.13b) when the emitter-base junction is forward-biased. Recalling that a voltage of

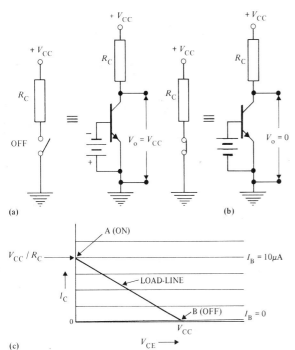

Fig. 6.13 The junction transistor in CE connection as an ideal switch

about 0.7V is required to forward-bias a silicon junction, it is seen that for an input voltage swing of about 1V (between the base and the emitter) the transistor can be switched from an ON to an OFF state.

Fig. 6.13c shows a load-line representing the equation

$$V_{CC} = I_C R_C + V_{CE}$$

drawn on ideal CE output characteristics. At the point A the transistor is conducting at saturation and the collector current I_C is limited only by R_C because the resistance of the transistor is ideally zero and $V_{CE} = 0$. No power is dissipated when it is operating as a closed switch. Conversely, at the point B on the load-line, the transistor is at cut-off, the collector current is zero and $V_{CE} = V_{CC}$. Once again, no power is being dissipated in the transistor. Hence, power is dissipated only during the transitions between the ON and the OFF state. Fig. 6.14 shows some actual output characteristics for a transistor in CE connection. Area 1 represents the saturated region, area 2 represents the active region and area 3 represents the cut-off region. The maximum power dissipation curve is shown in fig. 6.14. It is important that the three maximum values of power P_{max}, collector current $I_{C\ max}$ and collector-emitter voltage $V_{CE\ max}$ are not exceeded whether the transistor is operated in the linear region or as a switch. The values of P_{max}, $I_{C\ max}$ and $V_{CE\ max}$ are always quoted by the manufacturer in the specification of the device.

Point A (fig. 6.14) represents the transistor in saturation on the actual characteristics and V_{CE} is not zero. This voltage is referred to as the collector-emitter voltage at saturation and is denoted by $V_{CE\ sat}$; it has a value of approximately 0.3V for a silicon transistor and 0.1V for a germanium device.

Point B (fig. 6.14) represents the operating point when the switch is open. The base current is zero because the emitter-base junction is reverse-biased but the collector current is not zero: it has a value $h_{FE}I_{CBO}$, where I_{CBO} is the leakage current across the reverse-biased collector-base junction.

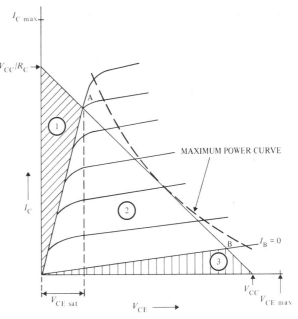

Fig. 6.14 Saturation and cut-off conditions on actual characteristics

For circuits in which the transistor is used as a switch, large signal or dc parameters apply. Hence, for the transistor in CE connection, h_{FE} and not h_{fe} is used for the current gain.

EXAMPLE 6.10a *The n-p-n transistor shown in the circuit of fig. 6.15 is assumed to behave as an ideal switch ($V_{BE} = 0$, $V_{CE\ sat} = 0$ and $I_{CBO} = 0$) and is guaranteed to have a minimum value of h_{FE} of 50. Calculate the value of V_{CC} and the components to be used in the switch if $v_i = 4V$ peak, $v_o = 20V$ peak and the saturation current $I_{C\ sat} = 20mA$.*

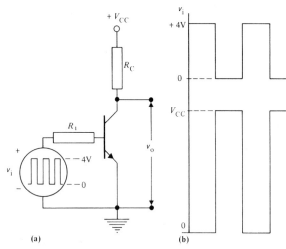

Fig. 6.15 The ideal switch of example 6.10 (a)

Because the switch is assumed to be ideal, $V_{CE} = 0$ when ON. Hence, V_{CC} determines the amplitude of the output voltage and

$$v_o = V_{CC} = +20V.$$

With a saturation current $I_{C\ sat} = 20mA$,

$$R_C = V_{CC}/I_{C\ sat} = 20\Omega/(20 \times 10^{-3}) = 1k\Omega.$$

If $I_{C\ sat} = 20mA$ and h_{FE} has a minimum value of 50,

$$I_B = I_{C\ sat}/h_{FE} = 20 \times 10^{-3} A/50 = 0.4mA.$$

$$v_i = I_B R_1 + V_{BE}$$

But $V_{BE} = 0$. Hence,

$$R_1 = v_i/I_B = 4\Omega/0.4 \times 10^{-3} = 10k\Omega.$$

Example 6.10a illustrates the calculations involved and fig. 6.15b shows the input and output voltage waveforms of this ideal switch. In a practical case of a transistor switch of the form shown in fig. 6.15a, the waveforms usually take the form shown in fig. 6.16.

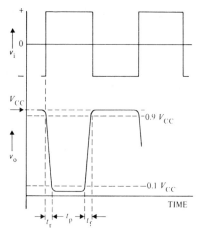

Fig. 6.16 The waveforms of the input (v_i) and output (v_o) voltages of a practical junction transistor switch.

We want the operation of a transistor switch to approach, as nearly as possible, that of the ideal switch. Consequently, the rise-time t_r (i.e. the time required for the collector current to rise from 10% to 90% of its saturation value $I_{C\ sat}$) should be as short as possible. When the transistor is OFF, the collector-base junction is reverse-biased and behaves as a capacitor with a junction capacitance, C_{CB}. When an input pulse forward-biases the emitter-base junction, the resistance of this forward-biased junction is very small. If the emitter-base junction capacitance can be neglected (because the emitter area is very small) the transistor switch of fig. 6.15a can be represented by the equivalent circuit of fig. 6.17. For I_C to rise from 10% to 90% of its saturation value, it must charge the collector-base capacitance C_{CB}. The growth of the collector current is represented by the equation

$$I_C = I_{C\ sat}[1 - \exp(-t/R_C C_{CB})]$$

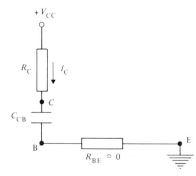

Fig. 6.17 An equivalent circuit for the ON state of the junction transistor switch of fig. 6.15 (a)

EXAMPLE 6.10b *Make use of the equivalent circuit of fig. 6.17 to express the rise-time t_r in terms of R_C and C_{CB}.*

There are three basic equations which govern the behaviour of the transistor switch of fig. 6.17.

(a) $I_{C\,sat} = h_{FE} I_{B\,min}$

where $I_{B\,min}$ is the minimum value of I_B required to provide $I_{C\,sat}$.

(b) $I_{C\,sat} = V_{CC}/R_C$

assuming that $V_{BE} = 0$

(c) $I_C = h_{FE} I_B [1 - \exp(-t/R_C C_{CB})]$

The rise time t_r is the time taken for I_C to rise from 10% to 90% of its saturation value. To simplify the analysis, assume that the time interval t_r is the same as that required for I_C to reach 80% of its saturation value starting at $I_C = 0$. Equation (c) can then be written:

$0.8 I_C = h_{FE} I_B [1 - \exp(-t_r/R_C C_{CB})]$

Writing $R_C C_{CB} = \tau$

$0.8 = 1 - \exp(-t_r/\tau)$

or $\exp(-t_r/\tau) = 5$,

so that,

$t_r = 1.15\,\tau = 1.15\,R_C C_{CB}$.

The value of t_r can be decreased by over-driving the base, i.e. by increasing I_B. Assume that the minimum base current specified by equation (a) in example 6.10b is doubled. Then,

$0.8 = 2[1 - \exp(-t_r/\tau)]$

$\exp(t_r/\tau) = 1.67$

$t_r = 0.51\,\tau$

Similarly, if three times the minimum base current value is used,

$t_r = 0.31\,\tau$.

These results are summarised in table 6.1.

TABLE 6.1

Relationship between rise time and over-driven base current

$I_{B\,min}$	$t_r = 1.61\,\tau$
$2 I_{B\,min}$	$t_r = 0.51\,\tau$
$3 I_{B\,min}$	$t_r = 0.31\,\tau$

To over-drive the base it is merely necessary to adjust R_1 (fig. 6.15a). This reduction in rise-time is achieved at a price. When the input pulse is removed, the excess charge carriers must be removed before I_C can decrease. A larger value of I_B means that the storage time and hence the decay time is increased resulting in a poorer trailing edge.

A very effective method of decreasing the rise-time is to use a 'speed-up' capacitor (fig. 6.18). This will enable the base to be rapidly saturated with charge carriers when the input pulse is applied, but subsequently the base current can be reduced to a value just sufficient to ensure saturation.

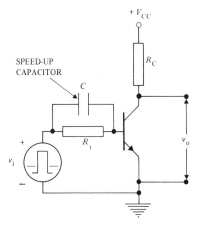

Fig. 6.18 A 'speed-up' capacitor C used to reduce t_r with a transistor switch

EXAMPLE 6.10c *In the switch circuit of fig. 6.18, $V_{CC} = 20V$, the transistor has $h_{FE} = 40$, $R_C = 2k\Omega$ and $R_1 = 8k\Omega$. When an input pulse of 2V peak is applied, without the capacitor C connected, the rise-time of the output pulse is 3μs. Calculate the value of C required to reduce this rise-time to 1μs.*

$I_{C\,sat} = V_{CC}/R_C = 20/(2 \times 10^3) = 10\text{mA}.$

$I_{B\,min} = I_{C\,sat}/h_{FE\,min} = 10 \times 10^{-3}/40 = 0.25\text{mA}.$

To make t_r one-third of its original value, the base current should be over-driven, i.e. $I_{B\,min}$ should be doubled (see table 6.1).

$I_{B\,total} = I_{B\,min} + I_{B\,overdrive}$

The capacitance C can provide this overdrive base current.

If Q is the charge on the capacitor

$I_{B\,overdrive} = Q/t_r$

But,

$Q = Cv_i$

where v_i is the input signal voltage. Hence,

$$C = \frac{I_{B\,overdrive}\, t_r}{v_i} = \frac{0.25 \times 10^{-3} \times 10^{-6}}{2}$$

$= 125\text{pF}.$

Thus, a 'speed-up' capacitor of capacitance 125pF can reduce the rise-time to $1\mu s$.

6.11 The bistable multivibrator

The three basic classes of multivibrator circuits are the monostable, the bistable and the astable or free-running. The bistable circuit is one which can exist indefinitely in either of two stable states. The circuit can be induced to make an abrupt transition from one state to the other by means of some external excitation, which usually takes the form of an input pulse. The bistable multivibrator is used extensively for counting and for storing binary information. Considerable effort has been devoted to producing reliable, fast, bistable circuits in integrated circuit form. These are now readily available at very low cost and decade counters on a single silicon slice are being used in large numbers in digital computers, digital frequency meters, digital voltmeters and many other instruments. The circuits to be described here make use of discrete components but serve to illustrate the principles of operation.

For obvious reasons, the bistable multivibrator is often called a 'scale-of-two', a 'flip-flop' or simply a 'binary'.

The bistable circuit employs positive (i.e. regenerative) feedback during the interval required for the switching to take place. This regenerative action will not occur unless initiated by an externally applied trigger voltage. The basic circuit is drawn in fig. 6.19a as a two-stage amplifier (each stage in this case being a junction transistor) with direct coupling

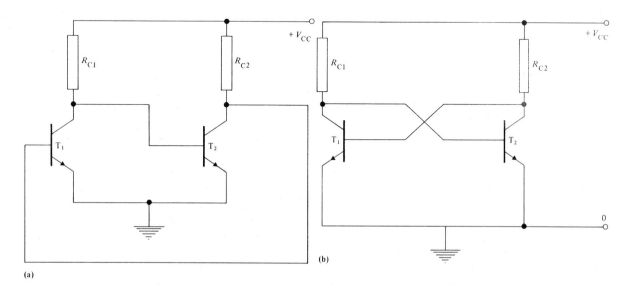

Fig. 6.19 Bistable multivibrator: **(a)** basic circuit as two-stage amplifier; **(b)** redrawn to show symmetry of arrangement

between the stages. In fig. 6.19b the same circuit is re-drawn to show the symmetry of the arrangement.

One might think it possible for the circuit to possess a state in which equal currents flow in T_1 and T_2. Such would be the case if both devices were biased to cut-off or to saturation. These extreme cases are of little practical interest.

Consider a small voltage fluctuation to occur at the base of transistor T_1. This fluctuation will be amplified and fed to the base of the transistor T_2 in an inverted state. Further amplified and inverted, this signal is fed-back to the base of T_1 to augment there the original voltage fluctuation. This regenerative feedback will ensure that one transistor is driven into saturation while the other one is cut-off. The circuit will remain in this stable state until an input pulse initiates the transition.

The capacitors C_1 (fig. 6.20) are termed commutating or speed-up capacitors (see also fig. 6.18). They serve to reduce the transition time, i.e. the time interval during which conduction transfers from one transistor to the other. Increasing the capacitance will reduce the transition time. However, a certain minimum time, called the *resolving time*, must elapse after one switching sequence before a succeeding trigger pulse will reliably induce the reverse state. This resolving time is the sum of the transition time and the settling time of the circuit. Increasing the capacitance increases the settling time; hence a compromise is required in the choice of C_1 for optimum behaviour. The resolving time of the bistable circuit is the reciprocal of the maximum frequency at which the binary will respond.

In the bistable multivibrator circuit of fig. 6.20, assume that initially transistor T_1 is conducting at saturation whereas T_2 is cut-off. The collector of T_1 is near earth potential. The steady voltage at the base of T_2 can be found from the values of R_1 and R_2 which form a potential divider network between earth and $-V_{BB}$. Thus, with the component values listed in fig. 6.20, the base of T_2 will be at approximately $-0.7V$, which ensures that T_2 is turned hard-off. The circuit will remain in this stable state until a triggering pulse is applied via the *steering diodes* D_1 and D_2.

When the negative pulse (indicated in fig. 6.20) arrives, D_1 will be forward-biased and will conduct whereas D_2 will be reverse-biased and will block the pulse. Hence, the input pulse will start the transition

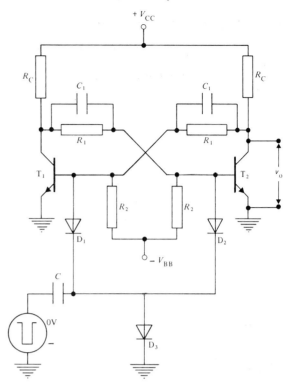

Fig. 6.20 A bistable multivibrator with symmetrical base triggering ($V_{CC} = 6V$, $V_{BB} = 1.5V$, $R_C = 220\Omega$, $R_1 = 2.2k\Omega$, $R_2 = 2.7k\Omega$, $C_1 = 150pF$, $C = 100pF$).

which turns T_1 OFF and T_1 ON. With the role of each transistor now reversed, the diodes will serve to steer the next negative pulse which arrives to the ON transistor — in this case T_2 — and the transition will be repeated. If a positive pulse, or an overshoot on a negative pulse, appears at the input, the diode D_3 is forward-biased and conducts this pulse to earth, so leaving the input circuit clear to receive the next genuine input pulse. Diode D_3 could be replaced by a resistor across which the input pulse would appear.

The divide-by-two function of this bistable circuit is now apparent: one output pulse is obtained for every two input pulses. To produce a counting circuit (a scaler) it remains to connect a number of binary stages in cascade and to use neon indicators or light-emitting diodes (section 8.11) to show the state of each stage. With 10 binary stages, a scaling factor of $2^{10} = 1024$ is achieved, and the output from the last stage could be connected via a power amplifier stage to a mechanical register.

It is obviously inconvenient to multiply the reading of the mechanical register by 1024: it would be preferable to work in scales of ten. The problem is therefore to convert binaries to a decimal system. Two methods of achieving this conversion will be outlined. The first makes use of a feedback system to convert four binary stages (forming a scale of $2^4 = 16$) into a basic scale of 10 (a decade counter). This method is shown diagrammatically in fig. 6.21. When the count reaches 8, the transistor in the left-hand half of the fourth binary stage is turned ON and the voltage pulse from its collector is fed to the input of both the second and the third binary stages. Hence $4 + 2 = 6$ extra pulses (generated within the scaler) have been added. When 10 pulses have arrived from the external source, therefore, a total of $10 + 6 = 16$ pulses have passed through the 4-stage unit, thus returning it to its initial state. The last binary stage of fig. 6.21 should now be labelled 2, not 8. So reading from right to left, the binary stages would indicate 2421. This system is therefore called a 2421 BCD (binary coded decimal).

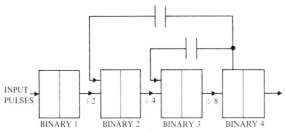

Fig. 6.21 A decade counter derived by the use of feedback from four binary stages.

The second method of converting four binary stages into a decade counter employs the re-routing of pulses by means of gates. At this stage it suffices to state that a gate is either open or closed dependent on the control voltage at the gate. If the gate is open, the input signal is transmitted whereas if the gate is closed the input signal is blocked. Fig. 6.22 shows a scale-of-ten produced by the use of two gates. In the initial condition of the counter shown in fig. 6.22, gate 2 is open and gate 1 is closed. Hence, the binary stages operate in the normal manner up to the count of 7. When the final binary stage changes at the count of 8, the two transistors in binary 4 change state; in so doing, gate 1 is opened and gate 2 is closed. The ninth pulse changes the state of binary stage 1 but produces no output pulse. The tenth pulse changes binary stage 1 to its initial state and, in so doing, produces an output pulse from binary stage 1 which is routed to binary stage 4, which returns it to its initial state. Thus, each binary stage is again in its initial state with gate 2 open and gate 1 closed, ready for the next sequence of 10. Because the binary stages are unchanged in operation, the system is called an 8421 BCD system, where the numbers refer to the four binary stages reading from right to left in fig. 6.22.

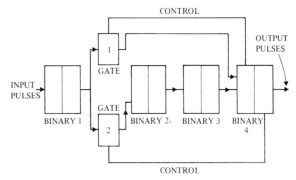

Fig. 6.22 A decade counter derived by the use of gates from four binary units

6.12 The monostable multivibrator

In the circuit of the monostable multivibrator, there is a stable state and a semi-stable state. When connected, it remains in the stable state until an externally applied pulse switches it to the semi-stable state. The time interval for which the circuit remains in the semi-stable state is determined by an internal RC network. At the end of this time interval, the circuit reverts to its stable state and remains there until caused to make another temporary transition by the next input pulse. This arrangement is often called a *one-shot multivibrator*. One important feature of the monostable circuit is that is shows true trigger action — the input signal can have a variety of forms provided that its amplitude is sufficient to initiate the transition. The output waveform is rectangular and is independent of the amplitude or duration of the input signal. Spurious noise pulses are ignored by the monostable serving as a *discriminator*, while each input pulse of amplitudes exceeding some critical value will produce a uniform rectangular output pulse of constant amplitude. The circuit can also serve to introduce a time delay, often of the order of milliseconds, between the input signal and the transition back to the stable state.

A monostable multivibrator circuit is shown in fig. 6.23. The stable state exists with transistor T_1

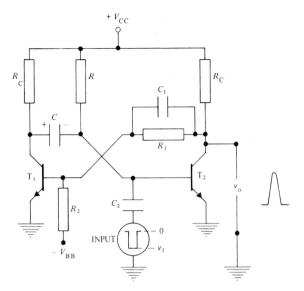

Fig. 6.23 A monostable multivibrator

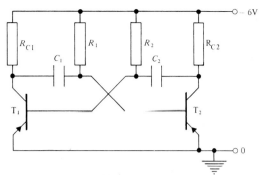

Fig. 6.24 An astable multivibrator

OFF and T_2 ON. The capacitor C is charged with the polarity indicated to a voltage V_{CC}. The potential divider R_1 and R_2 together with the voltage $-V_{BB}$ provide the reverse bias to T_1 which holds it to cut-off. When the negative pulse is applied to the base of T_2, this transistor is switched to cut-off and T_2 is turned ON. With the transistor T_1 conducting at saturation the positive end of capacitor C is connected to earth. The capacitor C now charges through resistor R and transistor T_2 is held at cut-off until C charges to 0V.

The interval during which T_2 is held at cut-off determines the duration of the output pulse.

6.13 The astable or free-running multivibrator

The astable multivibrator has two apparently stable or quasi-stable states and makes periodic transitions from one state to the other. One state is with transistor T_1 (fig. 6.24) full conducting (saturated) and the other transistor T_2 non-conducting (cut-off) while the other state is the reverse of this. Because, in this oscillator, each transistor switches rapidly from a fully conducting to a non-conducting state, the output waveform (normally observed between point A or B and earth) is approximately rectangular. Such a waveform can be shown (by Fourier analysis) to contain a large number of sinusoidal components covering a wide frequency spectrum – hence the name multivibrator.

With $R_{C1} = R_{C2}$, $R_1 = R_2 = R$ and $C_1 = C_2 = C$ in the circuit of fig. 6.24, the circuit is said to be symmetrical and each transistor remains in a conducting state for the same time. In this case, the ratio of the pulse duration to the time interval between pulses (known as the 'mark-to-space ratio') is unity and the time period T of the oscillator is given (as is shown below) by

$T = 1.38\,RC$ seconds

so the frequency $f = \dfrac{1}{1.38\,RC}$ Hz

where R is in ohms and C is in farads.

Because no two transistors are identical, when the circuit shown in fig. 6.24 is switched on, one transistor, say T_1, will conduct slightly more than the other. Immediately this occurs, transistor T_1 is driven to saturation while T_2 is cut-off. The p.d. across the capacitor C_1 which holds T_2 at cut-off, has a value of $+V_{CC}$. With T_1 at saturation, the left-hand plate of C_1 is effectively connected to earth while the right-hand plate begins to discharge through the resistor R from potential V_{CC} to $-V_{CC}$. When the potential difference across C_1 has fallen to zero, transistor T_2 will start to conduct and the rapid transition will occur. The time t_1 taken for the voltage across C_1 to fall to zero from $+V_{CC}$ when the applied emf is $-V_{CC}$, is given by

$1/2 = \exp(-t_1/R_1 C_1)$

or

$2 = \exp(t_1/R_1 C_1)$.

Hence,

$t_1 = 0.69\,R_1 C_1$.

After the transition, transistor T_2 is conducting at saturation and T_1 is being held-off by the p.d. across the charged capacitor C_2 for a time interval t_2 given by

$t_2 = 0.69 R_2 C_2$.

The time period of oscillation, T, is hence given by

$T = t_1 + t_2 = 0.69 (R_1 C_1 + R_2 C_2)$

If the multivibrator is symmetrical, i.e. $R_1 C_1 = R_2 C_2 = RC$,

$T = 1.38 RC$.

EXAMPLE 6.13a *The circuit of fig. 6.24 is set up using the following components:—*

$C_1 = C_2 = 100\mu F$ and $R_1 = R_2 = 10k\Omega$

(where each electrolytic capacitor needed to provide such a large capacitance as $100\mu F$ has its positive plate connected to the base of the transistor).
A 6V, 0.06A filament lamp is connected in series with a 47Ω resistor. This combination is used for R_{C1} and a similar combination for R_{C2}. Calculate the time period of the multivibrator. Comment on the use of electrolytic capacitors and suggest two other multivibrator circuits with a very long time period but in which capacitance values are limited to $1\mu F$.

The time period of the multivibrator is

$T = 1.38 RC = 1.38 \times 10^4 \times 100 \times 10^{-6}$ seconds
$\simeq 1.4$ seconds.

This time period is sufficiently long for the switching action to be observed by means of the filament lamps. Alternatively, a moving-coil voltmeter connected across the transistor to measure V_{CE} will be able to respond to this very slow switching action.
Electrolytic capacitors are used to enable this very long time period to be produced but the electrical leakage of these capacitors would mean that the period was not accurately reproducible.
If each transistor were replaced by a Darlington pair, very large values of R and relatively small values of C could be used. Fig. 6.25 shows such a circuit with a time period T of approximately 3 seconds.

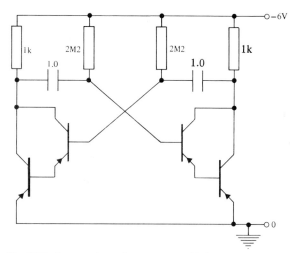

Fig. 6.25 A very low frequency multivibrator based on Darlington pairs

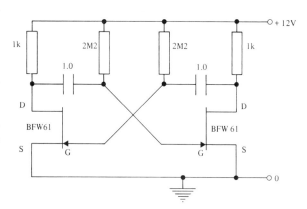

Fig. 6.26 A very low frequency multivibrator based on n-channel jgfets

Fig. 6.26 is the circuit of a very low frequency multivibrator based on the use of jgfets as the active devices. The high input resistance of the jgfet ($> 10^8 \Omega$) allows a long time period to be achieved without the need for large capacitance values.

6.14 Multivibrators as voltage-to-frequency converters

The time period T of a multivibrator can be changed without interchanging components by connecting R_1 and R_2 (fig. 6.24) to an auxiliary voltage $(-V)$, where the collector supply voltage remains constant at $-V_{CC}$. If the circuit is symmetrical, the time period of the oscillations is given by

$T = 2 RC \log_e (1 + V_{CC}/V)$ \hfill (6.10)

Hence, the frequency f is given by

$$f = 1/T = \frac{1}{2RC \log_e (1 + V_{CC}/V)} \qquad (6.11)$$

But,

$$\log_e (1 + V_{CC}/V) = V_{CC}/V - \frac{1}{2}(V_{CC}/V)^2 + \frac{1}{3}(V_{CC}/V)^3$$

and if V_{CC}/V is small, we can write equation (6.10) as

$$f = V/2 RC V_{CC}$$

or $f \propto V$.

The circuit of fig. 6.27 was used to investigate this relationship and the results are shown graphically in fig. 6.28.

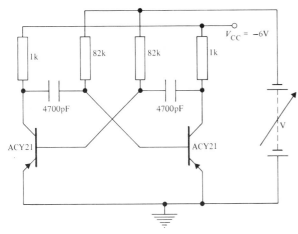

Fig. 6.27 A simple voltage-to-frequency converter

Fig. 6.29 shows the circuit of an astable multivibrator, the frequency of which is voltage controlled and extends over about five decades. Field-effect transistors, with their inherent high input resistance, are used as the switching elements. For each jgfet, a silicon junction transistor operating as a constant current source resets the voltage on the capacitor holding one of the jgfets in a non-conducting state. This constant current, determined by V_{EB}, hence determines the frequency. The voltage V_{EB}, provided by the potential divider network across the dc supply,

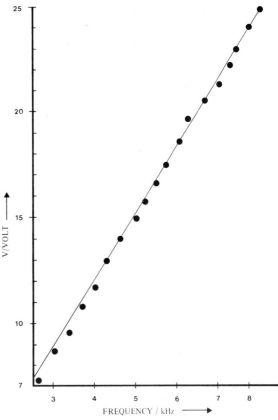

Fig. 6.28 Relationship between voltage and frequency obtained by means of the circuit of fig. 6.27.

is measured by a digital voltmeter with an input resistance exceeding 10MΩ. The curve of fig. 6.30 shows the relationship between the control voltage V_{EB} and the frequency of the multivibrator.

The constant current sources used in the circuit of fig. 6.29 are very temperature sensitive so the BCY 70 transistors used should be mounted on heat-sinks.

Multivibrators of this form have been used to design instruments for temperature measurements in the upper atmosphere or in outer space. Using constant circuit voltages, the temperature is expressed in terms of frequency, a very convenient form if it is required to be transmitted from space to a ground control station.

6.15 The Schmitt trigger circuit

Named after the inventor of the vacuum-tube version, the Schmitt trigger circuit is a special form of the emitter-coupled bistable multivibrator. The stable

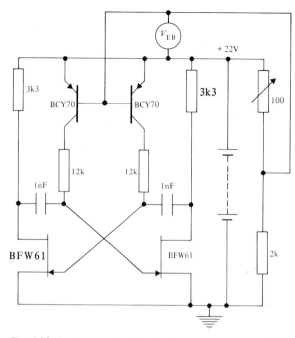

Fig. 6.29 An improved voltage-to-frequency converter which makes use of a constant current source

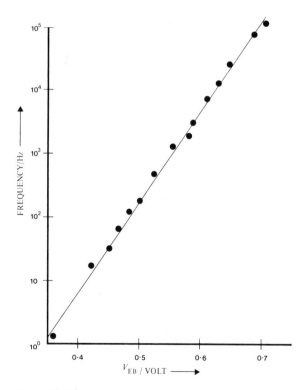

Fig. 6.30 Voltage-to-frequency conversion characteristic for the circuit of fig. 6.29

state of the Schmitt trigger circuit is determined by the amplitude of the input voltage. In the schematic circuit diagram of fig. 6.31, when no input voltage is applied, the transistor T_1 is cut-off and transistor T_2 conducts at saturation. Before the transistor T_1 can conduct, the input voltage must exceed the voltage drop (V_{E2}) across R_E. The amplitude of the input voltage required to cause T_1 to conduct is called the upper trigger potential (UTP). As T_1 starts to conduct, the forward bias on T_2 is reduced and T_2 is cut-off. This state will persist while the input voltage is above the UTP. If the input voltage is now reduced, the circuit will not switch back, so that T_2 is at saturation again until the input voltage has reached the lower trigger potential (LTP) which has a value below UTP. Fig. 6.32 shows the switching produced by a sinusoidal input signal and an input pulse with some damped oscillation on the flat top.

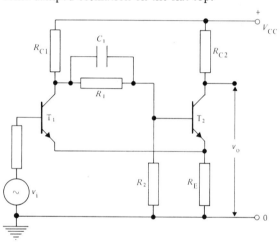

Fig. 6.31 The basic Schmitt trigger circuit

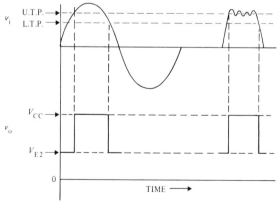

Fig. 6.32 Output from a Schmitt trigger circuit for various inputs

The circuit is said to exhibit *hysteresis* because the switching action is not initiated by the same input voltage: one voltage is required to switch-on T_1 but a different (lower) voltage is required before T_1 will turn-off. The hysteresis range V_H = UTP − LTP, can be controlled and serves to discriminate against noise, as shown in fig. 6.32.

An important application of the Schmitt trigger circuit is its use as an amplitude comparator to mark the instant at which some arbitrary waveform attains a particular reference level. In a second application, this circuit serves as a pulse shaper or a squaring circuit. Provided that the excursions of the input signal are large enough to carry it beyond the limits of the hysteresis range (fig. 6.32), the output is a square wave of the same frequency as the input but whose amplitude is independent of the amplitude of the input waveform.

A Schmitt trigger circuit in which the steady bias voltage on the transistor T_1 can be adjusted is shown in fig. 6.33. Having selected the bias voltage on T_1, the amplitude of the input signal required to turn-on T_1 is determined. The potential divider network providing the steady bias on T_2 ensures that T_2 is saturated before any input signal is applied.

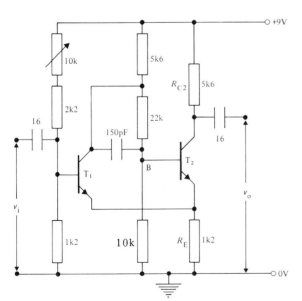

Fig. 6.33 A Schmitt trigger circuit in which the steady bias voltage on junction transistor T_1 can be adjusted

6.16 Digital instruments

It is becoming increasingly common to use moving-coil meters merely as indicators while measurements of voltage, current, resistance and frequency are made with digital meters which display the measured quantity in number form with the decimal point correctly positioned. All these digital meters make use of methods originally developed for digital voltmeters (DVMs) and now applied to a variety of other instruments.

Some of the advantages offered by a modestly priced DVM are:—

(a) Whereas a high-quality moving-coil meter, if used correctly, can provide an accuracy of ±1%, a DVM with an accuracy of ±0.1% is common and with a precision, more expensive instrument, an accuracy of ±0.01% is readily achieved. This accuracy can be readily checked by the use of stable reference voltages generated within the instrument.

(b) The input resistance of a DVM is generally in excess of 10MΩ, is often 100MΩ and an input resistance of 1000MΩ is possible.

(c) The voltage to be measured is sampled for a time interval of between 10ms and 100ms, depending on the choice of instrument.

(d) Filter networks and novel circuit configurations provide for very high rejection of common-mode voltages and also immunity from interference, particularly that at 50Hz.

(e) It is often arranged that the DVM prints out the recorded voltage each time it samples or when it is instructed by a print command signal.

(f) The output may be used to produce punched tape or may be fed directly to a digital computer.

It is apparent that the DVM is more than a measuring instrument: it is one unit in a system concerned with data recording, data processing and control. There are a number of different ways in which DVMs digitise the input signal, overcome interference and present their outputs.

One feature of the display system is important. Merely because an instrument has a display system containing five number tubes does not necessarily mean that it has a five-figure capability. A true five-figure instrument has a full-scale capability of 99999. In some instruments, the most significant digit has only two conditions, 0 or 1, and can only measure up to 11999; although the meter can display 19999, measurements above 11999 are inaccurate. This type of meter is best described as having a four-digit display with 20% over-range. This over-range facility is useful because it provides an overlap between ranges but it should not be confused with the circuitry and resolution necessary to justify the additional digit.

6.17 Ramp logic

This type of logic is used in low-cost digital instruments able to provide an accuracy of typically ±0.1%. The DVM has an internal generator of voltage which increases linearly with time (a linear ramp). When an appropriate command signal, arranged to occur at regular intervals, is received, the ramp starts. As it passes through zero volts, a voltage comparator opens a gate circuit and allows pulses from a stable oscillator to be counted. When the ramp voltage equals the input voltage (to be measured) a second comparator closes the gate and the final count is displayed. The sequence is illustrated by the block diagram of fig. 6.34, where t_c indicates the time interval during which the gate is open and pulses are counted.

By a suitable choice of ramp slope and oscillator frequency, the final count can be displayed directly as the measured voltage. A range switch selects an input attenuator and positions the decimal point. The measured voltage is sampled and displayed in about 200ms in this type of instrument. The overall accuracy is limited by the linearity of the voltage ramp and the quality of the comparators which control the gate.

6.18 Current-to-frequency converter DVMs

This type of voltage-to-frequency converter is used in another low cost DVM providing an accuracy of ±0.1%. The operating principles are of special interest because a student can readily assemble similar circuitry to construct a simple digital instrument or merely to become familiar with voltage-to-frequency converters.

In the commercial instrument, the voltage to be measured is applied, via an attenuator, to an operational amplifier which provides a constant current proportional to the input voltage. This current is used to charge a capacitor in a relaxation oscillator (similar in type to the UJT relaxation oscillator, section 9.11) which produces output pulses at a frequency proportional to the input voltage. These pulses are counted for the time period t_c chosen as the sampling time. If the charging current and the sampling time are chosen correctly, the total count can be displayed directly as a measure of the input voltage. The timing circuit determining t_c and the relaxation oscillator circuit are thermally and electrically coupled so tending to compensate for drift which exists in each separate unit. The unijunction transistor (UJT) and the programmable unijunction transistor (PUT) (section 9.13) in particular are ideal devices for use in the relaxation oscillator and the timing circuit.

6.19 Integrating logic

A DVM of higher quality with an accuracy of perhaps ±0.01% can be made by using a voltage-to-frequency converter based on the charging of a capacitor. In one form, the capacitor is charged by a constant current proportional to the input voltage V_i for a time interval t_1, controlled by a high quality crystal oscillator. At the end of this charging or sampling interval, a potential difference V_1 exists across the capacitor which is now connected to a reference voltage V_R (of polarity to discharge the capacitor) and allowed to discharge in a linear manner. The discharge process is terminated when the potential difference across the capacitor is zero. Such a charge and discharge sequence are illustrated in fig. 6.35. The time t_2 required to discharge the capacitor is proportional to V_1, the voltage across the capacitor when the discharge started, and is hence proportional to the input voltage V_i. The time interval t_2 is measured with the same crystal oscillator by counting

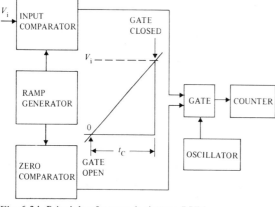

Fig. 6.34 Principle of a ramp-logic type DVM

the pulses from it for the discharge time t_2. Because the same oscillator is used to control t_1 and to measure t_2 and the same capacitor is involved in the charge and discharge sequence, any errors in the determination of the absolute value of the frequency or capacitance are eliminated.

The charging of the capacitor is an integrating process and the voltage recorded will be an average of the input voltage during the sampling time t_1. If some alternating interference is superimposed on the dc level to be measured, say, due to the 50Hz mains supply, this can be eliminated or rejected by arranging that $t_1 = nT$, where n is an integer and $T = 0.02$s.

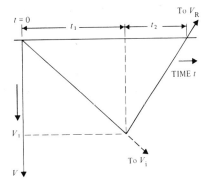

Fig. 6.35 The charge and discharge sequence in an integrating-type DVM

Exercises 6

1. Analogue data is often converted into digital form for storage, transmission, processing and display. Outline some of the advantages of providing data in digital form.

2. What is the *Barkhauszen criterion?*
   Comment on the statement: 'all oscillators are phase-shift oscillators'.

3. Illustrate the principles of a Wien bridge oscillator and derive an expression for the frequency of oscillation and the attenuation of the bridge network involved.

4. Draw an equivalent circuit for a mounted piezoelectric crystal.
   A quartz crystal has a Q value of 8.1×10^4 and a resonant frequency of 700kHz. If its effective inductance L is known to be 5.2H, calculate the effective resistance of this crystal.

5. Quartz has a density of 2.654×10^3 kg m^{-3}, and the Young modulus in the direction along the axis of a chosen cut is 8.0×10^{10} N m^{-2}. Explain why it is difficult to produce a crystal having a fundamental frequency above 10MHz.

6. Outline some uses of piezoelectric materials apart from crystal oscillators.
   Draw the circuit diagram of a form of crystal-controlled oscillator. What properties of quartz make it an ideal material to provide a frequency standard?
   Explain how you would measure the frequency of a crystal-controlled oscillator known to be of the order of 500kHz.

7. A junction transistor in CE connection is used as a switch.
   Draw an equivalent circuit to show the effect of the collector-base junction capacitance.
   Show, by calculation, that the turn-on time t_r of the switch can be reduced by providing a base current several times greater than that required to saturate the transistor.

8. Illustrate the effect of a 'speed-up capacitor' on the turn-on time of a junction transistor used as a switch in CE connection.

9. Draw the circuit diagram of a bistable multivibrator, explain the action of the components involved in it, and the role of the steering diodes.

10. With the aid of suitable diagrams, explain two methods by which four binary units can be used to construct a decade counter.

11. Draw the circuit diagram for and explain the action of an astable multivibrator.
    How can such a circuit be used as a voltage-to-frequency converter?

12. Define the terms *UTP*, *LTP* and *hysteresis* as applied to a Schmitt trigger circuit.
    Illustrate the value of hysteresis in providing clean rectangular output pulses.

13. Outline the various methods used in the design of digital voltmeters (DVMs).

14. What advantages does a digital voltmeter have over a moving-coil instrument?

15. Draw a circuit diagram of an UJT relaxation oscillator and explain the action of the components in it.

16. 'A junction transistor provides a better switch than a jgfet'. Comment on this statement.

17. Draw the circuit diagram of a symmetrical free-running multivibrator having a reliable time period of about two seconds.

7·Logic Circuits

7.1 Binary operation

A simple switch exists in either an 'on' or an 'off' state: it is correspondingly either conducting or non-conducting. As such, it performs a binary operation: the device exists in only two possible states. The 'on' state can be represented by 1 (called logical 1) and the 'off' state by 0 (called logical 0).

Mechanical switches and electromechanical devices such as relays are inevitably slow in operation. By the use of semi-conductor diodes and transistors (particularly in integrated circuit form, Appendix A) switching times from the 'on' to the 'off' position, or vice-versa (i.e. from 1 to 0 or from 0 to 1) can be a fraction of a microsecond or a few nanoseconds.

Such rapid switching enables the operation and control of a system to be achieved at such a speed that digital control systems and digital computers able to handle enormous numbers of digital inputs (representing information) have become a reality and have revolutionised the technology of control systems and the processing of numerical data.

An immediate advantage of binary procedures follows from the operation of a device such as a bipolar transistor in only one or other of two possible states, e.g. at cut-off or at saturation. It is that the operation is more readily made independent of changes of the characteristics of the device (e.g. the transistor) due to variations in ambient conditions (including those of voltage supplies) or due to replacement of the device by another of ostensibly the same characteristics. Thus, a transistor may be switched from a state where it gives across a load an output of, say, 4V to within ± 20% to one where the output is, say 0.2V to within ±50%. The higher voltage (4V nominally) corresponds to the 'on' or logical 1 and the low voltage to 'off' or logical 0. The system under control is correspondingly switched on or off. On the other hand, if the transistor were used as a voltage amplifier (i.e. used in an analogue instead of in a digital manner) to which the input was varying and the output was an amplified replica of the input, preservation of constant gain would become important, and this would not be so readily made independent of variations in ambient conditions, voltage supply and component replacement.

The logic specifications or binary-state terminology corresponding to 'on' and 'off' (i.e. logical 1 and 0) can be given as

On : 1 : voltage or current pulse output above (or below) a certain level

Off : 0 : voltage or current pulse output below (or above) a certain level.

Further elaboration of these ideas leads to table 7.1

Other possibilities also come to mind such as 'up' or 'down', 'excited' or 'de-excited', 'white or black', and so on.

Table 7.1

Logic specifications

One state	On	1	true	positive	negative	yes	north
Other state	Off	0	false	negative	positive	no	south

7.2 Binary arithmetic

In binary arithmetic only the two numerals 0 and 1 are used to represent any number.

The base (or radix) in the binary system is 2 whereas in the decimal system it is 10.

Consider a decimal number, say 71205. This is clearly interpreted as

71205
$= 7 \times 10^4 + 1 \times 10^3 + 2 \times 10^2 + 0 \times 10^1 + 5 \times 10^0$

A binary number consists of a sequence of 1's and 0's only. The base (or radix) is now 2. Consequently, a binary number such as 11001 is interpreted as

$11001 = 1 \times 2^4 + 1 \times 2^3 + 0 \times 2^2 + 0 \times 2^1 + 1 \times 2^0$
$= 16 + 8 + 0 + 0 + 1$
$= 25,$

i.e. 11001 in the binary system represents 25 in the decimal system.

To convert a decimal number into a binary one, the procedure is to break the number up into powers of 2. An equivalent procedure, often more readily undertaken, is as follows:

(a) divide the decimal number by 2 and place the remainder (0 or 1) to the extreme right of the binary number to be obtained;
(b) divide the quotient of the first operation (a) by 2 and place the remainder (0 or 1) in the next place to the left of the first binary digit in the sequence to be obtained;
(c) proceed in this manner until the quotient is zero (fractions are ignored); the binary number is then established.

EXAMPLE 7.2a *Obtain the binary number equivalent to 407*

Division of 407 by 2 gives 203 with a remainder 1
,,   ,, 203 ,,   ,, 101 ,,   ,,   ,, 1
,,   ,, 101 ,,   ,, 50 ,,   ,,   ,, 1
,,   ,, 50 ,,   ,, 25 ,,   ,,   ,, 0
,,   ,, 25 ,,   ,, 12 ,,   ,,   ,, 1
,,   ,, 12 ,,   ,,   6 ,,   ,,   ,, 0
,,   ,,   6 ,,   ,,   3 ,,   ,,   ,, 0
,,   ,,   3 ,,   ,,   1 ,,   ,,   ,, 1
,,   ,,   1 ,,   ,,   0 ,,   ,,   ,, 1

Arranging the digits in the right-hand column in the manner specified in section 7.2 gives

110010111, which is the binary number required.

In the *addition of binary numbers* it is clear that

$0 + 0 = 0$
$1 + 0 = 1$
$1 + 1 = 0$ with 1 to carry.

The 1 to carry is taken over to the next higher power.

EXAMPLE 7.2b *Convert the numbers 27 and 9 to their binary equivalents. Add the binary numbers obtained and show that the resulting binary number is equivalent to 36.*

Decimal		Binary
27	=	1 1 0 1 1
9	=	1 0 0 1
		1 0 0 1 0
carry		1 1
		0 0 0 0 0
carry		1 0 0 1 0 0
		1 0 0 1 0 0

$1 0 0 1 0 0 = 1 \times 2^5 + 0 + 0 + 1 \times 2^2 + 0 + 0$

$= 32 + 4 = 36.$

In the *subtraction of binary numbers:*

$0 - 0 = 0$
$1 - 0 = 1$
$0 - 1 = 1$ so borrow 1 from the next higher power
$1 - 1 = 0$

In the *multiplication of binary numbers:*

$1 \times 1 = 1$
$1 \times 0 = 0$
$0 \times 1 = 0$
$0 \times 0 = 0$

The term 'bit' is used as an abbreviation of *binary digit*. Thus an 0 or a 1 is a bit. A group of bits is called a 'word' or 'bite' (alternatively, 'byte'). Such a 'word' has a specific significance (e.g. 1 0 0 1 0 0 = 36, from example 7.2b).

In addition to the numbers, it is also necessary to have representation of the letters of the alphabet, of which in English there are 26 in A to Z. To represent by bits all of these 26 letters, clearly 26 different combinations of 0's and 1's are needed.

As the nearest integral power of 2 (i.e. 2^n) which just exceeds 26 is 5 ($2^5 = 32$ whereas $2^4 = 16$, which is inadequate), it is necessary to have 5 'bits' per bite to represent all the letters of the English alphabet. If to this are added the ten numerals 0, 1, 2, 3 ... 9, forming a total of 36 *alphanumeric characters*, then n must be 6, so a minimum of 6 bits per bite is needed because $2^6 = 64$ whereas $2^5 = 32$ is too small.

7.3 Logic gates

The three basic logic switching circuits are those of the AND gate, the OR and the NOT (or INVERTER) gate. The conventional symbols for these three basic gates are given in fig. 7.1.

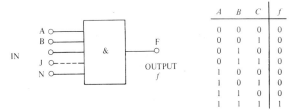

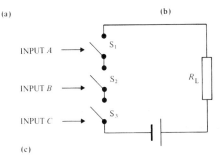

Fig. 7.2 **(a)** the AND gate **(b)** the corresponding truth table in the case where there are only three inputs A, B, C **(c)** a simple circuit illustrating that an AND operation involves essentially a series connection of switches

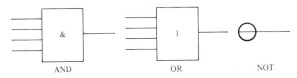

Fig. 7.1 The three basic logic gates

In all cases of logic gates the inputs will be referred to as A, B, C ... J ... N, where the number of inputs is N and where A, B, etc. can only represent 1 (usually known as 'logical 1') or 0 ('logical 0'). The output of the gate at terminal F will be denoted by f, which can also be only either 1 or 0.

The output of an AND gate is logical 1 if, and only if, ('if and only if' is to be abbreviated to 'iff') *all* its inputs are logical 1. Hence, in fig. 7.2a, iff the inputs A, B ... J ... N are *all* logical 1, the output f at F is logical 1. Thus, if any of these inputs is logical 0, the output is logical 0. The 'truth table' corresponding to the case where there are three inputs A, B, C is shown in fig. 7.2b.

The AND gate is essentially a series circuit arrangement: fig. 7.2c shows three switches S_1, S_2 and S_3 in series which are operated (opened or closed) by inputs A, B and C respectively; it is only when *all* these switches are closed that an output appears across the load resistance R_L.

By making use of the notation of Boolean algebra, the AND gate is represented mathematically by

$$f = A . B . C \qquad (7.1)$$

where there are three inputs A, B, C.

This expression (7.1) must be interpreted in the light of the fact that the only possible values of each of A, B, C and f are either 0 or 1. Furthermore, it reads that f equals A and B and C with the restriction that $f = 1$ iff $A = 1$, $B = 1$ and $C = 1$, whereas if one or more of these is 0, then f is 0.

The output f of an OR gate is 1 iff one or more of its inputs is 1. Hence, as in fig. 7.3a, if input A or B or C ... or J ... or N is 1, then the output f at F is 1. The corresponding truth table in the case where there are three inputs is shown in fig. 7.3b.

The OR gate is essentially a parallel circuit arrangement: fig. 7.3c shows three switches S_1, S_2 and S_3 in parallel which are operated by inputs A, B and C respectively. When one or more of these switches are closed, an output appears across the load resistance R_L.

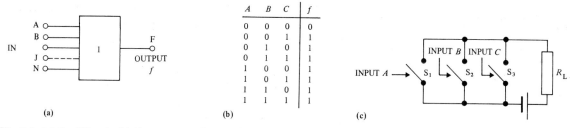

Fig. 7.3 (a) the OR gate (b) the corresponding truth table in the case where there are only three inputs A, B, C (c) a simple circuit illustrating that an OR operation involves essentially a parallel connection of switches

The Boolean representation of an OR gate is

$$f = A + B + C \qquad (7.2)$$

which is interpreted as meaning that $f = A$ or B or C with the restriction that f is logical 1 if any one of A, B or C is logical 1.

The output of a NOT gate is logical 1 *iff* the input is *not* logical 1. Hence the output is 1 if the single input is 0 whereas the output is 0 if this signal input is 1 (fig. 7.4).

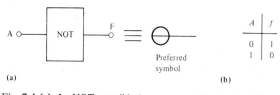

Fig. 7.4 (a) the NOT gate (b) the corresponding truth table

The Boolean representation of a NOT gate is

$$f = \overline{A} \qquad (7.3)$$

$\overline{A}$ means A negated, i.e. f is only present (as 1) if A is at 0, and vice-versa.

7.4 Logic gates based on diodes

For the purpose of discussing the use of a diode in a logic gate, the essential characteristics involved are:

(a) the reverse leakage current which may be regarded in an introductory account as negligibly small;
(b) the slope resistance of the forward characteristic ;
(c) the forward voltage V_γ across the diode, which must be exceeded so that the region of the characteristic is concerned where the slope resistance is a few ohms.

For a silicon planar junction diode, V_γ is about 0.6V, the forward slope resistance is about 3 ohm and the surface leakage current is as small as 1nA.

In logic gate circuits, the number of diodes equals the number of inputs so each diode has its separate input A, B or C (three are considered; many more can be provided, if required). The input is binary which demands that the input voltage changes from a value $V(0)$ representing the bit 0 to $V(1)$ to represent the bit 1.

If $V(1)$ is more positive than $V(0)$, a *positive logic system* is involved. If $V(1)$ is more negative than $V(0)$, the logic system is *negative*.

Usually, but not necessarily, $V(0)$ is at zero voltage (earthed) or near zero, so $V(1)$ is positive for positive logic and negative (for negative logic). In either case, the diode must be brought into its conducting region in its 'on' state by ensuring that a positive forward voltage exceeding V_γ is established across it.

In the OR gate circuit of fig. 7.5a, the input voltage level decreases in changing from $V(0)$ to $V(1)$: negative logic is concerned. The input terminals A, B and C are to the cathodes of the respective diodes D_1, D_2 and D_3. The anodes of these diodes are joined electrically and via the load resistance R_L to the supply voltage. This supply voltage is also $V(0)$, e.g. if $V(0)$ at the input corresponds to +4V and $V(1)$ to zero, the supply is also +4V.

If *all* the inputs (A, B and C) are at voltages $V(0)$, logic level 0, the voltage across each of the diodes is zero, no (or a vanishingly small) current flows through them so the voltage across the load resistance R_L is zero and the output voltage at F is also $V(0)$. If the input A or B or C is changed to $V(1)$, simulating logical 1, the voltage across the corresponding diode

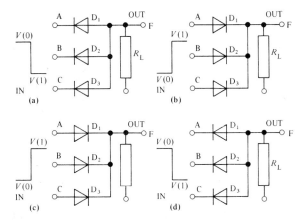

Fig. 7.5 Diode logic gates: **(a)** a negative logic OR circuit; **(b)** a positive logic OR circuit; **(c)** a negative logic AND circuit; **(d)** a positive logic AND circuit; (it is assumed in all four cases that the input device has a negligibly small internal resistance)

(D_1 or D_2 or D_3) is $V(0) - V(1)$ which is positive so current flows through the appropriate diode. There is then a voltage established across R_L and the output voltage at F becomes $V(1)$, assuming that R_L is much greater in value than the forward resistance of the diode and any internal resistance of the input device.

The logic involved is that if the input to A or B or C represents the bit 1, but the others represent logical 0, the output f at F represents logical 1. This is the OR operation.

If all the diodes in fig. 7.5 are reversed, i.e. the inputs A, B and C are to the respective anodes of the diodes D_1, D_2 and D_3, whilst their cathodes are electrically joined and lead via R_L to the supply voltage, the circuit is still an OR gate but now positive logic is involved (fig. 6.5b). The input is now such that $V(1)$ is more positive than $V(0)$.

Using diodes, an AND circuit for negative logic is the same as an OR circuit for positive logic *except that* the supply voltage has the value $V(1)$ instead of $V(0)$ (fig. 7.5c). If any one of the inputs to the diodes is at the voltage $V(0)$, logical 0, the diode involved conducts so the output voltage at F is $V(0)$ because there is a voltage drop across R_L (which is large) equal (or nearly equal for non-ideal diodes) to $[V(1) - V(0)]$. If, and only if, *all* the inputs are at the voltage $V(1)$, logical 1, will all the diodes have zero voltage across them so that the output voltage at F is $V(1)$ because no current flows through any of the diodes so that the voltage drop across R_L is zero. Hence the logic is that iff the inputs to A and B and C represent the bit 1, the output f at F represents 1. The AND operation is involved.

To change the AND circuit for negative logic to one for positive logic, the diodes are all reversed (fig. 7.5d). Note that fig. 7.5d is the same as that of fig. 7.5a. Thus the positive logic AND gate circuit and the negative logic OR gate circuit are the same. This is true not only for diode logic gates but also for any other device used in logic circuits, as is proved to be the case in example 7.7b.

The time taken for the full forward current to become established on applying a forward bias to the diode and the time taken for this forward current to decrease to zero on restoring the forward bias to zero (or a reverse value) are important factors concerning the speed with which a diode logic gate can be operated. Additional factors are the rates of charging and discharging of the voltage-dependent capacitance of the diode p-n junction.

7.5 A NOT gate based on an n-p-n junction transistor

The NOT gate (section 7.3) gives an output (simulating logical 1) when the input is representing logical 0 and, vice-versa, in that the output is logical 0 when the input is logical 1.

This function cannot be performed by a diode, but can be obtained from a simple bipolar junction transistor circuit.

The typical collector current (I_C) against collector-emitter voltage (V_{CE}) characteristics with the base current I_B as the parameter of the n-p-n junction transistor are shown in fig. 7.6. The forward current transfer ratio (current gain) of the junction transistor is h_{FE}. If V_{CC} is the supply voltage and the base current I_B exceeds $(V_{CC}/R_L)/h_{FE}$, the point P on the load-line represents the output and V_{CE} drops to $V_{CE\,sat}$, the voltage across the collector-emitter of the transistor when saturation current flows. If I_B is zero, the point Q on the load-line is concerned. Now the collector current is the small leakage current I_{CEO} and V_{CE} is almost equal to V_{CC}. For a typical small planar silicon n-p-n transistor, I_{CEO} is about 1nA and $V_{CE\,sat}$ is about 1V.

The NOT (or INVERTER) gate is achieved basically by the circuit of fig. 7.7a. If the input current I_i (= I_B) to the transistor is zero (or very small) the collector current I_C is very small so that the voltage f at the output terminal F equals (or is nearly equal to) the supply voltage V_{CC}, i.e. input

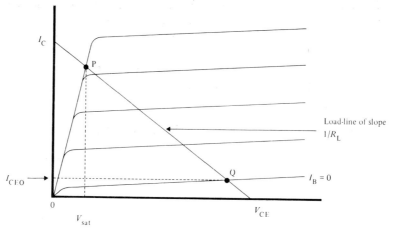

Fig 7.6. I_C against V_{CE} characteristics of an n-p-n junction transistor where I_B is the parameter and the load resistance in the collector circuit is R_L.

logical 0 gives output logical 1. If I_i (= I_B) equals or exceeds $(V/R_L)/h_{FE}$, the collector current I_C goes to the saturation value. This gives rise to a voltage drop across the load resistance R_L such that the output voltage f at terminal F falls to the small collector-emitter saturation voltage $V_{CE\,sat}$, i.e. input logical 1 gives output logical 0.

In practice, this NOT gate, operated with positive-going voltage pulses at the input, has to involve the resistances R_1 and R_2 (fig. 7.7b) to arrange an appropriate voltage on the base relative to the emitter. The transistor collector current goes to the saturation value when the input is positive; excess minority carriers are then stored in the base. To enable the transistor to respond without significant delay to a sudden charge of the input voltage level, this excess charge must be removed rapidly. To speed up this charge removal, the capacitance C (of value usually about 100pF) is connected in parallel with the resistor R_1 (cf section 6.10).

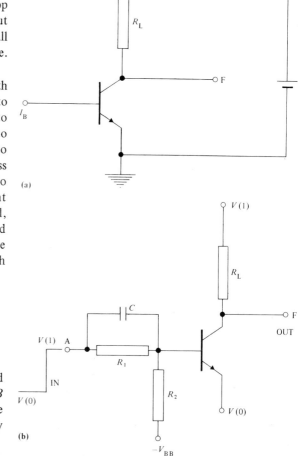

7.7 Boolean identities

On the basis that the Boolean expressions are related to switching operations and that in the following A, B and f can only have the values 0 or 1, consider the OR gate operation, where two inputs A and B only are involved:

$$A + B = f \qquad (7.4)$$

Fig. 7.7 NOT circuits based on an n-p-n transistor (a) the basic circuit (b) a working circuit for positive logic

i.e. in accordance with the definition of $+$, f is 1, iff A is 1 or B is 1. If B is 0,

$$A + 0 = 1 \quad (7.5)$$

Furthermore, it is readily seen that

$$A + 1 = 1 \quad (7.6)$$
$$A + A = A \quad (7.7)$$

and

$$A + \bar{A} = 1 \quad (7.8)$$

where $\bar{A}$ implies the negation of A (the NOT operation).

On the basis of the AND operation

$$A \cdot B = 1 \quad (7.9)$$

from which it clearly follows that

$$A \cdot 0 = 0 \quad (7.10)$$
$$A \cdot 1 = 1 \quad (7.11)$$
$$A \cdot A = A \quad (7.12)$$

and

$$A \cdot \bar{A} = 0 \quad (7.13)$$

The commutative laws of Boolean algebra are

$$A + B = B + A \quad (7.14)$$

and

$$AB = BA \quad (7.15)$$

The distributive law is basically

$$A(B + C) = AB + AC \quad (7.16)$$

The remaining identities of Boolean algebra which are of value in gate switching operations are the *dualisation laws*, based on de Morgan's theorem. This theorem may be expressed as:

The statement (as in AND operation) that if, and only if, all the inputs are 1, then the output is 1 is equivalent to the statement (as in OR operation) that if one or more of the inputs is 0, the output is 0.

In Boolean notation, this becomes for three inputs A, B and C

$$\overline{A \cdot B \cdot C} = \bar{A} + \bar{B} + \bar{C} \quad (7.17)$$

because the bar over the symbol implies negation (i.e. conversion of 1 to 0 or vice-versa).

It therefore follows that

$$\overline{A \cdot B \cdot C} = \bar{A} + \bar{B} + \bar{C} \quad (7.18)$$

and

$$\overline{A + B + C} = \bar{A} \cdot \bar{B} \cdot \bar{C} \quad (7.19)$$

EXAMPLE 7.7a *Verify the dualisation law in the form*

$$\overline{A \cdot B} = \bar{A} + \bar{B}$$

Put $\overline{A \cdot B} = f$

f is 1 iff $A \cdot B$ is *not* 1 ($A \cdot B$ is negated). But in order that $A \cdot B$ be 1, both A and B must be 1; therefore, $A \cdot B$ is *not* 1 if A is not 1 or B is not 1, which is tantamount to the statement that $A \cdot B$ is not 1 if $\bar{A}$ is 1 or $\bar{B}$ is 1. It follows that

$f = \overline{A \cdot B}$ is 1 iff $\bar{A}$ is 1 or $\bar{B}$ is 1

Therefore, $f = \overline{A \cdot B} = \bar{A} + \bar{B}$

EXAMPLE 7.7b *By the use of the dualisation laws in Boolean algebra, establish that an AND gate for positive logic is also an OR gate for negative logic*

For an AND gate with N inputs

$$f = A \cdot B \cdot C \ldots N \quad (i)$$

From equation (7.18)

$$\overline{A \cdot B \cdot C \ldots N} = \bar{A} + \bar{B} + \bar{C} \ldots + \bar{N} \quad (ii)$$

The OR gate with N inputs is represented by

$$f = A + B + C + \ldots + N \quad (iii)$$

From equation (ii) it is clear that equations (i) and (iii) are identical provided that $A, B, C \ldots N$ in (iii) are all negated and f is also negated. A positive logic AND gate is represented by equation (i). The inputs to this gate are all negated so that a 1 becomes a 0 and vice-versa if negative logic is used instead. Hence this positive logic AND circuit also functions as a negative logic OR circuit.

Fig. 7.8 Illustrating combinational switching

7.8 A combinational switching circuit

The requirement is an electronic system (a 'black box') which has a number of inputs $1, 2, 3, j, n$ and two outputs X and Y (fig. 7.8) and such that output X will initiate a different operation from output Y. For example, output X could (via a suitable relay) open a door or switch on a supply of cooling water or an electric lamp or heater whereas output Y respectively causes an alarm bell or buzzer to be sounded, or switches off the supply of cooling water or sets in operation a cooling-fan etc. The requirement is hence that output X is present (is logical 1) if a certain combination of inputs is present at $1, 2, 3, j, n$, whereas any other combination of inputs than this specified set results in an output at Y (i.e. Y is logical 1) and not at X (i.e. X is logical 0).

As a combination of digital inputs applied to 1, 2, 3 etc... n can represent numerical or written information on the lines of alphanumeric bits (section 7.2) such combinational switching leads to the possibility of the operation of a device provided that a certain set of instructions forms the input.

Table 7.2

Truth table for arranging that an output X is present iff the input combination actuated is 1 and 3; otherwise the output is Y

Inputs			Outputs	
1	2	3	X	Y
0	0	0	0	0
0	0	1	0	1
0	1	0	0	1
0	1	1	0	1
1	0	0	0	1
1	0	1*	1	0
1	1	0	0	1
1	1	1	0	1

*The required combination of inputs.

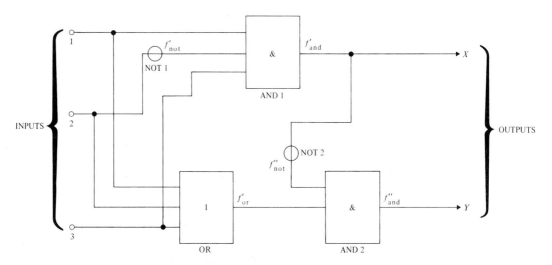

Fig. 7.9 An arrangement of logic gates which perform the functions required of the combinational switching circuit of fig. 7.8

A simple case is where there are only three inputs 1, 2 and 3 and the requirement is that an output is X iff inputs are present at 1 and 3. Any other combination of inputs at 1, 2 and 3 is to produce an output Y but not X. Recalling that the presence of a suitable signal is designated by 1 and the absence of any signal is designated by 0, a truth table can be constructed for this operation (table 7.2).

An arrangement of OR, AND and NOT logic gates which together operate in accordance with the truth table 7.2 is shown in fig. 7.9. Though this does not represent the minimal configuration of logic gates possible, it illustrates the ideas involved.

Let the outputs of the gates be f_{not} for a NOT gate, f_{or} for an OR gate and f_{and} for an AND gate and where f' is used for the first of such gates and f'' for the second. The truth table (table 7.3) is then constructed for the arrangement of logic gates in fig. 7.9.

Comparison of tables 7.2 and 7.3 shows that the outputs X and Y are identical.

Table 7.3

Truth table related to fig. 7.9

| Inputs | | | f'_{not} | $f'_{and} = X$ | f''_{not} | f'_{or} | $f''_{and} = Y$ |
1	2	3					
0	0	0	1	0	1	0	0
0	0	1	1	0	1	1	1
0	1	0	0	0	1	1	1
0	1	1	0	0	1	1	1
1	0	0	1	0	1	1	1
1	0	1	1	1	0	1	0
1	1	0	0	0	1	1	1
1	1	1	0	0	1	1	1

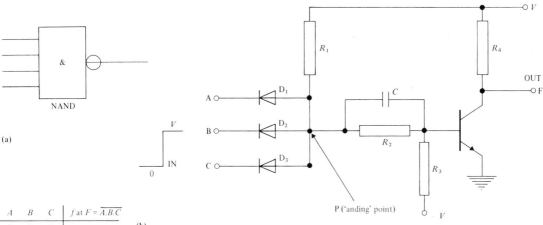

Fig. 7.10 (a) The logic symbol for a NAND gate (b) A positive logic DTL NAND gate circuit (c) Truth table for a 3-input NAND gate

7.9 AND and OR logic gates followed by a NOT circuit: NAND and NOR logic gates

The NAND or NOT-AND gate consists of an AND circuit followed by a NOT circuit. If the AND circuit is based on the use of diodes (section 7.4) and the negation (NOT circuit) is by a bipolar junction transistor (section 7.6) so-called diode-transistor logic (DTL) circuits are involved.

In fig. 7.10 are shown (a) the logic symbol for a

NAND gate, (b) the basic circuit diagram for a 3-input positive logic DTL NAND gate and (c) the truth table for a 3-input NAND gate

The NOR or NOT-OR gate comprises an OR circuit followed by a NOT circuit. In fig. 7.11, (a) is the logic symbol for a NOR gate and (b) is the truth table for a 3-input NOR gate.

The circuit for the positive logic DTL NAND gate (fig. 7.10) is the same as for a negative logic NOR gate, as is shown by the reasoning of example 7.7b.

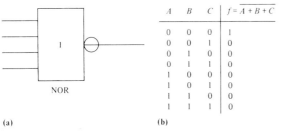

Fig. 7.11 (a) The logic symbol for a NOR gate (b) Truth table for a 3-input NOR gate

The positive logic DTL NAND gate circuit is usually with the emitter of the *n-p-n* bipolar transistor earthed. The voltage V is maintained across the circuit by the source of supply, the input voltage level rises from 0 to V and the maximum bias V_{BB} on the base relative to the emitter is equal to $-V$.

When one (or more) of the input terminals A, B, C (fig. 7.10b) is at 0V, the diode concerned conducts. The anodes of all the diodes are electrically joined at the so-called 'anding' point P. As R_1 is of much greater value than the forward resistance of any one of the diodes D_1, D_2, D_3, P is also set at or very near 0V. Assuming that the base current I_B of the transistor is negligibly small, the voltage V_{BE} applied across the base-emitter is then given by

$$V_{BE} = -V[R_3/(R_2 + R_3)]$$

By selection of appropriate values of R_2, R_3 and V, it is ensured that this negative voltage of magnitude $-V_{BE}$ will reduce the transistor collector current to zero. The output voltage at terminal F is therefore V. Hence the input of 0V at one or more of the inputs (representing logical 0) produces an output of V V (representing logical 1).

When *all* the inputs are at V V (representing logical 1), the output at F must be at or near 0V (representing logical 0). This requires that the transistor collector current is in the saturation region to ensure that the voltage drop across R_4 is much greater than that across the transistor. The transistor collector current will be at saturation provided that the input base current I_B equals or exceeds $(V/R)/h_{FE}$.

When the inputs are *all* at 0V, all the diodes are reverse-biased.

7.10 AND and OR logic gates preceded by a NOT circuit: enable gates

The AND gate operates in accordance with the Boolean statement (section 7.3)

$$f = A.\,B.\,C.\,.\,.\,N$$

where f simulates logical 1 provided that all inputs A, B and C . . . N simulate logical 1. The AND gate is a *coincidence circuit* in the sense that an output is obtained only when *all* the inputs coincide.

If a NOT gate is put before any one of the input terminals, say J of an AND gate (fig. 7.12a), the input at J will only simulate logical 1 if the input at S to this NOT gate represents logical 1, i.e. logical 0. Hence the NOT gate to J acts as an enabler in that the AND gate operation is only enabled provided that the input to S simulates logical 0. Conversely, if the input to S simulates logical 1, the AND operation is inhibited.

The truth table for a 3-input enable gate is given in fig. 7.12b.

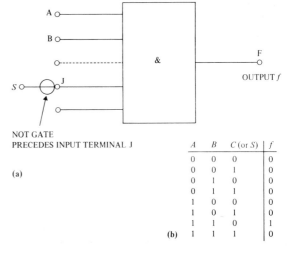

Fig. 7.12 (a) Symbol for an AND gate with a single enable input (b) Truth table for a 3-input enable AND gate

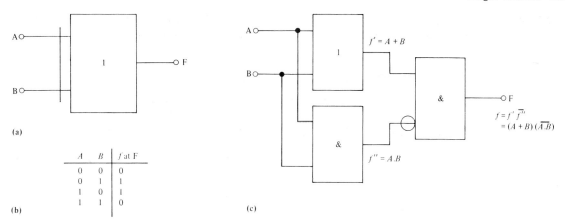

Fig. 7.13 (a) Symbol for a two-input EXCLUSIVE OR gate (b) The truth table for this gate (c) Block diagram of a two-input EXCLUSIVE OR gate, an inequality detector

More than one of the inputs to the AND gate can be preceded by a NOT gate giving enabling operation.

7.11 The exclusive OR circuit: inequality and equality detectors

Whereas the OR gate (inclusive OR gate) operates in accordance with the Boolean statement (section 7.3)

$$f = A + B + C + \ldots N,$$

the exclusive OR gate output f assumes the logical 1 state if one and *only* one input is put in the logical 1 state.

In fig. 7.13a for a two-input EXCLUSIVE OR, therefore, the output f at terminal F is 1 if $A = 1$ or $B = 1$, but not simultaneously. The truth table is then as in fig. 7.13b. Note that this truth table is the same as for a two-input OR gate *except that* f is 0 when A and B are both 1 whereas it is 0 in the OR (or INCLUSIVE OR) gate.

For a two-input OR gate the Boolean statement is

$$f = A + B$$

but for a two-input EXCLUSIVE OR gate, the Boolean statement is

$$f = (A + B)\,(\overline{A.B}) \tag{7.20}$$

because f must simulate logical 0 if A and B (i.e. $A.B$) are both 1. In other words, if A and B are 1 simultaneously, the 'anding' of A and B must be negated.

To arrange this the two inputs A, B must be fed to an OR gate to provide $(A + B)$ and simultaneously fed to an AND gate to provide $(A . B)$ where this AND gate is followed by a NOT gate so that $A + B$ and $A.B$ are the inputs to a succeeding AND gate (fig. 7.13c) for which the output at terminal F is $f = (A + B)(A.B)$. Other ways of arranging AND, OR and NOT gates can be alternatively adopted to give the same result.

As the output at F is logical 1 only if A is not equal to B in a two-input EXCLUSIVE OR gate, the inequality of two bits can be compared. Hence bites representing decimal numbers or alphanumeric statements can be compared so EXCLUSIVE OR configurations are used in the arithmetic sections of digital computers.

Whereas the EXCLUSIVE OR gate is essentially an inequality detector, an *equality detector* is also valuable. Here the concept is that the gate is able to detect when two bits are equal, i.e. both 0 or both 1. An output f simulating logical 1 is therefore desired if inputs A and B are both 1 or both 0 but f must simulate logical 0 if the bits at A and B are *not* the same. The Boolean statement for such an equality detector is therefore

$$f = A.B + \overline{A}.\overline{B} \tag{7.21}$$

This is seen to be the case on the basis that if A and B are logical 1 and 'anded', A and B will both be logical 0 so the OR operation (designated by +) will provide logical 1 but as this combination is negated the output is logical 0. Again, if A and B are both

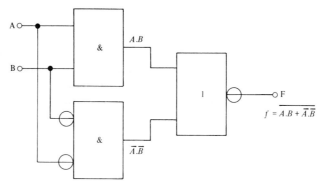

Fig 7.14. Gates arranged to provide an equality detector such that the output at F simulates logical 1 only if the inputs A and B both simulate logical 1 or both logical 0.

logical 0, $\bar{A}$ and $\bar{B}$ will both be logical 1 and f will be logical 0.

To provide an arrangement which undertakes the logic of equation (7.21) there is required an AND gate to which both inputs A and B are connected so as to give an output $A.B$ and where A and B are both negated (NOT gates used) before being also simultaneously presented to a second AND gate of which the output is $\bar{A}.\bar{B}$. $A.B$ and $\bar{A}.\bar{B}$ are then fed to the input of an OR gate to give $(A.B + \bar{A}.\bar{B})$ and where the output of this OR gate is negated to give finally an output f at F which represents $\overline{AB + \bar{A}\bar{B}}$. The block diagram of the arrangement of such gates is shown in fig. 7.14. This enables the equality of two bits to be detected.

7.12 General notes on the use of several gates in a logic system

The design of digital logic circuits involves generally combinations of several gates feeding into an output gate. The most useful configuration comprises inputs to several AND gates of which the outputs are combined by an OR gate or gates, followed by a NOT (INVERT) gate or several NOT gates. This AND-OR-INVERT arrangement is known as AOI.

As indicated by example 7.7b, it is seen that by suitable choice in the circuit of positive and negative logic operation of gates, AND gate operation can be undertaken by an OR gate circuit and vice-versa, as also can the operations of NAND and NOR. Indeed it is possible to construct an electronic calculator (or even the central processing unit of a digital computer) including the necessary storage (see section 7.13) based on NAND gate circuits alone.

The number of inputs to a logic gate is called the *fan-in;* the number of logic gates which a particular logic gate drives is known as the *fan-out* of that gate.

Clearly if the fan-out number is to be large, the load presented by the succeeding gates on a particular gate must be considered. Again, the power requirements of logic circuits are important, particularly as a considerable number of active devices must be included in a very small area in integrated circuit technology. Finally, logic gate circuits need to have significant *noise immunity* in that they should not be set in operation by random 'spikes', i.e. pulses due to the power supply, electric coupling between components, and, in some cases, electrical plant and apparatus in the vicinity. In all these respects, the mosfet is superior to the bipolar junction transistor (section 7.15).

7.13 The basic digital memory circuit

In digital systems it is often necessary to provide a memory circuit which receives a bit and stores it. The basic one-bit storage cell is a *flip-flop circuit*. This is a bistable circuit which can exist indefinitely in one of two stable states which simulate either logical 1 or logical 0 (i.e. $\bar{1}$).

This bistable circuit is based on cross-coupling two NOT gates in that the output of one NOT gate becomes the input to the other NOT gate and vice-versa. This storage cell or *latch* needs to be fed with information in the form of the bits 0 and 1. For this purpose, two 2-input NAND gates are used, NAND 1 and NAND 2 (fig. 7.15). There are two inputs A_1 and B_1 to NAND 1 whereas NAND 2 has the two inputs A_2 and B_2. Due to the cross-coupling, A_1 is the same as the output f_2 from NAND 2 and A_2 is the same as the output f_1 from NAND 1. So far, the NAND gates have thus each made use of only one input and consequently behave as NOT gates (a

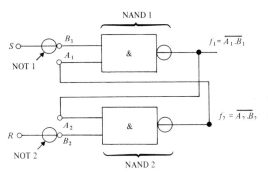

Fig. 7.15 A one-bit storage cell based on cross-coupled NAND gates with input NOT gates

single-input NAND gate is clearly the same as a NOT gate).

The other two inputs B_1 and B_2 to these gates NAND 1 and NAND 2 respectively are fed from NOT gates NOT 1 and NOT 2 respectively. The inputs to these NOT gates at S (for NOT 1) and R (for NOT 2) are the bits 1 or 0 which are to be stored in the latch.

If logical 1 (i.e. bit 1) is written into S and logical 0 (bit 0) into R, the input B_1 to NAND 1 is $\bar{1}$ (i.e. 0) and the input B_2 to NAND 2 is 1. The output from NAND 1 is $f_1 = \overline{A_1 . B_1}$ where A_1 is derived from the output from NAND 2. It is obvious that $f_1 = \overline{A_1 . B_1} = \bar{0} = 1$. The output from NAND 2 is $f_2 = \overline{A_2 . B_2}$, where A_2 is derived from the output of NAND 1, which is 1 and as B_2 is also 1, $f_2 = \bar{1} = 0$. Hence, the bit 1 input at S is stored as 1. The input of 1 at S is the *pre-set input*. To write 0 into this memory, the input at S is 0 and at R is 1. Then $B_1 = \bar{0} = 1$ and $B_2 = \bar{1} = 0$. The output f_1 from NAND 1 is $\overline{A_1 . B_1}$ whilst the output f_2 from NAND 2 is $\overline{A_2 . B_2}$. As B_2 is 0, f_2 is 1 whereas because B_1 is 1, f_1 is 0, so the latch flips over to the second of its bistable states. The input $S = 0, R = 1$ is the *re-set* or *clear input*.

7.14 Integrated circuit logic gates

Logic gate circuits are usually in integrated-circuit (IC) form. Integrated circuit technology (Appendix A) is unsuitable for the production of resistance values above about 30kΩ and of capacitance values exceeding about 100pF. Thus, to fabricate in IC form, for example, a positive logic NAND gate circuit comparable with that of fig. 7.10b, the capacitor C is eliminated and a single lower voltage supply is used. Furthermore, to ensure greater noise immunity, two additional diodes are included in the circuit between the 'anding' point P and the base of the output

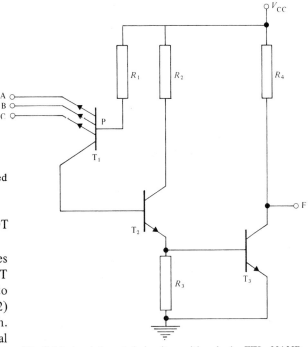

Fig. 7.16 An integrated-circuit positive logic TTL NAND gate

transistor. The voltage drop across these diodes is such as to exceed (and oppose) likely noise voltages.

However, in IC form, a widely used procedure is based on bipolar junction transistors: it is the transistor-transistor logic (TTL) gate. The input transistor T_1 to a TTL gate has a single base (and collector) but several emitters − as many as eight have been employed. Three emitters are shown in the positive logic TTL NAND gate circuit of fig. 7.16. There are consequently three emitter-base junctions which operate similarly to the input diodes (D_1, D_2 and D_3) of the DTL circuit (fig. 7.10b). The base of the transistor now acts as the 'anding' point P. As far as point P (from the input) the operation of this circuit is similar to that of DTL.

However, further features include the second bipolar transistor T_2 of which the base is joined directly to the collector of transistor T_1. The base-to-emitter junction of this transistor T_2 serves the function of a diode in improving the noise immunity of the circuit. The output transistor T_3 increases the fan-out of the logic gate.

Suppose that at least one of the inputs A, B, C to the multiple-emitter transistor T_1 is set at $V(0) = 0.2$V, say, simulating logical 0. One (or more) of the emitter-base junctions of T_1 conducts so that the voltage V_p at the 'anding' point P (the base of T_1) becomes, say 1V where V_{CC} is 5V. For the voltage across the collector-to-base of the transistor T_1 to become positive requires, say, 2V. Hence, with $V(0) = 0.2V$, transistors T_2 and T_3 do not pass collector current: they are 'off' so that the voltage at the output terminal F rises to V_{CC}, which is $V(1) = 5V$, say, simulating logical 1. The corresponding NAND operation is hence obtained. Simulation of logical 1 at *all* the inputs A, B, C requires a voltage $V(1) = V_{CC}$, i.e. +5V. The emitter-base junctions of transistor T_1 do not then pass current so that V_p becomes V_{CC}. This results in the switching-on to saturation of transistors T_2 and T_3 when the voltage at the output terminal drops to $V(0)$, simulating logical 0, again the NAND operation.

When T_2 and T_3 are in saturation (switched on and the output simulates logical 0) the voltage relative to earth in the collector of transistor T_2 (which is the same as that on the base of transistor T_3) is positive. Rapidly and subsequently the input changes (simulating logical 0) and the circuit has to revert quickly to the state where transistors T_2 and T_3 are cut-off. The switching time is delayed by stored charge in the base of transistor T_2. But this stored charge is quickly dispersed because the collector current of transistor T_1 is large when the input to one or more of its emitters is at $V(0)$. Thus, the first transistor T_1 (with its multiple emitters) does more than act as a set of diodes — its transistor action is important in ensuring the rapid switching action of this TTL gate.

Integrated-circuit logic gates based on TTL are widely used but they are strongly rivalled by increasing interest in integrated circuits based on mosfets and complementary mosfets, whose circuits are introduced in the next two sections.

7.15 MOS integrated circuit logic gates

The mosfets (chapter 3) are attractive instead of bipolar junction transistors in integrated-circuit construction because logic gates based on them have lower power dissipation, improved noise immunity, avoid the necessity of making passive components in IC form and, above all, enable several more active components to be incorporated within a given area of the silicon chip (*see* Appendix A). A disadvantage is that mosfet gate circuits are slower in action than the corresponding bipolar junction transistor circuits because of the relatively large capacitances involved due to the very thin silicon oxide films used.

In fig. 7.17 there are shown (a) a NOT gate, (b) a negative logic NAND gate and (c) a negative logic NOR gate. In all three circuits, *p*-channel enhancement type MOSFETs are used.

In the NOT gate circuit (fig. 7.17a) a mosfet T_2 has its drain D_2 and gate G_2 connected (so that $V_{GS2} = V_{DS2}$) and acts as a load in the drain circuit of mosfet T_1.

An immediate advantage of this circuit is that no passive components are used so that the fabrication in IC form of resistors and capacitors is not needed.

The input voltage V_i (in the form of a pulse representing a bit) is applied across the gate G_1 and the source S_1 (earthed) of T_1. With negative logic $V_i = V(0)$ is zero or near zero whereas $V_i = V(1) = -V_{DD}$, which is about $-20V$.

When $V_i = V_{GS1}$ (the gate-to-source voltage of T_1) $= -V_{DD}$, the drain current I_{D1} through the driver mosfet T_1 is large and the voltage V_{DS1} across its drain and source is small. The voltage at the output terminal F is therefore small and simulates logical 0, which is NOT or INVERTER action. When $V_i = V_{GS1} = 0$, the drain current I_{D1} is zero but drain current I_{D2} still flows through the load mosfet T_2 because its gate voltage is the same as its drain voltage. The voltage at the output terminal F therefore goes towards $-V_{DD}$, simulating logical 1, again the NOT action.

In the NAND gate circuit with negative logic (fig. 7.17b), three mosfets T_1, T_2 and T_3 are used. T_1 and T_2 are in series (as would be expected for AND operation, section 7.3) with their gates G_1 and G_2 respectively joined to the input terminals A and B respectively. The third mosfet T_3 has its drain and gate connected so that it is always conducting and acts as the load. Again, passive components are not used.

If either of the input voltages V_A or V_B to A and B respectively is at 0V (simulating logical 0) the corresponding mosfet is off so the voltage at the output terminal F goes towards $-V_{DD}$, simulating logical 1. With both V_A and V_B equal to $-V_{DD}$ (simulating logical 1), both T_1 and T_2 are on and the output voltage at F is zero, simulating logical 0. It is only when A and B are both $-V_{DD}$ that current passes; otherwise the current is zero, so this circuit is very economical in power consumption.

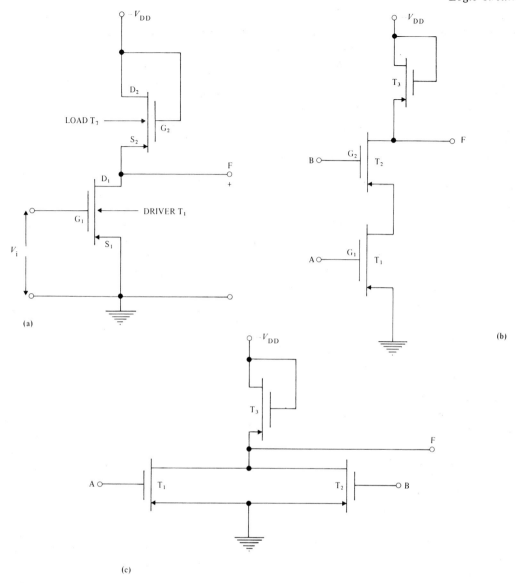

Fig. 7.17 (a) a mosfet NOT gate (b) a mosfet negative logic NAND gate (c) a mosfet negative logic NOR gate

The operation of the negative NOR gate based on two mosfets in parallel and with a common mosfet load (fig. 7.17c) can be readily deduced.

7.16 Complementary MOS integrated circuit logic gates

Complementary MOS (CMOS) integrated circuits are based on the use of complementary p-channel and n-channel enhancement mosfet devices on the same silicon chip. Logic gates based on CMOS devices reduce the power consumption per gate to levels of about 50nW.

To illustrate the further possibilities and advantages of CMOS, NOT gate circuits based on the popular TTL practice and on CMOS are compared (fig. 7.18).

The TTL NOT circuit (fig. 7.18a) utilizes a bipolar n-p-n transistor T_4 as an active pull-up circuit instead of a resistance in the collector circuit of the output bipolar n-p-n transistor T_3. If a resistor (giving so-called passive pull-up) were used as the load in the collector circuit of T_3 a prohibitively long time constant would be involved. The output transistor T_3 is used to increase fan-out. A capacitance load C_L is

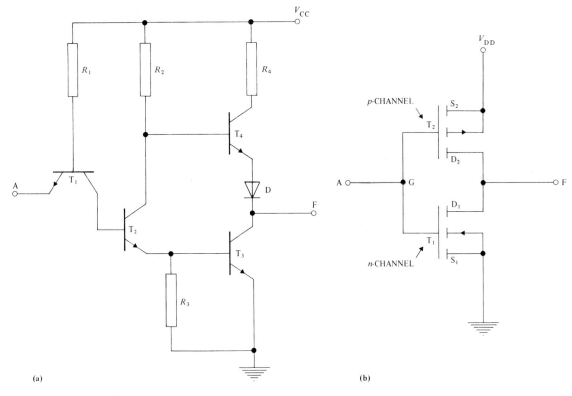

Fig. 7.18 NOT (INVERTER) circuits utilizing (a) TTL practice and (b) CMOS practice

present due to capacitances of the following gates and between leads. If this transistor T_3 were to utilize a load resistance R, during the change-over from input logical 1 to logical 0 the output would change from 0 to 1 and the capacitance C_L would charge exponentially from voltage $V_{CE\ sat}$ to voltage V_{CC}. RC_L would be too great as a time constant.

The TTL NOT gate therefore uses the transistor T_4 as the lead in the collector circuit of the output transistor T_3 and to which the bipolar n-p-n transistor T_2 acts as a phase-splitter. This arrangement is known as a *totem-pole output* driver. The input to transistor T_2 is from the initial NOT gate bipolar n-p-n transistor T_1. It is required that the output is at low voltage when transistors T_2 and T_3 are driven to saturation; in this state, T_4 should be non-conducting. To ensure that T_4 is off, the diode D is connected between the collector of transistor T_3 and the emitter of T_4.

At logical level 0, the input current to this TTL NOT gate is about 100μA, the output impedance is low in both states 0 and 1 and the output voltage change is about 60% of V_{CC}.

Using CMOS (fig. 7.18b), passive components are not needed, mosfet T_1 is an n-channel enhancement type and mosfet T_2 is a p-channel enhancement type. The input terminal A is to the gates (which are joined electrically at G) and the output is from the drains D_1 and D_2 which are joined electrically to the output terminal F. When the positive input voltage level (between G and earth) is high, T_2 is off and T_1 is on. The impedance of T_2 in the off (non-conducting) state is about $5 \times 10^9 \Omega$; the impedance of the conducting mosfet T_1 is only about 750Ω. Hence the output voltage at terminal F is very near to zero. When the input voltage level is low, T_1 is non-conducting whilst T_2 is conducting and the output voltage is very near to V_{DD}. The voltage change at the output terminal F is therefore substantially V_{DD}. Furthermore, when there is no input, the circuit is in a quiescent state with a power demand of only a few nW. The input impedance is some $3 \times 10^{10} \Omega$, so a particular CMOS gate can drive a considerable number of following gates: a large fan-out is possible. Another advantage is that the threshold voltage of a mosfet is about 40 to 50% of

together as a NOT gate whilst T_2 is the series load resistance which is either extremely high or comparatively extremely low; alternatively, T_2 and T_4 act as a NOT gate to T_1 as a series resistance which is either extremely high or low. Logical 0 will correspond to an input of 0V whereas logical 1 will correspond to an input of about 10V.

The CMOS NAND gate (fig. 7.19b) is clearly a simple re-arrangement of the NOR gate. Comparison of the circuits of fig. 7.19 with that of fig. 7.18b shows that the CMOS NOT gate circuit as a basic IC module (fig. 7.20 illustrates its construction) leads to combinations of NOR and NAND gates which enable any Boolean algebra function and data storage operation to be performed.

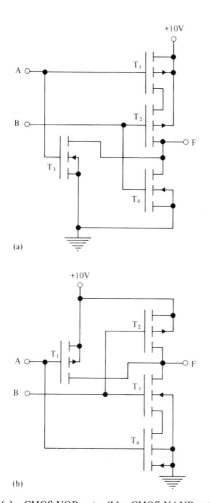

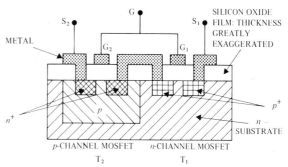

Fig. 7.20 Construction of a CMOS NOT gate in integrated circuit form

Fig. 7.19 (a) a CMOS NOR gate, (b) a CMOS NAND gate

V_{DD}, so the noise immunity exceeds considerably that of the TTL circuit.

Fig. 7.19 shows two-input CMOS gates where (a) is a NOR gate and (b) a NAND gate. In the circuit of the NOR gate, CMOS transistors T_1 and T_3 act

Rather large parasitic capacitances are associated with mosfet devices, particularly with the gate electrodes. The speed of operation of the gate is therefore limited. An IC technology in which mosfets are constructed by growing a silicon layer on top of an insulated substrate of sapphire or spinel enables excess silicon between diffusions to be removed completely by etching. Electrical leakage is thereby reduced and small capacitances are achievable.

Exercises 7

1. Obtain the binary numbers equivalent to 27, 333, 1041 and 13685.

2. Convert the numbers 365 and 52 into their binary equivalents. Subtract the binary number for 52 from that for 365 and show that the resulting binary number is equivalent to 313.

3. Show that it is necessary to have 6 bits in a word to represent all the alphanumeric characters in the English language. How many bits would be needed if, in a more complex language (such as Japanese) 48 characters were needed to form all the written language, in addition to the numbers 0 to 9?

4. Define and explain the operation of the following logic gates: AND, OR and NOT. Give the truth tables appropriate to each of these three gates, and also represent each of them by appropriate expressions in Boolean algebra.

5. Draw circuit diagrams of three-input AND and OR gates which make use of semiconductor diodes and explain their operation in the cases where (a) negative logic and (b) positive logic is involved. Explain why the circuit for a positive logic AND gate circuit is the same as that for a negative logic OR gate circuit.

6. Draw the diagram for and explain the operation of a working NOT circuit which is based on an n-p-n bipolar transistor. In your explanation of the operation, include an account of the necessity for a capacitor across the resistor used to connect the input terminal of the gate to the base of the transistor.

7. Write out logical arguments which justify the following identities in Boolean algebra:

   (a) $A + 1 = 1$ (d) $A.0 = 0$
   (b) $A + A = A$ (e) $A.1 = 1$
   (c) $A + \bar{A} = 1$ (f) $A.A = A$

8. State de Morgan's theorem, show that it leads to a dualisation law expressed in the form $\overline{A.B} = \bar{A} + \bar{B}$, and on the basis of this law, establish that a positive logic AND gate is equivalent to a negative logic OR gate.

9. Draw a block diagram of an arrangement of logic gates to which inputs 1, 2 and 3 are present and which enables a heater circuit to be switched on if a signal voltage is present at 1 and 3 but where any other combination of inputs to 1, 2 and 3 ensures that this heater circuit is off. Explain the operation of this combinational switching arrangement.

10. Draw circuit diagrams for (a) a positive logic DTL NAND gate and (b) a positive logic DTL NOR gate, where in each case three inputs are present. Explain the operation of one of these gates.

11. Explain the functions, principles of operation and main purposes of (a) enable gates and (b) exclusive OR circuits.

12. Write brief explanatory notes of the following terms:
    (a) AND-OR-INVERT (AOI), (b) fan-in, (c) fan-out, noise immunity of a logic gate system.

13. Show how 2-input NAND gates can be coupled together to form a flip-flop circuit and explain how this arrangement operates as a one-bit storage cell in a memory device.

14. Write an account of transistor – transistor logic (TTL), emphasising its importance in integrated-circuit form.

15. Give the advantages of the use of mosfets over bipolar transistors in logic gate circuitry. Draw circuit diagrams for negative logic NOT and NAND gates based on the use of mosfets, and explain their operation.

16. Explain the following terms: (a) totem-pole output driver and (b) complementary MOS (CMOS) circuitry.

17. Give a circuit diagram and explain the operation of a NOT circuit based on CMOS practice and outline the advantages and disadvantages of this practice compared with that of using bipolar transistors for designing a NOT integrated circuit.

8 · Opto Electronics

8.1 Photoconductivity (PC)

When photons of sufficient energy impinge upon and are absorbed by a semiconductor material, electrons may be raised from a lower level into the conduction band. If E is the energy which must be given to the electrons for this elevation to the conduction band to take place, the threshold wavelength λ_t which the wavelength λ of the incident radiation must not exceed is given by

$$\lambda_t = hc/E = \frac{1239}{E} \text{ nm} \qquad (8.1)$$

where E is in electron-volt.

This introduction is, however, an over-simplification in that it applies only to single crystal semiconductor materials having well-defined energy bands. If this is the case, for a pure material E is E_G, the width of the energy gap between the top of the valence band and the bottom of the conduction band (often called the energy gap between the valence and conduction band extreme). For a doped semiconductor E will normally be much lower than E_G for the host material. Thus, for an n-type semiconductor specimen, photoconductivity arises from the excitation of electrons from bound donor states into the conduction band whereas for a p-type semiconductor specimen, excitation of electrons from the valence band into the acceptor levels is required.

There are, however, important types of photocell which are based on semiconductor materials not in the single crystal form and where it is not uncommon for impurity additions to be added for other reasons than producing n-type or p-type materials, e.g. copper is added to cadmium sulphide (CdS) to enhance the probability of trapping holes being formed in the material and where the CdS may be in sintered powder form. The mechanics of the physical processes occurring on illumination by photons does not lead to such simple equations as (8.1).

A voltage is applied across suitable electrodes attached to the ends of the semiconductor specimen so that an electric field is established in which the current carriers move. On illumination of the semiconductor, the density of the photocurrent is determined by the total excess carrier concentrations and their mobilities.

8.2 Photoconductive cells for visible light

The two most used materials for making photoconductive cells sensitive in the visible and near infra-red regions of the spectrum are the II-VI semiconductor compounds, cadmium sulphide (CdS) and cadmium selenide (CdSe).

Cadmium sulphide is especially useful as by the incorporation within it of suitable materials its spectral response (fig. 8.1) can be made similar to that of the human eye. Cadmium selenide (CdSe) on the other hand, has a response which extends further into the infra-red than does that of CdS, so is more suitable for matching the light output from an incandescent lamp.

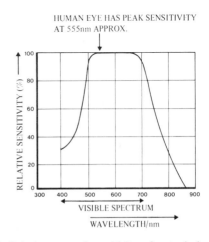

Fig. 8.1 Relative spectral sensitivity of a typical cadmium sulphide photocell

A photoconductive cadmium sulphide cell (fig. 8.2) consists of a pair of comb-like electrodes (e.g. of gold or indium) deposited on a tablet of sintered, suitably adulterated, cadmium sulphide powder, sometimes mounted on an insulating substrate. The arrangement adopted (ensured by suitable masking during deposition of the electrodes) provides a large sensitive area of the semiconductor between electrodes sufficiently closely spaced to enable an adequate electric field to be maintained by means of a suitable applied voltage. This structure is hermetically sealed in a housing with a suitable window for admitting the incident radiation and to inhibit the deleterious effects of moisture.

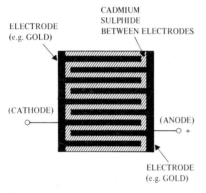

Fig. 8.2 'Comb-like' structure of deposited electrodes in a cadmium sulphide photoconductive cell

The cadmium sulphide used is always *n*-type and incorporates a small percentage of copper which results in the formation of copper ions.

This *n*-type CdS exhibits an important effect known as *charge multiplication*. This multiplication, which can attain values of 10^4 or more, arises because holes produced in the cadmium sulphide (on illumination) are much more readily trapped (especially by the copper ions) than are electrons. Indeed, the hole capture cross-section may well be 10^7 times the electron capture cross-section. Hence, when electron-hole pairs are produced, whereas electrons will travel virtually unhindered along the electric field lines to the anode, holes will not easily travel unhindered to the cathode. Much more likely is that the holes will only travel a short way towards the cathode and become trapped. These trapped holes (positive charges) attract electrons from the cathode which mostly pass to the anode; a few will neutralise the trapped holes. Those electrons which are produced initially by the incident photons are consequently greatly supplemented in number.

The quantum efficiency (yield) η of a photoelectric material is defined as the number of electrons made available (to form the photo-current) per incident photon. The value of η will depend on the nature of the material and the energy of the incident photons. In CdS (and CdSe) the multiplication effect results in a quantum efficiency which exceeds unity over a range of wavelengths in the visible region. Thus very high sensitivity to light in the visible region may be obtained.

Unfortunately cadmium sulphide photocells exhibit a time lag of about 0.1s. Thus, if the cell, initially in the dark, is subjected to a pulse of light it will take about 0.1s for the conductivity to reach $(1 - 1/e)$ (where e is the exponential factor 2.718), i.e. 63% of its final value and also to fall subsequently to 37% of this value after the light pulse has ceased. Cadmium selenide has a much shorter time constant of about 0.01s. These time constants depend on the level of illumination and on the previous use of the cell as a light-sensitive device.

The conventional circuit symbol for a photoconductive cell is shown in fig. 8.3a. The Mullard cadmium sulphide photoconductive cell ORP 12 (fig. 8.3b) has a sensitive area of $0.6cm^2$, operates at a

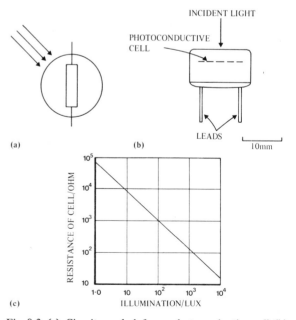

Fig. 8.3 (a) Circuit symbol for a photoconductive cell (b) Housing of the Mullard ORP12 cadmium sulphide cell (c) Characteristic of resistance against illumination for the ORP12

maximum pd of 110V and has a maximum allowable power dissipation at 25°C of 0.2W. A plot of resistance against illumination (fig. 8.3c) is for light from a lamp at a colour temperature of 2700K. The illumination is in lux (i.e. lumen/m^2).

Three applications of photoconductive cells of the cadmium sulphide type are shown in fig. 8.4 where (a) is a simple circuit, (b) is a twin-cell circuit for use in a photographic exposure meter and (c) utilises a Mullard RPY 20 cell in a twilight switching circuit.

Amongst the several applications of cadmium sulphide, photoconductive cells of various electrical characteristics may be mentioned: exposure meters in photography and particularly for automatically operated 35mm cameras; automatic contrast and brightness control of television receivers; flame indicating and control equipment, as in gas-fired central heating plant; automatic car head-lamp dipping devices and automatic control of street lighting.

8.3 Photoconductive cells for infra-red radiation

As indicated by equation (8.1) photoconductive materials must have small values of E to be sensitive in the infra-red region. The energy required to elevate the electrons into the conduction band begins to become not much greater than thermal energies. As a consequence, thermally generated noise becomes important. Such noise results within the semiconductor material itself (i.e. excluding noise of associated electronic circuitry) when current carriers are generated thermally and also when they recombine. To reduce this noise, infra-red semiconductor detecting devices are often cooled to a low temperature, 77K being convenient as it is obtained readily by the use of liquid nitrogen in a Dewar flask. The boiling point at atmospheric pressure of liquid nitrogen is 77K, i.e. $-196°C$.

To consider infra-red photodetectors more fully, specific terms need to be defined:—

Noise equivalent power (NEP): the root mean square (rms) radiant power P incident on the detector which produces at the output an rms signal voltage V_s which is equal in magnitude to the rms detector noise voltage V_n.

As an amplifier of frequency band width Δf is usually employed to enhance the output signal from the photodetector head, the appropriate relationship is

$$\text{NEP} = PV_n / [V_s (\Delta f)^{1/2}]$$

because the extent of the noise will be limited by narrowing the band width.

Detectivity D^*: the reciprocal of NEP multiplied by $A^{1/2}$, where A is the area of the sensitive surface of the photodetector. The $A^{1/2}$ is included because the NEP is proportional to $A^{1/2}$ for almost all photodetectors. Hence

$$D^* = \frac{V_s A^{1/2} (\Delta f)^{1/2}}{P V_n}$$

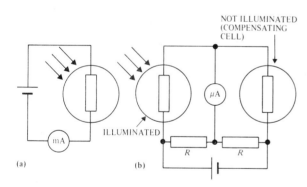

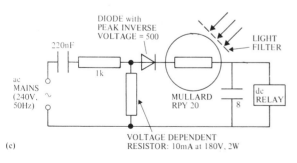

Fig. 8.4 (a) A simple operating circuit for a cadmium sulphide photoconductive cell; (b) a twin-cell circuit for a photographic exposure meter; (c) circuit for twilight switching which makes use of a Mullard RPY20 photoconductive cell

As V_s/V_n is a dimensionless ratio, the appropriate unit for D^* is that of $\dfrac{A^{1/2} (\Delta f)^{1/2}}{P}$. The unit is therefore cm W^{-1} Hz$^{1/2}$.

In quoting the detectivity D^* of a photodetector, two practices are usually adopted. The first is to give D^* (TK, f,1) where the detectivity is for radiation from a black-body at TK (T is commonly 500K) at a chopping (or modulation) frequency of fHz, where f is usually 800 and square-wave modulation is practised, and the bandwidth is 1Hz. Thus the detectivity is commonly for D^* (500, 800, 1). The second way is to quote D^* (xμm, f, 1) in the case where a collimated beam of monochromatic radiation of wavelength xμm is incident on the cell.

The detectivity D^* will be also clearly dependent on the geometry of the photodetector and its housing in relation to the window through which the radiation is admitted. D^* is inversely proportional to the sine of the half angular entrance aperture.

The smaller the value of E in equation (8.1) the more likely is noise to be significant compared with the signal so the lower the temperature to which the semiconductor material must be cooled. A useful working result is that the operating temperature T should be below $(E_G/25k)$K, where E_G is in eV and the Boltzmann constant $k = 8.6 \times 10^{-5}$ eV K^{-1}.

To indicate the importance of cooling, for indium antimonide (InSb) D^* at 77K is about 100 times that at 193K; but the response time is increased by lowering the temperature.

For a peak spectral response in the wavelength region round about 2μm, lead sulphide photoconductive cells are convenient. They have a detectivity D^* (λ at peak response, 800, 1) of about 5×10^{10} cm W^{-1}(Hz)$^{\frac{1}{2}}$ for evaporated lead sulphide with a cell of sensitive area 6 × 6mm and give at the peak wavelength an output of about 10^5 V per watt of incident radiation, i.e. the typical responsivity at the peak wavelength is about 10^5 V/W. Chemically deposited lead sulphide cells of sensitive area 1 × 1mm have a detectivity D^* (λ at peak response, 800, 1) of about 10^{10} cm W^{-1} (Hz)$^{\frac{1}{2}}$ and a typical responsivity of 200mA/W. Evaporated lead sulphide cells of the dimensions noted have a minimum resistance of about 1MΩ and a maximum of about 4MΩ, whereas chemically deposited cells have smaller resistances of a few hundred kΩ.

For a peak spectral response in the region of 5μm, the III-V semiconductor compound indium antimonide at 77K is useful. A cell with a sensitive area of 6 × 0.5mm has a detectivity D^* (λ at peak response, 800, 1) of 5 to 6 × 10^{10} cm W^{-1} (Hz)$^{\frac{1}{2}}$, a responsivity of 3.5×10^4 V/W, and a resistance of 60kΩ (max) to 20kΩ (min). At room temperature the detectivity is less: about 2×10^8 cm W^{-1}(Hz)$^{\frac{1}{2}}$ at a peak response wavelength which is somewhat longer at 6.0 to 6.5μm.

Indium antimonide photoconductive cells are often operated with a bias current which causes ohmic heating. An optimum value of this current is chosen to give the highest signal-to-noise ratio.

Fig. 8.5 shows typical spectral curves for indium antimonide photoconductive cells.

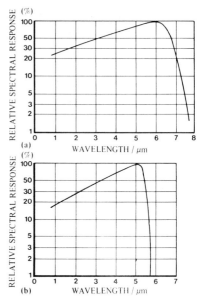

Fig. 8.5 Spectral response curves for indium antimonide photoconductive cells: (a) for the Mullard ORP10 at 293K; (b) for the Mullard ORP13 at 77K

The typical arrangement of an indium antimonide photoconductive cell operated at 77K is shown in fig. 8.6.

At room temperature, the response (per photon) of indium antimonide photoconductive cells is roughly constant for wavelengths up to about 7μm but falls off rapidly at longer wavelengths. The technology of indium antimonide (which can be produced in an extremely pure state) has been so advanced that their sensitivity to infra-red is limited by fluctuations in the radiation from the surroundings at room temperature.

Extrinsic photoconductive devices make use of the group IV elements germanium or silicon (and usually germanium) doped with donor (group V) or acceptor (group III) elements. The energy E involved in

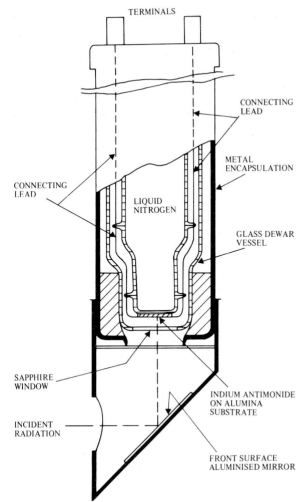

Fig. 8.6 Arrangement of an indium antimonide photoconductive cell at 77K for detecting infra-red radiation in the wavelength region from 1 to 5.5 μm

equation (8.1) is the impurity ionisation energy E_i, which is given approximately by assuming that it is the same as for a hydrogen atom (E_i = 13.6eV) immersed in the host material (e.g. germanium) of relative permittivity ϵ_r. Thus,

$$E = E_i = [13.6\, m^* / (m\epsilon_r^2)]\ \text{eV}$$

where m^* is the effective mass of the electron (for n-type germanium) and m is the mass of the free electron.

Assuming that $m^* = m$, so

$$E_i = 13.6/16^2$$

on substituting ϵ_r = 16 for germanium.

Hence E in equation (8.1) is about

13.2/256 = 0.053eV.

In fact, because $m^* < m$, energies E are about 0.008eV for group V donors in germanium and about 0.05eV for group III acceptors in silicon.

Hence, from equation (8.1) for n-type germanium λ_t the threshold wavelength, can be as long as

$$\frac{1239}{0.008}\ \text{nm} = 1.55 \times 10^5\ \text{nm} = 155\ \mu\text{m}.$$

A problem is to incorporate donor or acceptor impurities within the host germanium or silicon at levels much above 10^{15} to 10^{16} atoms per cm^3. Higher doping is not only difficult but is likely to lead to impurity band formation instead of levels. The absorption of the infra-red radiation in the doped material is consequently very low. To overcome this poor absorption in a practical device, the photoconductor element is mounted in a cavity which 'integrates' the radiation, i.e. radiation not initially absorbed is reflected back onto the detector.

8.2 Photovoltaic Effects (PV)

A photovoltaic effect is the generation of a voltage on the absorption of photons by a material. The immediate practical advantage of a photovoltaic cell is that no external source of emf is needed to produce an electric field in the material. The electric field required to cause oppositely charged carriers to move in opposite directions is created by a built-in potential barrier brought about either by a p-n junction or an interface between a metallic contact and the semiconductor.

Photovoltaic devices based on the action of a p-n junction include the photodiode, the phototransistor, and large area p-n junctions made by alloying, diffusion, epitaxial growth or ion implantation (*see* Appendix A) which are used as infra-red detectors and in the construction of solar cells, i.e. devices which generate electricity on the incidence of sunlight.

In fig. 8.7, p-type semiconductor material in region P is separated by a junction from n-type semiconductor material N. An electrode A is attached to N and an electrode B to P. In the dark, positive donor ions exist within N and electrons are the majority carriers whereas negative acceptor ions exist within P and holes are the majority carriers. A pd becomes established across the junction between N and P where N is positive (because of the donor ions) with respect to P (in which there are acceptor ions). Let this junction pd of N with respect to P be V_{np} — which is positive. Furthermore, N is positive with respect to the attached electrode A and P is negative with respect to the attached electrode B. No current will pass through any external load connected across AB because the pd established across the junction forbids any transfer of carriers across it and the pd across AB is zero.

On irradiation with photons having energies exceeding the energy gap, electron-hole pairs are created in the semiconductor material. Electrons so created will move across the junction to the positive n-side where they increase the number density of majority carriers within N. Holes so created will move across to the negative p-side and increase the number density of majority carriers within P. Consequently, within N a further number of positive donor ions become neutralised by the additional electrons whereas within P more negative acceptor ions become neutralised by the additional holes. The pd V_{np} across the junction is therefore reduced. The interface potential differences between electrode A and N and the electrode B and P remain unchanged. Hence the absorption of the photons causes the electrode A (attached to N) to become lower in potential. If electrode A is earthed, the irradiation thus causes the potential of electrode B to become positive. This is the photovoltaic emf.

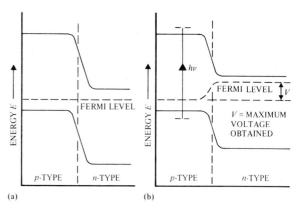

Fig. 8.8 Energy-band structure of a p-n junction (a) in the dark; (b) irradiated with photons of sufficient energy

A current will thus pass through the external load resistance across AB on irradiation of the semiconductor. This will cause the photovoltaic emf to decrease slightly. The maximum emf obtainable is equal to the difference between the Fermi levels (fig. 8.8).

Photovoltaic devices based on the action of an interface junction between a semiconductor and a metal include two types of *barrier layer cells:* the silicon photovoltaic cell and the cuprous oxide/copper photovoltaic cell. The latter is not much used so the former only will be considered.

A selenium barrier-layer photovoltaic cell (fig. 8.9a) consists of a polycrystalline layer of

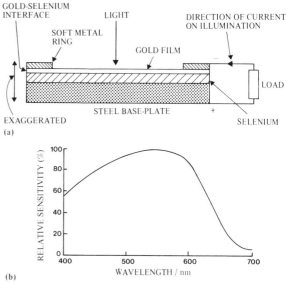

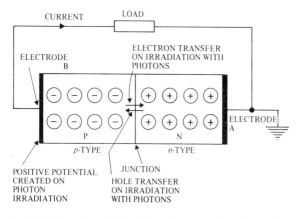

Fig. 8.7 The photovoltaic effect when a p-n junction exists between semiconductor materials

Fig. 8.9 (a) A selenium barrier-layer photovoltaic cell (b) Spectral response of a typical iron-selenium barrier-layer photovoltaic cell

selenium formed by temperature-curing of the molten element on a base-plate, usually of steel. The surface of this selenium is coated with a thin, transparent layer of gold, prepared by evaporation *in vacuo*. The contacts to the cell are, one, to a soft metal contact ring bearing on the gold coated surface and, two, a contact to the steel base-plate.

There will be established a pd between the conducting gold film and the selenium with the gold film positive because it loses some of its plentiful free electrons to the semiconducting selenium. In the dark, this potential will be equal and opposite to that between the selenium and the steel base. The pd across the electrodes of the cell in the dark is therefore zero.

On illumination of the gold-coated surface, usually with visible light, a significant fraction of this light will be transmitted by the semi-transparent gold film, which exhibits a green filter action. The light reaching the selenium will create electron-hole pairs there. Some of the electrons generated will move to the positive gold film so the potential difference across the gold-selenium interface is reduced. This causes the potential at the contact to the gold to become negative with respect to the steel base-plate.

Iron-selenium photovoltaic cells of this type cause an electric current to flow through a load resistance (usually that of a suitable microammeter) and where the photoelectric current varies with the illumination (fig. 8.10).

Such iron-selenium cells are convenient in a range of applications but their response decreases if the incident light fluctuates at frequencies above about 1kHz because of their significant resistance and capacitance.

8.3 Silicon photovoltaic cells; solar cells

The most commonly used silicon photovoltaic cell of the *p-n* junction type comprises a slice of *n*-type silicon into the surface of which is diffused a *p*-type acceptor element to a depth of about 5μm. Incident light is able to penetrate such a thin layer and reach the *p-n* junction. The current generated through a low external load resistance across such a cell is proportional to the intensity of the incident light. As would be expected, this current is proportional to the area of the exposed surface of the cell, but the pd established is independent of this area. Response times are a few microseconds, much shorter than for iron-selenium cells of the type described in section 8.2.

Tiny photovoltaic cells of this type can be arranged as required in the surface of a silicon chip or slice. A number of applications of such arrays of cells have been explored: two examples are automatic coin selection and photoelectric read-out devices for computer punched card or tape.

The conversion of the energy of sunlight into electrical energy is the function of a solar cell where the chief application is to provide a source of electric power in an artificial satellite. A typical arrangement of an element of a junction type of solar cell (fig. 8.11) comprises *n*-type material diffused into or epitaxially grown on the surface of *p*-type material. As indicated, current flow to the contact strips occurs. The energy gap of the device needs to be an optimum at about 1.4eV, a compromise between energy which is wasted because incident photons of high energy in the solar radiation will simply cause heat losses consequent upon lattice vibrations set up in the semiconductor and the fact that photons of low energy will not be absorbed.

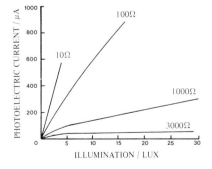

Fig. 8.10 Characteristics of current against illumination for an iron-selenium photovoltaic cell. The parameter is the load resistance across the cell

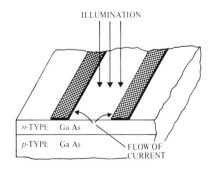

Fig. 8.11 Outline of a typical solar cell element

The surface layer thickness is an optimum at about one-third of the electron diffusion length. The majority carrier flow is mostly transverse. As these transverse currents will introduce heating losses in the surface layer it is important that this layer should be of high conductivity, more easily obtained with *n*-type than *p*-type material. Generally, gallium arsenide (Ga As) is the preferred host semiconductor material: power conversion efficiencies η of as much as 20% are feasible, where

$$\eta = \frac{\text{the maximum electrical power output from the cell}}{\text{the total radiation power incident upon the cell}}$$

8.4 Photodiodes and phototransistors

If a reverse bias is set up across a *p-n* junction diode, a small leakage current is produced due to thermally generated current carriers.

When radiation is incident in the region of the *p-n* junction, if the absorbed photons have energies exceeding the band-gap energy, electron-hole pairs are produced. On the *n*-side of the junction, electrons are the majority carriers and the fractional increase in their number is small due to the radiation-generated electron-hole pairs; but the fractional increase in the relatively few holes (the minority carriers) is large due to the holes provided in the electron-hole pairs created. On the *p*-side of this junction, the opposite is the case: the number density of the electrons (the minority carriers) is significantly increased but that of the holes (the majority carriers) is not.

With the reverse bias provided, the minority carriers generated by the radiation will be transferred across the *p-n* junction. The photocurrent which passes through a suitable load resistance increases in proportion to the light intensity.

A typical *silicon photodiode* with a sensitive area of 12mm^2 in the *p-n* junction region has a peak spectral response at 920 nm and provides a photocurrent of 32μA when the illumination is 2000 lux. The photodiode exhibits linearity between the illumination and the photocurrent and exhibits smaller time lag than the phototransistor.

Avalanche photodiodes make use of a reverse bias which is near the breakdown voltage. This gives rise to a multiplication effect because electron-hole pairs generated by absorbed photons are sufficiently accelerated to acquire energies which enable additional electron-hole pairs to be created on collision with atoms in the host material. They give greater sensitivities than normal photodiodes, exhibit less noise and smaller time lag.

p-i-n photodiodes have an intrinsic layer of semiconductor material sandwiched between *p*-type material and *n*-type (fig. 8.12). On the application of a fairly high reverse bias, an approximately uniform electric field of high intensity is set up in the intrinsic region. This region is considerably wider than the depletion region associated with a reverse-biased straightforward *p-n* junction. It consequently has a much smaller capacitance and a greater volume within which incident photons may be absorbed. The intense electric field provided enables fast transport of the current carriers produced. Response times of about 1ns are obtainable.

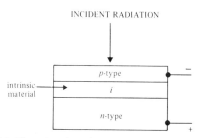

Fig. 8.12 The *p-i-n* photodiode structure (schematic)

In a phototransistor, frequently adopted practice is to utilise an *n-p-n* bipolar transistor in common-collector connection and where the light-sensitive region is the *p-n* junction between the base and the collector. A positive potential of V_{CE} is maintained on the collector with respect to the emitter but the base is not connected directly to any voltage supply: it is 'open-circuit'. There is then a reverse bias across the base-collector junction and a forward bias across the emitter-base junction, so that transistor action is enabled.

In the dark, a few thermally generated electron-hole pairs are created in the *p*-type base. The electrons of these pairs will travel to the positive collector. The remaining holes will attract electrons from the base region but most of these electrons will go through to the collector region.

When photons of incident radiation are absorbed near the collector-base junction, additional electron-hole pairs are created. The electron current to the collector is then considerably enhanced. This photocurrent output at the collector equals h_{FE} times the

Opto Electronics 181

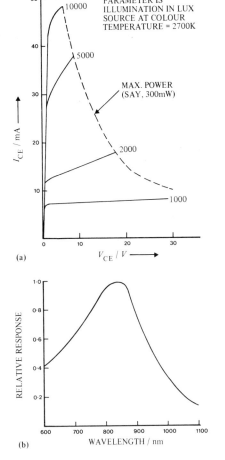

(a)
(b)

Fig. 8.13 (a) Current-voltage characteristics of a phototransistor (b) Spectral sensitivity characteristics of a silicon planar phototransistor

current which would be obtained from an equivalent photodiode, where h_{FE} is the transistor forward current transfer ratio or current gain (section 4.3) which can readily be about 100. The current-voltage characteristics and spectral response curves of a silicon planar phototransistor are shown in (a) and (b) respectively of fig. 8.13.

On the absorption of photons in the collector-base junction region, electron-hole pairs are formed at various distances from the junction itself. The transit times of electrons from their point of origin to the collector are therefore various. Clearly to obtain satisfactory light sensitivities, a significant area around the junction concerned has to be irradiated. When a sharp pulse of photons is absorbed over this defined area, electrons generated (multiplied in number by the transistor action) produce a collector

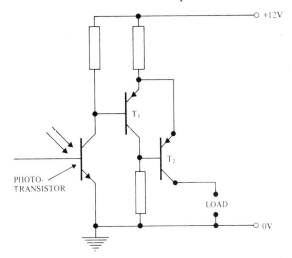

Fig. 8.14 Amplification of the output from a phototransistor

current pulse which is less well-defined than the light pulse: in particular, it has a significantly longer rise time. Consequently, silicon planar phototransistors exhibit a marked fall-off in response at frequencies of modulation of the incident light greater than about 50kHz.

Fig. 8.14 shows a typical circuit in which the output from a photo-transistor is amplified.

Instead of basing a photo-sensitive device on a bipolar junction transistor, the field-effect transistor (section 3.1) is also used, giving the so-called *photofet*.

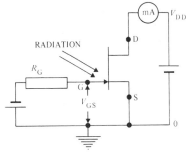

Fig. 8.15 The photofet circuit

In the circuit of fig. 8.15, the input voltage to the gate G relative to the source S is via a gate resistor R_G and the connection of the drain D to its voltage supply V_{DD} is via a milliammeter.

To obtain photofet action, radiation is incident in the region near the gate junction (between p-type and n-type silicon) where absorption of the photons creates electron-hole pairs. The holes so generated flow to the negative gate causing an increase ΔI_G of the gate current from a negligible value and a corresponding change of the gate-source voltage of ΔV_{GS} which depends on $\Delta I_G \cdot R_G$. The drain current I_D is therefore altered by $g'_m \Delta I_G \cdot R_G$, where g'_m is the dynamic mutual conductance of the FET.

The sensitivity to light of the photofet is much greater than for the phototransistor of the bipolar type. Being based on silicon, its spectral response is similar to that of a silicon planar phototransistor. This sensitivity can also be conveniently varied over a wide range by alteration of R_G which can be made variable from zero to as much as 500MΩ (though this high value results in a long time constant of the circuit) resulting in a drain current change from say 10nA to 10mA (i.e. over a ratio of 10^6) for illumination of 10 lux by light at the peak wavelength. Relay action is also readily obtained by means of a photofet in that drain current can be switched-on by a pulse of incident light which relaxes the gate-source bias from the cut-off value.

Light-actuated relays or switches for the control of power are also based on silicon controlled rectifiers (SCRs). Absorption of photons in the silicon of a specially constructed SCR (a LASCR, or light-actuated SCR) produces electron-hole pairs which are arranged to cause a gate current which switches the device to the conducting state.

Another phototransistor device of interest is the *Darlington-connected phototransistor* (fig. 8.16) in which the two n-p-n bipolar transistors are incorporated. The collector of transistor T_1 is connected to the collector of transistor T_2 whereas the emitter of T_1 is connected to the base of T_2. High sensitivity to light incident on the base region of T_1 is obtained because the emitter current from T_1 is amplified by the transistor T_2.

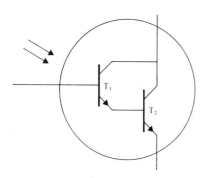

Fig. 8.16 The Darlington-connected phototransistor

8.5 The photoemissive effect (PE)

The photoemissive effect or *surface photoemission* is that phenomenon in which the incidence of radiation on a surface causes the ejection from this surface of electrons. This surface is either in a vacuum to enable the electrons from the surface (a photocathode) unobstructed passage to an anode or in an inert gas at low pressure so that gas amplification results from ionisation of the gas molecules when struck by sufficiently energetic electrons.

The basic equation involved, due to Einstein, is

$$h\nu = \phi + \frac{1}{2} m v^2_{max} \qquad (8.2)$$

where h is the Planck constant, ν is the frequency of the incident radiation, ϕ is the work function energy of the photocathode material, m is the mass of the free electron and v_{max} is the maximum speed of the electrons ejected.

Putting $v_{max} = 0$ gives

$$h\nu_t = \frac{hc}{\lambda_t} = \phi = Ve \qquad (8.3)$$

where ν_t and λ_t are the threshold frequency and the threshold wavelength respectively and V is the work function potential ($= \phi/e$, where ϕ is the work function energy) of the surface. Clearly for a surface to exhibit the photoemissive effect, the wavelength of the radiation incident upon it must be shorter than λ_t.

Equation (8.2) only applies to true *surface* photoemission from an element. The emission of electrons from an illuminated semiconductor results from the generation of electron-hole pairs *within* the semiconductor, as in photoconductivity. The electrons involved subsequently have to reach the surface and there be liberated into the surrounding space (usually a vacuum) and where, to attain low values of the work function, special kinds of composite surface layers are used.

Photons are absorbed close to the semiconductor surface. To be capable of causing emission of electrons from this surface, the energy required of the photon must be at least $(E_G + E_A)$ where E_G is the energy gap between the top of the valence band and the bottom of the conduction band and E_A is the electron affinity energy of the material. This energy

E_A is the difference in energy between that at the bottom of the conduction band and the ionisation energy, because the ionisation energy corresponds to the case where the electron is free of the molecule.

The electrons produced within the semiconductor near the surface will lose energy in trying to reach the surface. This energy loss is due chiefly to scattering in the crystalline lattice and also to electron-hole pair production, whereby an electron raises another electron into the conduction band and inevitably creates also a hole. The cross-section for such pair production by an electron is very small (indeed, negligible) unless the electron energy exceeds $2E_G$ by an appreciable amount.

If the energy of the electron produced within the semiconductor by sufficiently energetic photons incident on the surface is less than $2E_G$, electron-hole pair production can be neglected. Then the loss of energy is due to lattice scattering, which is so small as to enable electrons from depths of about 30nm to escape. In some important photosensitive semiconductor materials these conditions apply. The electrons produced with energies appreciably in excess of $2E_G$ will not be able to escape from such depths because they suffer rapid energy losses as they produce electron-hole pairs; the depths from which such energetic electrons escape do not exceed about 2nm.

It is also clearly necessary for the electron affinity energy E_A of the material to be small if the quantum efficiency is to be significant for radiation of the longer wavelengths. To put this in an alternative way, the potential barrier at the surface must be small. This is achieved by coating the surface with a thin film of an electropositive metal of low work function: caesium is frequently used. Such a film is particularly effective on an atomically clean surface of a p-type semiconductor because the very thin electric dipole layer produced at the surface enhances electron release. Caesium oxide (CsO) has been shown to be superior to caesium alone in this respect.

Semiconductor surfaces for photocathodes with quantum efficiencies as high as 0.3 have been produced.

Photocathodes are classified in accordance with their spectral response numbers. Some of those frequently used in commercial photocells are given in table 8.1. The spectral response will depend on the material from which the window of the photocell (through which the incident radiation is transmitted to the photocathode) is made: soda-lime glass, Pyrex, fused quartz and a high silica content ultra-violet (uv) transmitting glass are all used. Soda-lime glass transmits light of wavelength exceeding about 330nm; Pyrex glass has a wavelength cut-off rather shorter than soda-lime glass; fused quartz will transmit satisfactorily down to wavelengths of 200nm.

Table 8.1

Some photoemissive materials for photocathodes

Spectral response number	Photo-emissive material	Window material	Wavelength (in nm) at which response is a maximum
S 1	Ag – O – Cs	Lime glass	800
S 3	Ag – O – Rb	Lime glass	420
S 5	Cs_3 Sb	uv transmitting glass	340
S 8	Cs_3 Bi	Lime glass	365
S 10	Bi – Ag – O – Cs	Lime glass	450
S 11	Cs_3 Sb	Lime glass	440
S 13	Cs_3 Sb	Fused quartz	440
S 20	Sb – K – Na[Cs]	Lime glass	420

The relative spectral sensitivities for photocathodes of types S1 (Ag – O – Cs implies caesium on oxidised silver), S11 (Cs_3 Sb is an alloy of caesium and antimony) and S13 (as S11 but with a quartz window which transmits uv radiation at wavelengths down to 200nm) are shown in fig. 8.17.

S3 (Ag – O – Rb implies rubidium on oxidised silver) has a spectral sensitivity curve more like that of the human eye than does S1 in which caesium is used. The photoemissive material with a spectral sensitivity curve which approximates best to that of the human eye is, however, S10 (caesium on oxidised bismuth and silver).

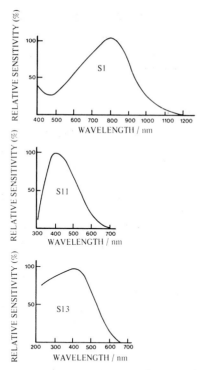

Fig. 8.17 Relative spectral sensitivities of photocathodes S1, S11 and S13

8.6 Vacuum photoemissive cells

Vacuum photoemissive cells of various constructions are made (fig. 8.18). A typical commercial vacuum photocell is the Mullard 90AV (fig 8.18c) which has a caesium antimony film on the cathode surface of projected area 4.0cm^2. This cell is particularly sensitive to daylight and in the blue end of the visible spectrum. When the whole of this cathode area is illuminated by light from a lamp at a colour temperature of 2700K, the luminous sensitivity S is 45μA/lumen when the pd between the anode and cathode is 100V and a series resistance R_L of 1MΩ is in the anode circuit.

The characteristic curves shown in fig. 8.19 are typical of a vacuum photocell. For a given level of illumination I of the photocathode, an increase from zero of the anode voltage relative to that of the cathode causes the anode current (the photocurrent) to increase at first until an anode voltage is reached beyond which further increase of the anode current is very small. This 'saturation' current arises because all the electrons emitted from the cathode travel directly along electric field lines which terminate at the anode.

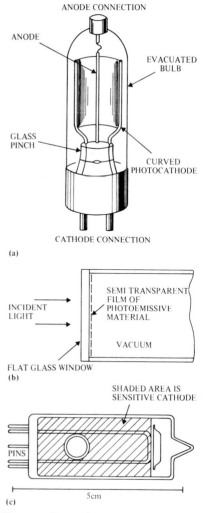

Fig. 8.18 Vacuum photoemissive cells (a) Half cylinder of metal coated with photoemissive material; anode rod on axis (b) Semitransparent cathode of photoemissive material on inside surface of flat glass window where incident light reaches cathode by passing through the glass window (c) A Mullard vacuum photocell, model no. 90AV

Vacuum photocells are useful for measuring amounts of light in photometry because, at a given anode voltage (in the saturation region) the anode current is directly proportional to the number of photons incident on the cathode per second. Another advantage is the very short time lag (due to electron transit time between cathode and anode) of about 1ns between incidence of a photon on the cathode and the arrival at the anode of the electron released. Thus, vacuum photocells can be used with light modulated at a frequency of up to about 1000MHz,

8.7 Gas-filled photoemissive cells

An inert gas, usually argon, at a pressure of a few pascal (1 torr = 1mm of mercury = 133 pascal) within the phototube increases significantly the luminous sensitivity compared with the corresponding vacuum photocell provided that the anode voltage exceeds considerably the ionisation potential of the gas. This increase arises from the phenomenon of *gas amplification* whereby electrons released from the cathode are accelerated to sufficient energies to ionise the gas atoms. Such ionisation results in the production of an ion pair (an electron and a positive ion). The extra electron enhances the anode current and, moreover, the positive ions created cause still further release of electrons from the photocathode. This impingement of positive ions on the cathode may, however, cause the luminous sensitivity of the gas-filled photocell to fall significantly during some hundreds of hours of operation at the maximum rated anode voltage.

The Mullard 90AG gas-filled photocell has the same geometry as the vacuum cell 90AV. Its luminous sensitivity under the same conditions of illumination is 130μA/lumen as against 45μA/lumen for the vacuum cell. The typical anode current — anode voltage characteristics of a gas-filled photocell (fig. 8.21) do not exhibit the plateau region typical of the vacuum photocell.

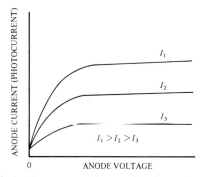

Fig. 8.19 Anode current against anode voltage characteristics of a typical vacuum photocell for various levels of illumination I of the photocathode ($I_1 > I_2 > I_3$)

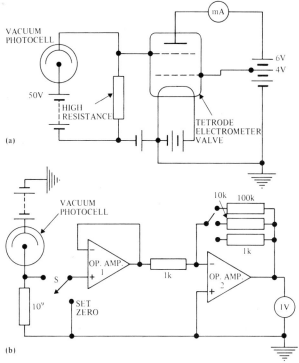

Fig. 8.20 (a) Amplification of the output from a vacuum photocell (b) Operational amplifier 1 has dual fet input; input resistance $> 10^{12}\,\Omega$; used as voltage follower, i.e. unity gain. S is high quality ceramic mounted switch

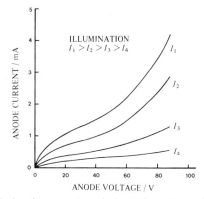

Fig. 8.21 Anode current against anode voltage characteristics of a typical gas-filled photocell for various levels of illumination of the photocathode

the limit being set by the circuit time constant brought about by the anode-cathode capacitance (0.7pF for the Mullard 90AV photocell) stray capacitance and the resistance in the circuit.

Amplification of the output from a vacuum photocell is often practised: two circuits for such a purpose are that of fig. 8.20a, utilising an electrometer valve, and that of fig. 8.20b which makes use of an IC module amplifier.

Unlike the vacuum photocell, the anode current at a given anode voltage of a gas-filled photocell is not directly proportional to the illumination of the photocathode. Indeed, the current increases more than the illumination increase because increase of the number of photons incident on the photocathode causes more electrons to be produced, so more ion pairs and the comparatively slow-moving positive ions create a positive space charge around the cathode. A further consequence of the important effect of the less mobile positive ions is the significant time-lag experienced in the use of such cells whereby the maximum frequency of operation is about 10kHz with a fall-off of response at a frequency of about 1kHz.

Gas-filled photocells are much used in film sound recording and with relays or a thyristor whereby a light beam is used to switch-on an electrically operated device.

8.8 Photomultiplier tubes

Within a vacuum tube, electrons emitted by the photo-emissive cathode on the incidence of photons, are directed by a suitable electric field to a first dynode. This dynode is an electrode having a surface with a secondary emission coefficient (number of electrons ejected divided by a number of electrons incident per second) which exceeds unity. Suppose this secondary emission coefficient is m. The secondary electrons from the first dynode D_1 are then guided by accelerating electric fields to successive dynodes D_2, D_3 etc. At each dynode, the number of electrons per second is increased n times. With n dynodes, the overall increase is therefore m^n. Hence, if 10 dynodes are used and m is 5, the overall factor by which the original number of electrons is multiplied is 5^{10} or 10^7 approx.

In practice with photomultiplier tubes, eleven dynodes are frequently used and an electron amplification of 10^8 or even 10^9 is obtained, where the restriction on large amplification is set, in practice, by thermally generated electrons (noise).

Typical constructions are of the 'venetian blind' type (fig. 8.22a), and the 'box-and-grid' type (fig. 8.22b). More compact arrangements of electrodes are also used*

The photocathode is often of the semi-transparent type (figs. 8.22a and b) consisting of a thin layer of the photoemissive material deposited on the inside of an optically flat end window to the tube. The photoemissive material is one of those used for vacuum photocells, chosen depending on the spectral

---

*See, for example, p. 418 of *Atomic and Nuclear Physics*, Yarwood (U.T.P., Cambridge, 1973).

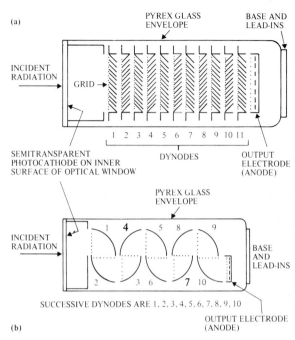

Fig. 8.22 Photomultiplier tubes (a) The 'venetian blind' construction (b) The 'box-and-grid' arrangement

response characteristics required. The coating on the dynode surfaces has to have a high secondary emission coefficient and composite materials similar to those for high photoemission are employed. For example, caesium antimonide is widely used.

A 10-stage photomultiplier tube suitable for photometry in the visible region is the Mullard XP1002. This has an S20 type photocathode for which the spectral response is shown in fig. 8.23b. The S20 is a semi-transparent photocathode formed from an antimony-potassium-sodium-caesium preparation designated by Sb − K − Na [Cs] (Table 8.1). The secondary emitter material with which the dynode surfaces are coated is an oxidised silver-magnesium alloy activated with caesium which has a maximum secondary emission coefficient of about 6. The useful diameter of the photocathode is 44mm. The overall luminous sensitivity (known as the 'anode sensitivity') when the total voltage across the tube is 1.8kV, is 400A/lumen, which shows the enhancement achieved by ten-fold secondary emission multiplication in that the average luminous sensitivity of the photocathode alone is 150μA/lumen.

The voltage divider circuit arrangement used with such a photo-multiplier tube (fig. 8.23a) is rated so

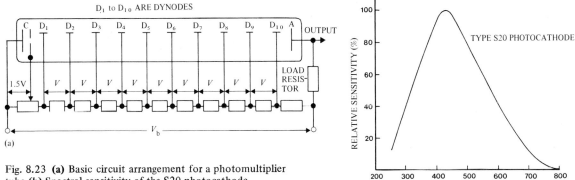

Fig. 8.23 (a) Basic circuit arrangement for a photomultiplier tube (b) Spectral sensitivity of the S20 photocathode

that the supply voltage V_b across the whole tube (i.e. on the anode A relative to the photocathode C) is 1.8kV, the pd V across neighbouring dynodes is 160V and the voltage on the electron accelerator aperture immediately beyond the photocathode is 240V.

In use photomultiplier tubes are commonly shrouded in a mu-metal shield to inhibit stray magnetic fields from affecting the electron paths, and the incident light intensities are restricted so that the output current at the final anode does not exceed about 100μA in continuous operation.

Photomultiplier tubes are also much used in scintillation counting in X-ray and gamma-ray spectrometry and other applications in nuclear physics. Infra-red sensitive types are available commercially.

Channel electron multipliers (channeltrons) make use of a continuous film on the inside wall of a small glass tube which is curved, although straight and even parallel-plate types have been made. This film has a secondary emission coefficient greater than unity for incident electrons of energies exceeding about 50eV. The film is usually of metallic lead having an overall resistance of about $10^9 \Omega$ or of a vanadium phosphate glass with a higher resistance of 10^{10} to $10^{11} \Omega$. The principle of operation is indicated by fig. 8.24, though it must be emphasised that this diagram is greatly simplified and not to scale. A pd V of about 3kV is applied across the resistive film coating on the inside wall where the input end (the 'cathode') is negative with respect to the output end.

A sufficiently energetic photon incident on the resistive film near the cathode end causes an electron to be emitted. This electron follows a path determined by its ejection speed and the distribution of the electric field lines within the region surrounded by the resistive film. This path will be towards the output end of the channel because of the continuous increase of positive potential across the resistive film from zero (at the cathode end) to $+V$ (at the output end). The electron will strike the film at such a point as A (fig. 8.24) where it will produce secondary electrons provided that the electron has gained sufficient energy (exceeding about 50eV) in the electric field. These secondary electrons will then be accelerated along paths towards the location B where further secondaries are emitted and then this increased number of electrons will be further multiplied on subsequent incidence in location C. In the diagram (fig. 8.24) many more collisions with the resistive film will occur during the paths from the

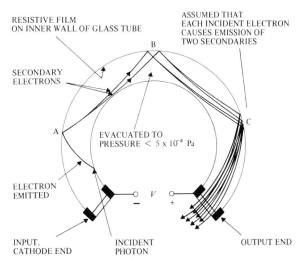

Fig. 8.24 Principle of operation of the channel electron multiplier (not to scale)

cathode to the output than are indicated. Indeed, depending on the magnitude of V, the output pulse may contain as many as 10^8 electrons for a single electron created by the initial incident photon.

Such channel electron multipliers will respond to incident electrons and ions of sufficient energies as well as to uv photons and X-ray photons. They have been used in space research for studying ionising radiations, in which case a window is not needed at the input or the output as the vacuum in space (provided it is at a pressure below 5×10^{-2} Pa) is satisfactory. For operation in the laboratory, however, input and output windows are needed and the channel multiplier is evacuated by conventional vacuum practice and then sealed-off.

The length/diameter of the channel is important. Because of the continuous resistive coating (which, in effect, provides a continuous dynode), several different electron paths are possible within the evacuated tube so the number of stages of electron multiplication is not definite. However, for a given value of the length/diameter (where the length is the mean length between cathode and output and the inside diameter of the wall is concerned), the electron multiplication remains the same and the gain is the same irrespective of the channel length. The values of length/diameter used have to exceed 30 and are usually between 50 and 100.

Although a straight tube could be used it is liable to give rise to problems of spurious pulses because positive ions produced by electron collisions with any residual gas present within the tube will be accelerated back towards the cathode end. If these positive ions gain sufficient energies and hit the tube wall they will generate secondary electrons which are confused (give rise to spurious pulses) with the desired secondaries generated by the electrons. The pd through which a positive ion needs to be accelerated to gain enough energy to generate secondary electrons on collision with the resistive coating is considerably in excess of the 50eV or so needed for an electron. If the channel is straight, such high ion energies may be produced in the direct paths possible from output to input; if the channel is curved, on collision with the resistive coating, the ions will not have traversed sufficient backward path lengths to acquire sufficient energy to release secondary electrons on collision with the inner wall.

A typical channel electron multiplier circuit for pulse counting (fig. 8.25a) produces a pulse of charge of duration less than 15ns which is fed into the capacitance C coupled to a pulse amplifier. The pulse amplitude fed to this amplifier is $1.6 \times 10^7 \, G/C$ volt where the output pulse contains G electrons and C is in pF. The typical variation of the gain G of the multiplier with applied voltage V (fig. 8.25b) is where

$$G = \frac{\text{charge in the output pulse}}{\text{the charge of the electron}}$$

$$= \frac{\text{charge in pulse in coulomb}}{1.6 \times 10^{-19}}$$

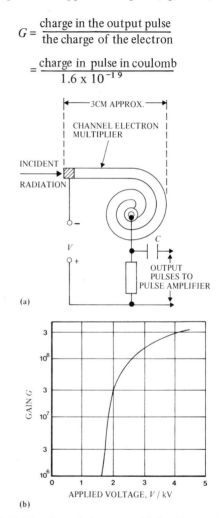

Fig. 8.25 The channel electron multiplier (a) a typical pulse counting circuit; (b) a typical characteristic of gain G against applied voltage V

8.9 Image converter and image intensifier tubes

Certain photoemissive materials are sensitive to infrared radiation. In particular, the silver-oxygen-caesium photocathode (Ag–O–Cs) prepared by activating silver oxide with caesium, which is photocathode type S1 (*see* table 8.1 and fig. 8.17) has sensitivity in the near infra-red extending up to wavelengths of 1000nm. If, therefore, photons of near infra-red radiation in the wavelength region from about 700 to 1000nm are incident on the back of a semitransparent S1 photo-cathode, electrons are emitted. These electrons can be accelerated and focused to form an image at a fluorescent screen beyond the photocathode. This fluorescent screen is of a material (e.g. willemite) which produces green light (in the region of the spectrum where the human eye has high sensitivity) on impact by electrons. Hence green visible light is produced by invisible infra-red incident radiation.

An object will emit infra-red radiation if it is at a temperature above that of the surroundings but may otherwise be invisible or only discerned with difficulty. An infra-red image of this object is formed by a suitable lens on the flat window of the image-converter tube where the opposite side of this window (in the vacuum tube) is coated with a semi-transparent layer of the silver-oxygen-caesium. At any given location on this photocathode surface, the number of electrons emitted will depend on the intensity of the infra-red radiation there. The gradations in the infra-red image are therefore 'represented' in the electrons emitted. Focusing of the emitted electrons and hence of this electron pattern is by an electrostatic field, an electrostatic electron-lens or a magnetic electron lens so that the pattern is preserved at the fluorescent screen to which the electrons are accelerated and where is produced a visible picture of the object.

A useful form of such image converter tube (fig. 8.26) employs a fluorescent screen parallel to a plane photocathode at a separation of about 0.5cm where this screen is at 3 to 6kV positive in potential with respect to the photocathode. The electrons emitted by the photo-cathode (illuminated by infra-red radiation) travel rapidly along the electric field lines to the fluorescent screen; the transit times of these electrons are so short that spread of the electrons in directions perpendicular to the field line paths is very small. Hence a good image is formed, although it is not sharply focused.

Image converter tubes able to produce a visible image from invisible infra-red radiation have been used to detect the presence of aircraft by virtue of the infra-red radiation from hot engine parts and exhaust; to 'see' in the dark an object (a human being or an animal, for example) which is 'illuminated' by radiation from an infra-red lamp; to examine aspects of furnace construction and in the study of inks and pigments in documents and paintings.

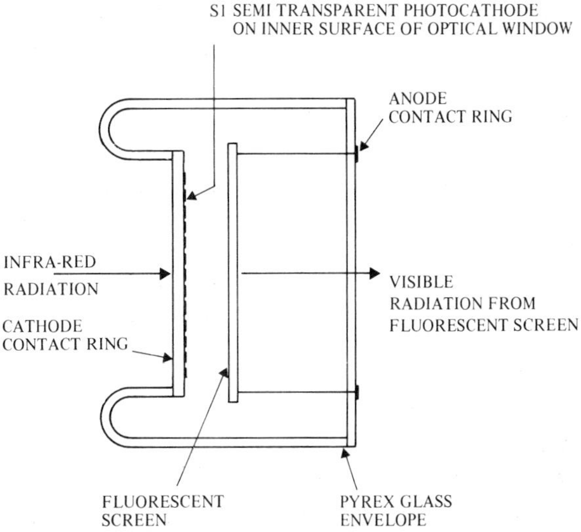

Fig. 8.26 An image converter tube with direct focusing utilising a fluorescent screen in close proximity to the photocathode and at high potential

An *image intensifier tube* has resemblances to the image converter tube but now the photocathode is sensitive to visible radiation and the electrons emitted are accelerated to a fluorescent screen (a phosphor material) at such a potential and made of such a phosphor that each electron incident produces a number of photons. Then the number of photons per second given out by any element of the image formed on the fluorescent screen can be considerably greater than the number of photons per second which form the corresponding element in the image focused at the photocathode. This enhancement of image intensity will also depend on the relative areas of the final and initial images: the final image area can be made larger or smaller than the initial image or of the same size. Multi-stage phosphor-photocathode image intensifier tubes have also been produced in which electrons emitted from the first photocathode are accelerated and focused to a phosphor on the near side of a very thin supporting membrane of which the other side is coated with photosensitive material. Multi-stage image intensifier tubes have been made in which the final image is 10^6 times as intense as the initial image.

The chief field of application of image intensifier tubes is to night vision systems.

8.10 Electroluminescence

The emission of light can be obtained from semiconductor materials on excitation by electric fields or currents in contradiction to the case where a material emits light on being bombarded by particles or as a consequence of a chemical reaction.

In such electroluminescence, charge carriers gain energy from the electric field and light is emitted when the charge carriers recombine.

There are several kinds of electroluminescence. Two important ones are (a) injection electroluminescence and (b) 'Destriau effect' luminescence.

(a) Injection electroluminescence occurs when minority charge carriers are injected into a semiconductor. The commonest case is that of a forward-biased *p-n* junction, as in the light-emitting diode (section 8.11). Electrons from the *n*-side of the junction (in which they are majority carriers) are transported across the junction under the action of the electric field into the *p*-side, in which they are minority carriers; holes are transported in the opposite direction. In the *p*-type material, for example, the electrons injected as minority carriers combine with holes. In doing so, light is emitted; such injection electroluminescence can be of high efficiency.

(b) In 1936 Destriau made a 'capacitor' from large plane electrodes separated by a small gap filled with an insulating oil in which was suspended specially prepared zinc sulphide powder. On applying an alternating voltage across these plane electrodes, light was emitted from the phosphor.

This 'Destriau effect' (still not properly understood) forms the basis of modern electroluminescent light panels. Zinc sulphide adulterated with small amounts of copper or manganese depending on whether the light to be emitted is to be predominantly green or yellow respectively (and to a lesser extent other phosphors such as cadmium sulphide and the selenides of zinc and of cadmium) is dispersed in a thin sheet of plastic (usually polystyrene or Perspex) or glass or ceramic maintained between two plane parallel electrodes, one of which is transparent. When a suitable alternating pd is maintained across these electrodes, the light emitted is transmitted through the transparent electrode.

8.11 Light emitting diodes (LEDs)

The light emitting diode is based on a forward-biased *p-n* junction. Electrons within the conduction band (to which they are readily transferred at room temperature from the donor level) of the *n*-type semiconductor side of the junction are transported by the advancing electric field brought about by the forward bias into the *p*-side. Within the *p*-side some of these electrons may combine with holes in the valence band. In such recombination, these electrons lose an amount of energy given by $(E_C - E_V) = E_G$ where in the host semiconductor material, E_C is the energy level at the bottom of the conduction band, E_V is the energy level at the top of the valence band and E_G is the band gap energy. This loss of energy by the electron appears in the form of radiated photons of frequency ν and wavelength $\lambda = c/\nu$, given by

$$\nu = \frac{c}{\lambda} = \frac{E_C - E_V}{h} = \frac{E_G}{h}$$

where h is the Planck constant.
Hence,

$$\lambda = \frac{hc}{E_G} = \frac{4.15 \times 10^{-15} \times 3 \times 10^{10}}{E_G} \text{ m}$$

$$= \frac{1.245}{E_G} \ \mu\text{m} \qquad (8.4)$$

where E_G is in eV.

The radiation emitted is from near the *p-n* junction because the majority of the electrons and holes are within a diffusion length of this junction.

Energy is also lost by the electrons because they cause lattice vibrations on colliding with molecules of the semiconductor. Such phonon loss gives rise to heating. Clearly materials are best for LED devices in which lattice vibration (phonon production) is small compared with photon production brought about by direct transfer from conduction to valence band.

The visible spectrum has a long wavelength limit at 720nm = 0.72μm. From equation (8.4) it is seen that E_G must therefore be greater than $(1.245/0.72) = 1.73$eV for visible light to be generated. Germanium

(E_G = 0.72eV at 300K) and silicon (E_G = 1.1eV at 300K) are hence unsuitable for making light-emitting diodes.

Gallium arsenide (Ga As) a III-V compound, has E_G = 1.41eV and emits efficiently infra-red radiation of wavelength exceeding (1.245/1.41) = 0.88µm = 880nm. Gallium phosphide (Ga P), also a III-V compound, has a band gap energy of 2.25eV and will therefore emit radiation in the visible region but unfortunately it is not very efficient in producing photons because considerable phonon production results from lattice vibration excitation. Solid solutions of gallium arsenide and gallium phosphide give a compound gallium arsenide phosphide, Ga $As_x P_{1-x}$ where variation of x in the compound enables the band gap energy E_G to be arranged at any value between 1.41eV for Ga As and 2.25eV for Ga P. Ga As P is consequently the most used III-V compound for making light-emitting diodes. As x is varied from 0.25 to 0.6, the wavelength limit decreases from 725nm to 600nm, i.e. from red light at the long wavelength end of the spectrum (to which the human eye is not very sensitive) to orange light (to which the human eye has good sensitivity). The human eye has a peak sensitivity at about 550nm but gallium arsenide phosphide has greater electron-photon conversion efficiency in the far red. The compromise to give the best visibility of the light emitted is to use a gallium arsenide phosphide compound in which x = 0.4 (i.e. the formula is Ga $As_{0.4} P_{0.6}$) which gives peak light output at 650nm.

Clearly it is desirable to shift nearer to the 550nm value in the green region of the visible spectrum at which the light output is at a peak. This can be done by including nitrogen into the gallium arsenide phosphide lattice. Indeed, red, yellow and green semiconductor lamps (light-emitting diodes) have been produced based on this material.

The light output increases with the current through the forward-biased diode. It is specified in lumen/ampere.

Light-emitting diodes of small sizes can operate with a forward bias of about 2V and with a current of about 10mA, and have a luminance of about 1500 candela per square metre, which is readily seen under normal viewing conditions. These electrical characteristics are convenient for operation by transistor logic circuits. Furthermore, small LEDs in

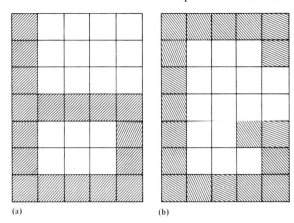

SHADED SQUARE DENOTES LED 'ON'

Fig. 8.27 A 7 x 5 matrix of LEDs in which selected diodes are arranged to be 'on' so that **(a)** the number 6 is displayed and **(b)** the letter G is displayed

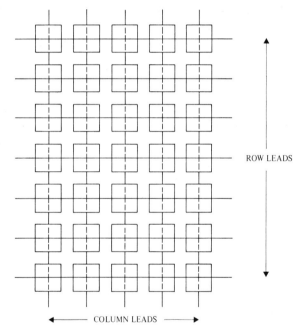

Fig. 8.28 Coordinate array of LEDs in a matrix showing connecting leads

an array such as that shown in fig. 8.27, enable alphanumeric displays to be set-up conveniently.

In a matrix of LEDs, a coordinate array is usually employed (fig. 8.28). The LEDs are in rows and columns in the matrix. All the anodes in a row are electrically connected as are also all the cathodes in a column. This array is operated row by row or column by column so as to display a character and avoid energising unwanted diodes.

Exercises 8

1. Write brief notes on the following:
   (a) the principle involved in photoconductivity in the simple case of a single-crystal semiconductor material in which the energy-bands are well defined;
   (b) charge multiplication, as experienced in the use of n-type cadmium sulphide adulterated with copper as a photoconductive material;
   (c) the quantum efficiency of a photoelectric material;
   (d) some applications of photoconductive cells.

2. In relation to the use of photoconductive cells for infra-red radiation, explain why it is usually desirable to operate the cell at a low temperature of about 77K or below.
   Define the terms 'noise equivalent power' and 'detectivity' as applied to photoelectric detectors.

3. It is generally stated that the operating temperature T of a semiconductor material should be below the value given by $(E_G/25k)$ K, where E_G is in eV and k is the Boltzmann constant = 8.6×10^{-5} eV K^{-1}, in order that a satisfactory signal-to-noise ratio be obtained in the use of this semiconductor material in photoconductivity detection practice.
   Give a justification for this statement.

4. On the basis of a calculation which makes use of the Bohr theory to estimate the ionisation energy of n-type germanium, show that this semiconductor material can form the basis of a device for detecting infra-red radiation of wavelengths above 100μm.
   What is the chief problem in the use of n-type germanium for the construction of a photoconductive cell for use in the infra-red?

5. Define in precise terms what is meant by the 'photovoltaic effect'.
   Describe with a suitable diagram the construction and method of operation of an element of a solar cell.

6. Write explanatory notes aided by suitable diagrams on the following:
   (a) the barrier layer photovoltaic cell;
   (b) the photodiode, including the p-i-n type;
   (c) the phototransistor;
   (d) the photofet.

7. Write an account of the photoemissive effect, including the relevant theory, and explain why the vacuum photoemissive cell is still widely used despite the fact that its sensitivity to radiation is significantly less than that of the corresponding photoconductive and photovoltaic devices.

8. Describe in some detail three types of photomultiplier device (a) of the 'venetian blind' type, (b) of the 'box-and-grid' type and (c) of the channel electron multiplier type.
   Explain the relevant advantages and disadvantages in various applications of these three types of photomultiplier.

9. Describe in detail with a suitable diagram an image converter tube which enables a visible image to be obtained of an object which emits infra-red radiation but is otherwise invisible.

10. Write an account of the use of image intensifier tubes in X-ray practice.

11. Explain the operation of a light-emitting diode. Why has the compound $GaAs_xP_{1-x}$ become important in the manufacture of light-emitting diodes?
    Describe briefly in outline how a matrix array of light-emitting diode elements is used to display the alphanumeric characters.

9 · Power Control Utilising Thyristors

9.1 Introduction

The term **thyristor** is used to identify one member of a family of semiconductor devices which have characteristics like those of a thyratron. A thyratron is a hot-cathode gas-filled tube in which a signal on a control electrode initiates the anode current but does not normally limit it. The commonest member of the family is the *reverse blocking triode thyristor,* better known as the *silicon controlled rectifier* (abbreviated to SCR).* Other members of the family (abbreviations are given in brackets after the full name of the device) are the *bidirectional trigger diode* (DIAC), the *bidirectional triode thyristor* (TRIAC) and the silicon controlled switch (SCS). A bidirectional triode thyristor is also available with a built-in trigger diode, the integrated package being called a QUADRAC.

To avoid the confusion which often arises when any one of the family is simply called a thyristor, we shall refer to the most common member as an SCR and similarly use to identify the other devices the convenient abbreviations DIAC, TRIAC, SCS and QUADRAC.

In many of the circuits in which these devices are used, it is common to produce a power gain of 10^6. Thus, an input control signal of a few milliwatts can switch power of the order of kilowatts. Although the small size and low power consumption of semiconductor devices is rightly frequently emphasised, thyristors can be used for such large power requirements as the speed control of heavy electric locomotives, the temperature control of electric furnaces and, indeed, larger SCR's are designed to carry currents of hundreds of amperes for control equipment in electric power stations.

Notice that we refer to power control and power gain because the sole function of the input signal to the thyristor is to perform the switching action. The term 'power amplifier' is reserved for the different case where an amplifier has an output which reproduces the essential characteristics of the input signal, as in the output stage of an audio-frequency amplifier used in, for example, a public address system.

9.2 The silicon controlled rectifier (SCR)

The SCR is a three-terminal, four-layer device of which the circuit symbol and outline structure are shown in fig. 9.1.

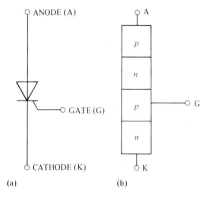

Fig. 9.1 **(a)** the circuit symbol and **(b)** the outline structure of an SCR

The circuit symbol (fig. 9.1a) indicates satisfactorily that the SCR is a diode with a third electrode used for control and called the *gate* (G). Essentially, the SCR is a power diode of which the conduction periods can be controlled. As for any ordinary silicon rectifier, it will not conduct when reverse biased. Unlike the ordinary rectifier, it will not conduct when forward biased until triggered into a conducting state by a pulse of current applied to the gate. Thus the very efficient principle arises of blocking the unwanted power in contrast to the inefficient method of dissipating power in the control

---

*Sometimes called a controlled silicon rectifier (CSR).

device which occurs if a series variable resistance or a series control transistor is used.

Alloy-diffusion and all-diffusion methods are used to produce this four-layer device (Appendix A) which is mounted between discs of molybdenum or of tungsten — chosen because these metals have thermal expansion coefficients which match those of silicon. A modest temperature increase will not cause sufficient mechanical stress to fracture the silicon slice. Fig. 9.2 shows the structure of an SCR in more detail and also two typical encapsulations for low-cost, low-power devices.

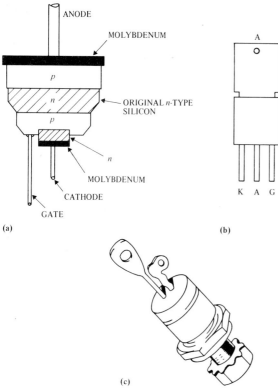

Fig. 9.2 (a) Typical structure of an SCR; (b) and (c) are typical encapsulations

9.3 Theory of operation and voltage-current (V-I) characteristics of an SCR

In fig. 9.3 the three junctions within the SCR are denoted by J_1, J_2 and J_3. When the device is reverse-biased (i.e. the anode A is negative w.r.t. the cathode K) junctions J_1 and J_3 are both reverse-biased, the depletion region at each of them is wide and only a small leakage current can flow. The reverse characteristic (fig. 9.4) is similar to that for an ordinary silicon rectifier.

When the SCR is forward-biased (i.e. A is positive w.r.t. K) junctions J_1 and J_3 are forward-biased whereas J_2 is reverse-biased. Because most of the applied voltage across AK is dropped across junction J_2, the depletion layer at this junction is wide. The SCR is in the forward blocking state. If a current pulse is applied to the gate which makes its potential positive with respect to the cathode K, electrons will flow across junction J_3 into the region P_2. The three regions N_2, P_2 and N_1 (fig. 9.3) may be regarded as comprising an *n-p-n* junction transistor where N_2 corresponds to the emitter, P_2 to the base and N_1 to the collector. By transistor action, some of the 'emitter' electrons which cross junction J_3 will cross junction J_2 into the collector. This flow of electrons across junction J_2 will reduce the depletion layer width there and the forward bias voltages at junctions J_1 and J_3 are increased slightly. The three layers P_2, N_1 and P_1 may be considered to constitute a *p-n-p* junction transistor. Holes from the emitter (P_1) flow through the base (N_1) into the collector (P_2). In so doing, the depletion width at junction J_2 is further decreased. The action is cumulative: once initiated by the positive-going current pulse at the gate, the action builds-up rapidly until the depletion layer at J_2 disappears. The effective anode-to-cathode resistance then drops very rapidly. This ON or forward-conducting state can be initiated in about 2-3μs.

Once the SCR is on, the gate signal can be removed without affecting the conducting state

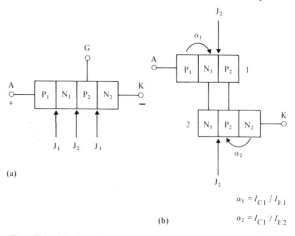

Fig. 9.3 (a) Junctions concerned in the operation of an SCR; (b) the two-transistor analogy in the theory of its operation

provided that the anode-to-cathode current is above a certain minimum level called the 'holding current' (I_H). When the firing pulse applied to the gate triggers the SCR into a forward-conducting state (i.e. it is ON), any signal applied to the gate no longer has any control over the conduction. To switch-off the SCR either the anode current has to be reduced below the holding current (I_H) or the device has to be subject for an instant to reverse bias.

Referring to fig. 9.3b, illustrating the two-transistor analogy of the *p-n-p-n* operation, the current I_{J2} crossing junction J_2 can be written

$$I_{J2} = I\alpha_1{}^* + I\alpha_2 + I_{CO}$$

where $I\alpha_1$ is the hole current from the P_1 region, α_1 being the fraction of the hole current injected at emitter 1 that reaches collector 1 in the *p-n-p* structure.

$I\alpha_2$ is the electron current from the N_2 region, α_2 being the fraction of the electron current injected at emitter 2 which reaches collector 2 in the *n-p-n* structure and I_{CO} is the leakage current across the junction J_2, when the SCR is in a forward blocking state.

Hence the current I through the SCR, which must equal I_{J2}, is given by

$$I = I_{CO}/(1 - \alpha_1 - \alpha_2)$$

If $(\alpha_1 + \alpha_2) = 0.9$, $I = 10 I_{CO}$ and the device is in a forward blocking state because I_{CO} is very small for silicon. It is apparent that low values of α for the *n-p-n* and *p-n-p* parts are an essential requirement of the forward blocking state. High values of α are obtained by utilising thin base regions, low values demand wide base regions. The SCR is thus a relatively high voltage device.

If $(\alpha_1 + \alpha_2)$ tends towards 1, the current through the SCR is limited only by the external circuit impedance. The current pulse on the gate creates this condition because α increases as the emitter current of a junction transistor increases. This ON state will exist when the N_1 and P_2 regions are saturated with current carriers and the reverse bias initially across J_2 is caused to disappear. The voltage drop across the SCR in this ON state is about that across one forward-biased *p-n* junction.

The heavily-saturated regions N_1 and P_1 affect greatly the turn-off mechanism. To turn off the SCR in a minimum time it is necessary to apply a reverse voltage. When the reverse voltage is applied, holes and electrons in the regions of junctions J_1 and J_3 will diffuse to these junctions and a reverse current pulse will flow in the external circuit. With the carriers removed in the regions of junctions J_1 and J_3, the junctions will assume a blocking state. However, recovery is not complete because a high carrier concentration exists in the region of junction J_2. This concentration decreases mainly by recombination which is largely independent of the external circuit. The turn-off time of an SCR is about $12\mu s$.

In usual operation with an ac input, the SCR is operated with a peak ac voltage below the breakover voltage (V_{BO}) (*see* fig. 9.4), and triggering is

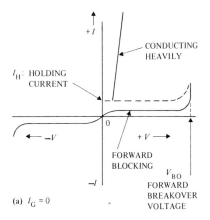

(a) $I_G = 0$

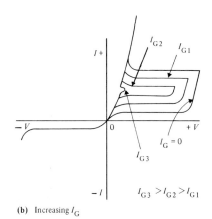

(b) Increasing I_G

*α is used conventionally as the large signal current gain of the junction transistor in the CB configuration; it is not to be confused with α used as the phase-angle, in section 9.4.

Fig. 9.4 The voltage-current (*V-I*) characteristic of an SCR

196 Electronics

accomplished at the desired time in the forward-biasing cycle by supplying a triggering current pulse to the gate electrode.

9.4 AC phase control with SCRs

Phase control is the process of rapid ON-OFF switching whereby an ac supply is connected across a load for a controlled fraction of each half-cycle. This is a very efficient method of controlling the average power supplied from an ac source to a load such as a lamp, heater or motor. The SCR is an ideal device for providing such control.

Fig. 9.5a shows a half-wave circuit with control by one SCR, whilst fig. 9.5b illustrates the meaning of the terms 'phase angle' and 'conduction angle'. It is assumed here that the control device shown provides current pulses which trigger the SCR into a conducting state at a phase angle α of 120°; the SCR will conduct for the remainder of that half-cycle, i.e. between 120° and 180°. The control device can be varied to select the value of the phase angle α and hence select the fraction of the half-cycle during which current passes through the load. Thus, the power supplied to the load can be controlled between a maximum (α near zero) and a minimum value (α near 180°). The voltage waveform across a resistive

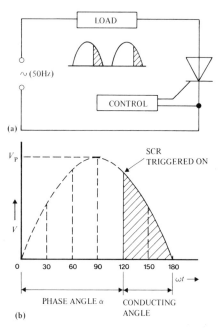

Fig. 9.5 ac phase control by means of an SCR: **(a)** controlled half-wave circuit; **(b)** illustrating the terms 'phase angle' and 'conducting angle'

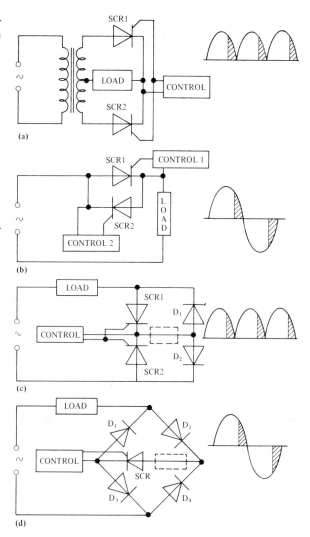

Fig. 9.6 Full-wave rectified ac phase-control circuits utilising SCRs: **(a)** two SCRs with a centre-tapped transformer; **(b)** two SCRs connected in inverse parallel; **(c)** two rectifiers and two SCRs; **(d)** one SCR and four rectifiers, the current through a load in the dotted position would be unidirectional

load (shaded portions shown in figs. 9.5 and 9.6) can readily be observed with a suitably connected cathode-ray oscillograph.

Fig. 9.6 shows four different arrangements for full-wave ac phase control.

In the circuits (a) and (c) of fig. 9.6, one control unit serves to provide firing pulses for two SCRs. This gives a simplified control or trigger circuit. Whichever SCR is forward-biased when the gate pulse arrives will be triggered into conduction and will conduct for the remainder of that half-cycle.

Only one SCR is required for the control in the circuit of fig. 9.6d. Some loss of efficiency is then caused by the sum of the voltage drops across the rectifiers and the SCR. Turn-off of the SCR is sometimes a problem because the voltage across the SCR never drops below zero (section 9.16). The SCR will turn-off if the time for which the current is less than the holding current exceeds the turn-off time.

An SCR can be switched to a forward conducting state by the application of a steady source of gate current; this would require continuous power supply from the dc control device and continuous power dissipation at the gate, which is undesirable. In circuits for ac control by means of SCRs, the almost universal practice is to trigger the SCR with a train of current pulses, then no power is dissipated in the gate circuit in the intervals between the pulses. Moreover, the first pulse is intended to switch the SCR but if this fails, succeeding pulses in the train will succeed. The power dissipated in the gate circuit by this pulse train is minimal.

9.5 Introductory experiments with SCRs

The circuits of figs. 9.7 and 9.8 enable certain basic features of an SCR to be examined. An SCR capable of carrying 1.5A and withstanding a peak reverse voltage of 100 is ideal for this purpose. The anode plate of the SCR is bolted to a small heat sink. Although the power dissipated in the SCR during the following experiments does not demand a heat sink, it is a worthwhile precaution to attach one with experimental equipment and it is good design practice to assess the heat sink requirements when the control devices are being selected.

Using the dc circuit of fig. 9.7 one can readily examine:—

(i) *The forward blocking state:* with resistor R_3 set at its maximum value and R_2 set at approximately 1kΩ, the forward voltage across the SCR can be increased to some chosen value — say 20V initially. Negligible current will flow through the SCR which will be in the forward blocking state. If resistor R_3 is now gradually reduced in value, the gate current will increase until it is of sufficient magnitude to switch the SCR into a conducting state. This minimum gate current requirement is related to the forward voltage and can be determined for a number of values of the forward voltage. Hence the essential features of the characteristics shown in fig. 9.4b can be obtained.

(ii) *Effect of the gate current when the SCR is conducting:* once the SCR is conducting and the current passing through it is above some minimum value, the gate no longer has any control function. The gate current can be reduced or the gate circuit disconnected without affecting the current through the SCR.

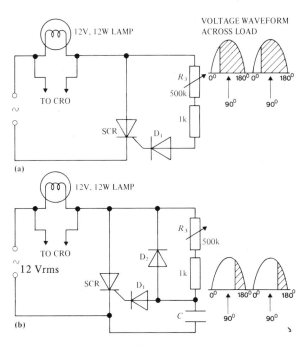

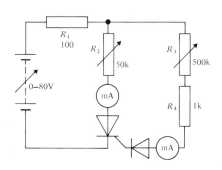

Fig. 9.7 Circuit for investigating the holding current and the role of the gate in an SCR

Fig. 9.8 (a) Circuit to enable half-wave phase control to be examined: phase-angle between 0° and 90° (b) inclusion of a capacitor and a diode so that the phase-angle can be selected at a value exceeding 90°

(iii) *The holding current* (I_H): having switched the SCR into a conducting state, the gate circuit is disconnected. The resistance R_2 is gradually increased in value so that the current through the SCR decreases. As this current falls, a critical value will be reached below which the SCR cannot remain ON. This critical value is the holding current (I_H).

The circuit of fig. 9.8a enables simple phase control to be examined, although the phase angle can only be varied between 0 and 90°. A CRO connected across the load (e.g. a 12V, 12W lamp) enables the fast switching action of the SCR to be observed. The inclusion of a capacitor C as shown in fig. 9.8b enables the phase angle to be controlled beyond 90°. The diode D_1 in the circuit serves to prevent current flowing in the gate circuit during the half-cycle for which the SCR is reverse-biased; diode D_2 serves to discharge the capacitor C at the end of each half cycle.

9.6 Problems associated with ac phase control

When power is delivered to a resistive load by means of ac phase control, the connection of the ac supply for a chosen fraction of each half-cycle creates a number of problems:—

(a) Considerable radio-frequency interference (r.f.i.) is generated which can disturb neighbouring electronic equipment. This is because the current waveform (as in fig. 9.9) contains several harmonic components of high frequency.
(b) There is a significant effect of the phase control on the ac generator. If the load is resistive the current through it and the voltage across it are in phase. While the generator is providing a sinusoidal output voltage, current may be demanded from it only during the last part of each half-cycle. Hence the current and the voltage provided by the generator are out-of-phase: the fundamental component of the current lags on the supply voltage. It is as if the generator were connected to a very inductive load. The Electricity Authority have always specified a minimum value allowable for the power factor of any load connected to the ac mains supply because it is demanded that the energy loss in the distribution lines is kept to a

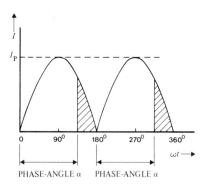

Fig. 9.9 Current wave-form for average current calculations

minimum (*see* example 9.7)
It has been proposed that phase control should only be used with domestic equipment having a power rating of less than 2kW. Thus, while SCR phase control modules are readily available to dim room lighting, the power demanded by electric cookers is sufficiently high to require zero voltage switching (section 9.7).
(c) Because the voltage and current waveforms in a phase-control circuit are non-sinusoidal, measurements cannot be made with several types of ac instrument as they are calibrated for inputs of sinusoidal waveform, e.g. those instruments which utilise rectifiers and a moving-coil meter.

9.7 Current and power measurements in circuits utilising phase control

(a) The vacuum thermocouple meter can be used to measure the root-mean-square value of an alternating current whatever the waveform and it is therefore useful with phase-control circuits. The current to be measured is passed through a short fine platinum wire mounted in an evacuated glass envelope (fig. 9.10). To the mid-point of this wire, but electrically insulated from it, is attached a fine thermocouple which enables the temperature of the platinum to be measured. The output from the thermocouple, normally measured with a sensitive millivoltmeter, is a function of I_{rms}, the root-mean-square current through the platinum wire. The instrument is calibrated by means of known direct currents. A number of plug-in thermocouple units can be used which cover between them a wide current range; the low thermal inertia of each unit ensures rapid response.

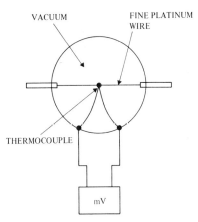

Fig. 9.10 The vacuum thermocouple ammeter

(b) To measure the average current with controlled rectification, it has to be considered that a moving-coil ammeter connected in series with the load in the circuit of, for example, fig. 9.6a would be subjected to unidirectional current pulses of the form shown in fig. 9.9. The current flows through the load between the times $\alpha T/2\pi$ and $T/2$ and again between the times $\alpha T/\pi$ and T during a complete cycle of period T. The total charge Q passing during one cycle is given by

$$Q = 2 \int_{\alpha T/2\pi}^{T/2} i \, dt = 2 \int_{\alpha T/2\pi}^{T/2} I_P \sin \omega t \, dt = \frac{2 I_P}{\omega} \int_{\alpha}^{\pi} \sin \omega t \, d(\omega t)$$

$$= -\frac{2 I_P}{\omega} \left[\cos \omega t \right]_{\alpha}^{\pi} = -\frac{2 I_P}{\omega} \left[-1 - \cos \alpha \right]$$

$$= \frac{I_P T}{\pi} \left[1 + \cos \alpha \right] \qquad (9.1)$$

where $\omega = 2\pi f = 2\pi/T$, f being the frequency of the sinusoidal alternating supply of peak current I_P.

Hence, the average current I_{dc} as recorded by the moving-coil ammeter is given by

$$I_{dc} = Q/T = \frac{I_P}{\pi} \left[1 + \cos \alpha \right]$$

For the special case of $\alpha = 0$, $I_{dc} = 2I_P/\pi$, as would be expected for a full-wave rectified supply.

(c) To evaluate the power available in terms of the phase angle, for a resistive load, R_L, in a phase-control circuit, the energy E' dissipated in the load during the time interval $\alpha T/2\pi$ to $T/2$, can be written

$$E' = \int_{\alpha T/2\pi}^{T/2} R_L \, i^2 \, dt = \frac{R_L I_P^2}{\omega} \int_0^{\pi} \sin^2 \omega t \, d(\omega t)$$

In the full-wave case (rectified or not) the energy dissipated in R_L is twice this value, i.e. is given by

$$E = \frac{2 R_L I_P^2}{\omega} \int_{\alpha}^{\pi} \frac{1 - \cos 2\omega t}{2} \, d(\omega t)$$

$$= \frac{R_L I_P^2}{\omega} \left[\omega t - \frac{\sin 2\omega t}{2} \right]_{\alpha}^{\pi}$$

$$= \frac{R_L I_P^2}{\omega} \left[\pi - \alpha + \frac{\sin 2\alpha}{2} \right]$$

The power is therefore

$$\frac{E}{T} = \frac{R_L I_P^2}{2\pi} \left[\pi - \alpha + \frac{\sin 2\alpha}{2} \right] \qquad (9.2)$$

For the special case of $\alpha = 0$,

$$\text{Power} = \frac{R_L I_P^2}{2} = R_L I_{rms}^2 \text{ as would be expected.}$$

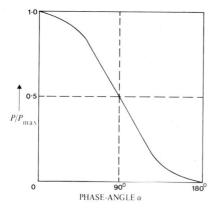

Fig. 9.11 Fraction of total power available plotted against the phase-angle α

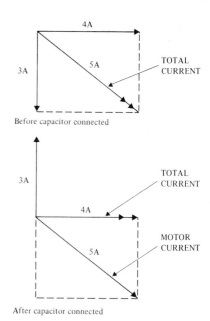

Fig. 9.12 Phasor diagrams for example 9.7

Equation (9.2) is used to plot the graph of fig. 9.11 showing how the power available varies with the phase angle α. The curve is approximately linear over a relatively large range of values of α.

EXAMPLE 9.7. *A 250V ac motor operating on a 500Hz supply takes 1.0kW at a power factor of 0.8, the current lagging on the applied voltage. Calculate (a) the total current taken from the supply and (b) the capacitance required to bring the power factor to unity. Show that as the power factor increases towards unity, less power is lost in the distribution lines.*

(a) $1000 = EI \cos\phi = 250 \, I \times 0.8$
where I is the total current.
Therefore, $I = 5.0\text{A}$.

(b) To bring the power factor to unity, a capacitor must be connected in parallel with the motor.
The in-phase component of the total current of 5.0A is
$I \cos\phi = 5.0 \times 0.8 = 4.0\text{A}$.

From the phasor diagram (fig. 9.12) it is seen that for a power factor of unity, the capacitor must take a current 3.0A, 90° ahead of the applied voltage.
The reactance of this capacitor = $\dfrac{250}{3}\,\Omega$
Therefore, $1/\omega C = \dfrac{1}{2\pi \times 500\,C} = \dfrac{250}{3}$,
where C is the capacitance.
Therefore, $C = (12/\pi)\,\mu\text{F} = 3.8\,\mu\text{F}$.

The total current to the circuit has been decreased from 5.0A to 4.0A as the power factor was increased from 0.8 to 1.0. The power loss in the distribution lines must therefore decrease as the power factor of the load increases.

9.8 Zero voltage switching of SCRs

Radio-frequency interference is avoided and no distribution or generator problems arise if the SCRs are switched at the beginning of each half-cycle during which they are to conduct. The firing can be arranged so that the SCRs conduct for a whole number of cycles and remain off for a whole number of cycles. Fig. 9.13 illustrates such an arrangement in which 60%, 50% and 20% of the available power is utilised. Whole cycle control of this type is unsuitable for loads of low thermal inertia such as filament lamps when flicker of the light output is created by the ON and OFF duty cycles. A zero voltage control system resembles that of a thermal vacuum relay. However, while the instant at which the relay contacts are opened or closed is not related to the input waveform, in the case of zero voltage switching the conduction is initiated at the beginning of a half-cycle and terminated at the end of a half-cycle.

One circuit which may be used for zero voltage switching is shown in fig. 9.14. The control is entirely

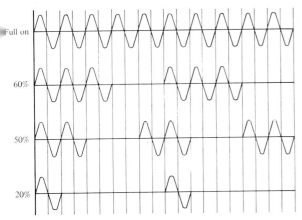

Fig. 9.13 Proportional control for zero voltage switching

Zero voltage switches are available commercially in integrated circuit form. These circuits are often supplied from a low voltage transformer. An oscillator provides a train of current pulses of which the first pulse occurs at the beginning of each half-cycle of the mains supply. Whether these pulses are communicated to the gate electrodes of the SCRs and power delivered to the load depends on the state of a gate controlled by a dc voltage level, provided perhaps from a sensing element.

Pulse triggering offers the advantage over dc gate firing that a wide spread in gate sensitivity can be tolerated. The gate of the SCR is over-driven to guarantee the conducting state. In addition, the power requirement in pulse triggering circuits can be quite low because the triggering energy can be stored slowly and discharged quickly.

dependent on the state of SCR1. If SCR1 is conducting no power is delivered to the load whereas if SCR1 is non-conducting the load receives maximum power.

Consider the behaviour of this circuit (fig. 9.14) in more detail. When line A goes positive with respect to B, SCR2 is forward-biased. If SCR1 is now in a forward blocking state the component values ensure that SCR2 is fired very early in the cycle. Whilst current flows through the load during this half-cycle, capacitor C_2 is charged through resistor R_3 and diode D_2. During the next half-cycle when SCR3 is forward-biased the discharging of C_2 through R_4 turns on SCR3 to supply the second half of the load current. SCR3 is said to be fired by a slave circuit which is primed during the half-cycle that SCR2 is conducting.

9.9 Firing circuits for ac phase control by SCR

A simple resistor-capacitor (RC) network with R variable can provide a convenient arrangement for manually controlling the phase-angle in a simple ac phase-control system (fig. 9.8b).

Firing circuits often provide a train of current pulses to the gate electrode. The first pulse of this train is intended to fire the SCR. A blocking oscillator or a unijunction transistor (UJT) relaxation oscillator is arranged to provide a pulse train during the interval for which the SCR is forward-biased. If the first pulse is made to occur immediately the SCR is forward-biased, the phase-angle α will have a value near zero; if the first pulse is delayed for 5ms, with a 50Hz ac supply, α will have a value of 90°. This phase-angle control is often conveniently determined by the level of a dc voltage signal at the input of the firing module. The silicon unijunction transistor and also the programmable unijunction transistor (PUT) are ideal devices for use in SCR firing circuits. The characteristics and some applications of the UJT and the PUT are considered in sections 9.10 to 9.13 whereas in section 9.14 there is described a typical commercial firing unit based on a blocking oscillator.

9.10 The unijunction transistor (UJT)

This type of transistor (fig. 9.15), sometimes called a *double-ended diode*, consists of a bar of *n*-type silicon with ohmic contacts B_1 and B_2 at the ends. A third contact created by alloying between an aluminium

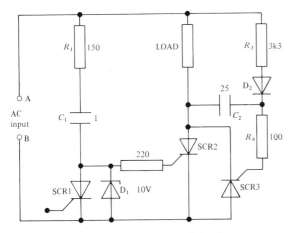

Fig. 9.14 Zero voltage switching control circuit

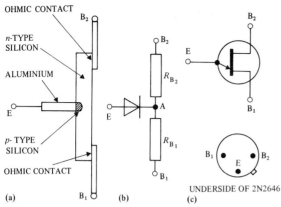

Fig. 9.15 The unijunction transistor (a) structure, (b) equivalent circuit (c) circuit symbol and base connections

wire (trivalent) and one side of the silicon rod is a p-n junction; this aluminium wire forms the emitter electrode E.

The resistance between the base contacts B_1 and B_2 (the interbase resistance R_{BB}) is usually between 5kΩ and 10kΩ. In normal practice, a voltage V_{BB} of less than 30V is maintained across these base contacts with B_2 positive with respect to B_1. A small current therefore flows through R_{BB} and the point A (fig. 9.16b) acquires a potential with respect to B_1 of $[R_{B1}/(R_{B1} + R_{B2})]V_{BB}$, where the constant $R_{B1}/(R_{B1} + R_{B2})$ is called the *intrinsic stand-off ratio*, denoted by η.

If the emitter voltage V_E is gradually increased the emitter junction will at first be reverse-biased and only a very small leakage current I_{EO} will flow. When V_E reaches some critical value V_p – termed the emitter peak point voltage – and if the emitter current exceeds the peak point current I_p, the UJT will turn on. In this ON state, the resistance between the emitter and base 1 is very low and the emitter current is limited chiefly by the external resistance connected in series with the base one. The peak point voltage is related to the interbase voltage by the equation

$$V_P = \eta V_{BB} + V_D$$

where V_D is the equivalent junction voltage, which is about 0.5V at 25°C. The value of η lies between 0.51 and 0.82.

It is found that the peak point voltage V_p decreases by about 3mV/°C, this negative coefficient being due primarily to the variation of V_D with temperature. It is possible to compensate for this effect by making use of the positive temperature coefficient of R_{BB}. If the resistance R_2 in series with base B_2 (see, for example, fig. 9.19) has a value given by

$$R_2 = (10^4/\eta V_1)\Omega$$

good compensation is achieved within the temperature range −40 to +100°C.

The emitter input characteristics of a General Electric 2N 2646 UJT are shown in fig. 9.16.

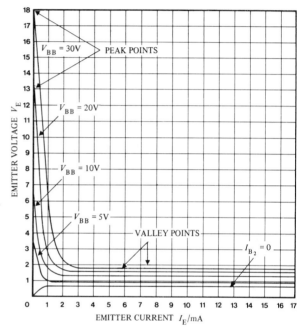

Fig 9.16. The emitter input characteristics of the 2N 2646 unijunction transistor

When $V_E = V_p$, the junction is forward-biased and holes are injected into the n-type silicon. Because B_2 is positive with respect to B_1 these holes are swept into the base B_1 region and the resistance R_{B1} falls rapidly. This resistance change causes the voltage at A to fall and so increases the forward bias at the junction. The characteristics show this interesting region in which the emitter current I_E increases whilst the emitter voltage falls. The effective resistance of the device over this region is negative; this feature makes the UJT valuable as a basis for a relaxation oscillator.

Values of η and V_D may be determined experimentally by use of the circuit of fig. 9.17. For a particular value of the interbase voltage (V_{BB}), the emitter voltage is gradually increased until the peak point value (V_p) is reached. At this value the voltage recorded by the voltmeter V_E will fall as the UJT turns on. The $2\text{k}\Omega$ resistor in the emitter lead serves to limit the emitter current and therefore protect the emitter junction. Typical results obtained from such an experiment are shown in fig. 9.18. By use of the equation $V_p = \eta V_{BB} + V_D$, the curves yield values of $\eta = 0.64$ and $V_D = 0.5\text{V}$.

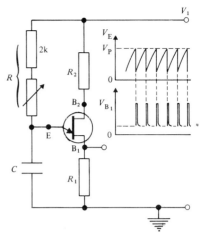

Fig. 9.19 Basic UJT relaxation oscillator with typical waveforms.

Typical component values: $V_1 = 25\text{V}$, $R = 2\text{k}\Omega + 50\text{k}\Omega$ (variable); $C = 0.1\mu\text{F}$; $R_2 = 470\Omega$; $R_1 = 56\Omega$

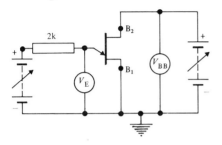

Fig 9.17. Circuit used to determine values of η and V_D

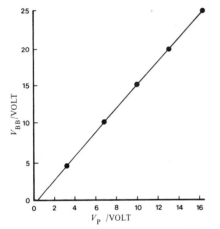

Fig. 9.18 Curve relating experimental values of V_p and V_{BB} (From $V_p = \eta V_{BB} + V_D$, $V_D = 0.5\text{V}$ and $\eta = 15/24 = 0.64$)

9.11 The UJT relaxation oscillator as a firing circuit

The basic UJT relaxation oscillator (fig. 9.19) is frequently used in phase-control systems.

The capacitor C is charged through the resistor R until the emitter voltage reaches V_p, when the UJT turns on and discharges C through R_1. When the emitter voltage drops to about 2V, the emitter ceases to conduct, the UJT turns off, and the cycle is repeated. The time period T of the oscillations is given by

$$T = RC \log_e [1/(1-\eta)] = 2.3 RC \log_{10}[1/(1-\eta)].$$

Note that for $\eta = 0.63$ (a realistic value) $T = RC$.

The frequency of this relaxation oscillator is virtually independent of small changes in supply voltage or temperature.

The resistance R may be varied between about $5\text{k}\Omega$ and $3\text{M}\Omega$. Below $5\text{k}\Omega$, the UJT will not turn off and current flows continuously through the forward-biased emitter junction. The upper limit of about $3\text{M}\Omega$ is set by the requirement that the emitter current must exceed I_p for the UJT to be able to turn on. The output pulse appearing across R_1 has a fast leading edge (approximately 100 ns) and may be coupled directly to the gate of an SCR.

If the output pulses from the oscillator of fig. 9.19 are to be used to fire SCRs in a phase control system, some arrangement is required to synchronise the oscillator with the alternating supply across the SCRs. One method of achieving this synchronisation is to discharge C at the end of each half-cycle. The values of R and C will then determine the time interval which lapses until the first firing pulse occurs. In this type of synchronisation (fig. 9.20) the UJT relaxation oscillator is supplied from a full-wave rectified supply with a Zener diode limiting the interbase voltage V_{BB} to 22V. Because this interbase voltage drops to zero

204 Electronics

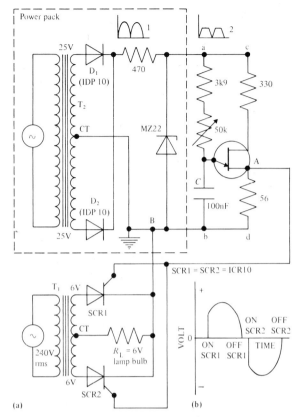

Fig. 9.20 (a) a lamp-dimming circuit, (b) the voltage waveform across the lamp R_L

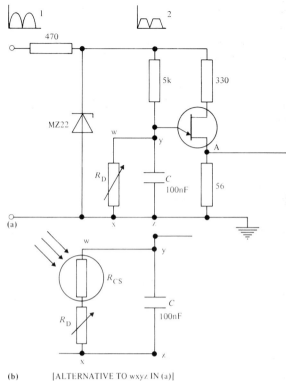

Fig. 9.21 (a) An alternative method of controlling the phase angle; (b) A cadmium sulphide photoconductive cell to provide negative feedback in a circuit which maintains constant illumination from the lamp

at the end of every half-cycle, any voltage on C will forward-bias the emitter junction and the capacitor will be discharged. The values of R and C can be selected to determine the time interval for the emitter to reach peak point voltage and produce the first firing pulse.

The lamp dimming circuit of fig. 9.20 provides a good demonstration of phase control. The output pulses from B_1 are coupled directly to the gate of each SCR. Whichever SCR is forward-biased when the first firing pulse occurs will be triggered into a conducting state and will conduct for the remainder of that half-cycle. With the whole circuit isolated by transformers from the mains supply, it is particularly convenient to use a CRO to observe the voltage waveforms.

9.12 Negative feedback to provide for automatic control of the phase angle

In the lamp dimming circuit of fig. 9.20 it is possible to control the phase angle by connecting a variable resistance R_D across the capacitor C. In fig. 9.21a, a small part of the circuit shown in fig. 9.20 has been isolated to show this alternative control method.

If the resistance R were low (about 5 kΩ) and R_D were infinitely large, the first firing pulse would occur early in each half-cycle and maximum power would be delivered to the load. As R_D is decreased, some of the charging current to the capacitor C is by-passed through R_D so the time taken for V_E to reach the peak point voltage is increased. Hence, by draining away some of the charging current to C the phase angle can be increased. In the limit, as R_D is further decreased, V_E will never reach the peak point voltage during any half-cycle so no power will be delivered to the load.

A cadmium sulphide (CdS) photoconductive cell connected in series with R_D (fig. 9.21b) and illuminated by light from the filament lamp R_L can provide a simple automatic control system. At constant temperature, the resistance R_{CS} of the CdS cell is a function of the light flux incident on its active surface. In this application, only a small portion of the active surface is exposed to illumination from the lamp. With no light falling on this photoconductive cell, its resistance (R_{CS}) is very high so its shunting effect on the capacitor C is negligible and the lamp receives maximum power. When light falls on the photoconductive cell, its resistance (R_{CS}) falls, the phase angle is increased and the system will stabilise at a light level determined by the value of R_D and the characteristics of the photocell. Any deviation from the selected behaviour immediately creates a signal which serves to correct the deviation and return the system to the chosen stable state. Such a system is an attractive demonstration of negative feedback and provides a straightforward method of providing a constant illumination. One shortcoming of this system is obvious if a careful examination is made of the light from the filament lamp. Power is supplied to the filament lamp in pulse form at a frequency of 100Hz so that a 100Hz flicker exists superimposed on the constant illumination. This fluctuation results from filament temperature variations and to some extent is determined by the thermal inertia of the filament.

A thermistor bead with a large negative temperature coefficient of resistance can be used in series with R_D instead of the CdS photocell. If this bead is maintained in contact with the glass bulb of the filament lamp, a negative feedback system is again obtained. In this case, it is unlikely that a steady state will be established because the temperature of the glass bulb and so of the thermistor will change relatively slowly. A time delay is introduced whereby the feedback signal will over-compensate resulting in 'overshoot' as the lamp is switched on and off.

The resistance R_D of fig. 9.21a can be replaced by an n-channel field-effect transistor operating as a voltage controlled resistance. Control is then available merely by the application of a dc voltage between the gate and source electrodes. An n-channel fet connected across the capacitor C of the UJT relaxation oscillator (fig. 9.19) and operated as a voltage-controlled resistance enables the circuit to function as a voltage-to-frequency converter. The behaviour, stability and linearity of such a system provides an interesting experimental exercise.

9.13 The programmable unijunction transistor (PUT)

The programmable unijunction transistor is a planar passivated four-layer (p-n-p-n) device with three terminals called the anode, the anode gate and the cathode (fig. 9.22). Whereas its structure is quite different from that of a UJT its electrical behaviour is similar and because it often replaces a UJT, the same terminology is used to describe its characteristics.

The four-layer structure is reminiscent of an SCR, the difference being that an anode gate constitutes the third electrode instead of a cathode gate. The device is used in SCR trigger circuits, oscillators and particularly in long duration timer circuits. Among the advantages afforded by the PUT are:

(a) low leakage current (perhaps 10nA at 25°C);
(b) low leak point current, I_p (perhaps 1µA compared with 5µA for a corresponding UJT);
(c) the facility whereby external resistors can be arranged to determine the values of the intrinsic stand-off ratio η, the interbase resistance R_{BB} and I_p.

The two transistor equivalent circuit (fig. 9.22c) shows that, when forward-biased (anode positive with respect to cathode) the device will switch rapidly from a blocking state to its ON state if the gate is made negative with respect to the anode. When operated as a UJT, an external reference voltage is maintained at the gate terminal by means of a potential divider R_1 and R_2 (fig. 9.24) and the anode voltage is increased until it reaches the peak point voltage V_p. Immediately the anode-gate junction is forward-biased and conducts, the regeneration inherent in a p-n-p-n device causes the PUT to switch ON. The anode characteristic (fig. 9.23) exhibits a negative resistance region as does the emitter characteristic of the UJT. The peak point voltage V_p is essentially the same as the external gate reference

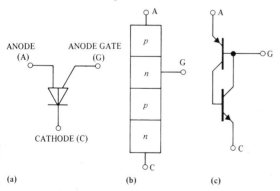

Fig. 9.22 The programmable unijunction transistor: (a) the circuit symbol; (b) the structure; (c) the two transistor equivalent

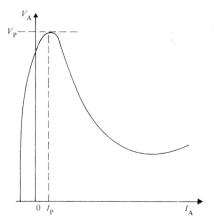

Fig. 9.23 The anode characteristic of the PUT

voltage, the only difference being the junction voltage drop. Because the reference voltage is not device dependent but may be varied by means of the two external resistors (R_1 and R_2 of fig. 9.24) the peak point voltage is programmable.

The ease with which the peak point voltage V_P can be selected is well demonstrated by means of the circuit of fig. 9.24.

In the circuit of fig. 9.24, V_{CC} could initially be set at 20V and the anode voltage gradually increased until the peak point voltage is reached when the PUT switches ON. No current flows through the PUT until this critical value is reached. The experiment can be repeated using different values for R_1 and R_2 whilst maintaining $(R_1 + R_2) \geqslant 10\text{k}\Omega$. Also V_{CC} can be varied and the values recorded of V_{CC}, R_1, R_2, V_P, and $\eta = V_P/V_{CC}$ as measured, compared with $\eta = R_2/(R_1 + R_2)$ as calculated.

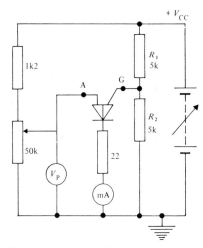

Fig. 9.24 Experiment to select the η value

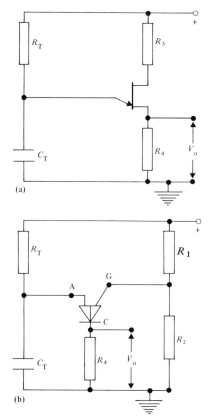

Fig. 9.25 Typical relaxation oscillator circuits: (a) utilising a UJT; (b) a PUT replaces the UJT

The use of a PUT in place of a UJT in a typical relaxation oscillator is shown in fig. 9.25.

The frequency f of the oscillation is given by

$$f = \frac{1}{R_T C_T \log_e [1/(1-\eta)]}$$

where R_T and C_T are the variables which determine the time period whereas the intrinsic stand-off ratio, η, is programmed by choice of the values of the resistors R_1 and R_2. The circuit of fig. 9.26 is ideal for examining the behaviour of a PUT relaxation oscillator. Of particular interest is the very fast rise time of the output pulses, approximately 40ns, compared with perhaps 100ns for a UJT. Even fast oscilloscopes (e.g. with a band-pass of 100MHz) degrade the direct measurement of this rise time.

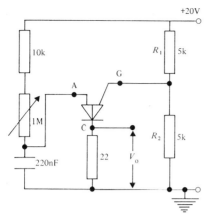

Fig. 9.26 An experimental PUT relaxation oscillator

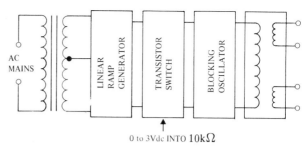

Fig. 9.28 Block diagram of MC16 SRC firing unit

When a conventional UJT relaxation oscillator is used to trigger an SCR the resistor R_4 (fig. 9.25) serves to by-pass the standing current through R_{BB} ($R_{BB} = R_{B1} + R_{B2}$) which would otherwise trigger the SCR. Using a PUT, no standing current flows and the cathode can be connected directly to the gate of the SCR (fig. 9.27). Resistor R_3 (fig. 9.25a) is normally used for temperature stabilisation of the UJT working point against changes in R_{BB}. Because R_{BB} is composed of linear fixed resistors in a PUT circuit, the stability of the working point is inherent so R_3 is not required.

9.14 A commercial SCR firing unit

A number of compact and versatile firing units are available at relatively low cost for use in single-phase or (in groups of three) three-phase motor control, power regulation or other applications which require

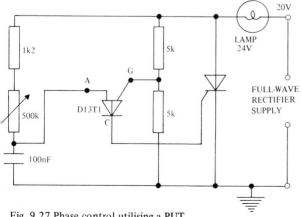

Fig. 9.27 Phase control utilising a PUT

a smooth control of SCR firing angle by means of an external low voltage source — typically 0 to 3V dc. One such unit (the MC16) was designed by the International Research and Development Co. Ltd. and is manufactured by Reyrolle Parsons Ltd. The operation of this firing unit is illustrated by the block diagram of fig. 9.28. Power to the unit is provided by an external 25 – 0 – 25V, 50Hz, 60mA transformer.

As this unit is isolated by its external and internal transformers it may be 'floated' at any reasonable voltage relative to earth. The metal casing of the unit is earthed at all times. An input voltage signal of 0 to 3V dc into a 10kΩ load controls the phase angle between approximately 0° and 165°. The unit comprises a transistor ramp generator driving a transistor switch which controls the operation of a blocking oscillator. The linear ramp has a duration of one half-cycle of the mains supply or 10ms and is initiated at the beginning of each half-cycle. When the ramp voltage exceeds a selected voltage reference level, the transistor switch turns on and firing pulses from the blocking oscillator appear at the output. The input dc signal lifts the ramp and therefore controls the instant at which the reference level is exceeded and the first firing pulse occurs at the output. The unit is free from spurious outputs due to mains-borne interference which can cause the SCRs to fire in a random manner. The pulse train of output is simple to generate with a blocking oscillator and yet provides an ideal waveform to fire SCRs even under difficult conditions, e.g. inductive loads or with low anode-to-cathode voltages. The output pulses available at the two isolated outputs have the following characteristics:

(a) a pulse width of not less than 12μs;
(b) a repetition frequency of 2.4kHz;
(c) a pulse amplitude ⩾ 5V;
(d) a rise time of approximately 3μs.

These features ensure the firing of all commercially available SCRs up to 150A rms rating.

An internal pre-set adjustment enables the matching of firing angle-input voltage characteristics between units required for three-phase working.

With such units readily available in modular or integrated circuit form it is generally uneconomical and therefore rare for firing circuits to be designed and built around discrete components.

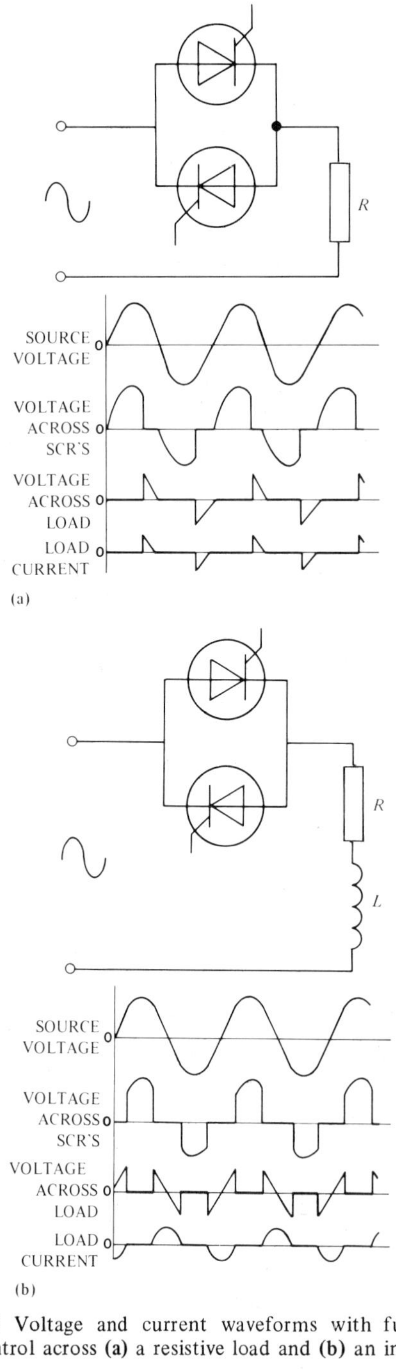

9.15 AC phase control: form factors with resistive and inductive loads

So far phase control has been described only for resistive loads for which the voltage across the load is in-phase with the load current. Voltage and current waveforms for a resistive load are shown in fig. 9.29a. Typical loads such as electric motors, solenoids and even some 'resistive' heaters have inductive components as part of their impedance. The effect of this inductive reactance on the voltage and current waveforms is shown in fig. 9.29b. The load current waveform is non-sinusoidal and the question arises as to the selection of SCRs with adequate rating for such a circuit.

The required rms current rating of an SCR, as shown on the specification sheet, is necessary to prevent excessive heating in the resistive elements of the SCR such as joints, leads, interfaces, etc. By definition, the rms and average values are identical for direct steady current but not for half- and full-wave rectified currents. In the specification sheet of each SCR there appears a series of curves which show the maximum average current as a function of case temperature and conduction angle (conduction angle = 180° − phase angle α). A series of such curves, based on a sinusoidal current waveform, is shown in fig. 9.30 for a 380A rms. SCR [General Electric C380 series − maximum case temperature = 125°C]. The characteristics of non-sinusoidal waveforms (and this includes portions of sinusoidal waveforms as shown in fig. 9.29a) are conveniently expressed in terms of a *form factor F* defined as

$$F = \frac{\text{rms value}}{\text{average value}}$$

An expression for the form factor F for a resistive load for a half-wave sinusoidal waveform can be obtained from equations (9.1) and (9.2), bearing in mind that each SCR in the inverse parallel connection of fig. 9.29 is concerned only with the current waveform on one side of the zero,

Fig. 9.29 Voltage and current waveforms with full-wave phase control across (a) a resistive load and (b) an inductive load

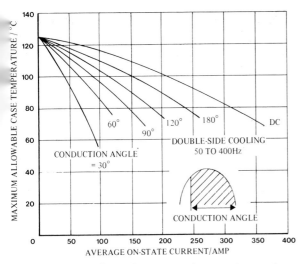

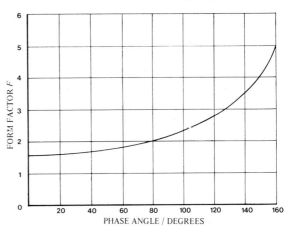

Fig. 9.31 Form factor against phase angle for a phase-controlled resistive load

$$I_{av} = \frac{I_P}{2\pi}(1 + \cos \alpha)$$

and $$I_{rms}^2 R_L = \frac{R_L I_P^2}{4\pi}[\pi - \alpha + (\sin 2\alpha)/2]$$

Hence, the form factor

$$F = \frac{\dfrac{I_P}{2\sqrt{\pi}}[\pi - \alpha + (\sin 2\alpha)/2]^{\frac{1}{2}}}{\dfrac{I_{pk}}{2\pi}(1 + \cos \alpha)}$$

Therefore,

$$F = \frac{\sqrt{\pi}\,[\pi - \alpha + (\sin 2\alpha)/2]^{\frac{1}{2}}}{1 + \cos \alpha}$$

For the special case of $\alpha = 0$, i.e. full conduction,

$F = \pi/2 = 1.57$.

Fig. 9.31 shows the variation of this form factor F for other values of the phase angle α for a resistive load. With the C380 SCR considered above, the average current must be limited to $380/1.57 = 242$A, when operating in a half-wave circuit with phase angle zero.

Fig. 9.30 Average forward current against case temperature for the C380 series SCRS (courtesy of General Electric, USA)

For a phase angle firing of 150° the form factor is approximately 4 (fig. 9.32), so the maximum average current with a resistive load must not exceed 380/4 or 85A. If the load is slightly inductive, resulting in a current waveform as shown in fig. 9.30b, the form factor is changed. The change is not obvious merely from the appearance of the waveform. In fact, the form factor is always reduced when inductance is present in the load. Fig. 9.32 shows the variation of the form factor with phase angle for loads of different values of lagging power factor.

At a phase angle of 150° a 25% reduction in the form factor is realised by using a load of power factor 0.9 and the maximum average current could now be $380/3 = 125$A instead of the 80A calculated for a resistive load. In conclusion, it is apparent that SCRs selected using the form factor for a resistive load

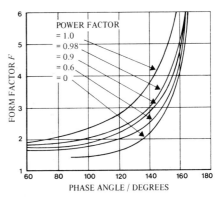

Fig. 9.32 Variation of the form factor with phase angle for different power factors

would have a more than adequate rating for use with a load containing inductance. When high current, high voltage SCRs are being used with inductive loads, the additional current capability can provide substantial additional power handling capacity. Design procedures for making use of this saving are given in the manufacturers' literature.

9.16 Using SCRs with high power incandescent lamps

When a high power (e.g. 1000W) filament lamp is switched on there is a large current surge through the 'cold' filament. This current surge may extend over five to ten cycles of the supply voltage with the initial current as much as 20 times the final operating current. SCRs are typically capable of withstanding a current surge of ten times their operating current, so are particularly suitable for this and similar applications provided that due care is taken in selecting adequate ratings.

9.17 Commutation of SCRs in ac circuits

Commutation (the switching from a conducting to a non-conducting state) of the SCR in an ac circuit usually presents no problem because of the normal periodic reversal of the supply voltage. Difficulties can be encountered as a consequence of insufficient time for turn-off or excessively large time rate of variation of the re-applied forward voltage (i.e. $\frac{dV}{dt}$).

Supply voltage, frequency and inductance in the load are often responsible for failure to commutate properly. Consider the two SCRs (fig. 9.33) connected in inverse parallel with an inductive load. When SCR1 turns off the decaying magnetic field induces a voltage which will forward bias the SCR while the applied voltage is zero. The addition of resistor R_1 and capacitor C_1 (shown dotted in fig. 9.33) can reduce the value of dV/dt to acceptable limits so as to allow SCR1 to turn off. An alternative and preferable solution is to select SCRs capable of turning off in a short time with a high dV/dt and a high voltage.

When an inductive load is used in a full-wave rectifier circuit the inductance causes a holding current to flow through the SCR during the time that the supply voltage drops to zero. Hence commutation is inhibited unless a 'free-wheeling' diode D is

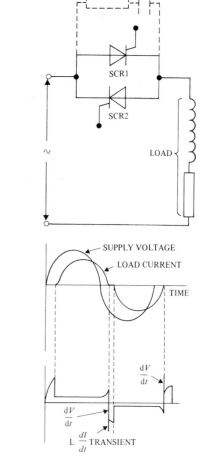

Fig. 9.33 Suppression of $\frac{dV}{dt}$ for inductive load

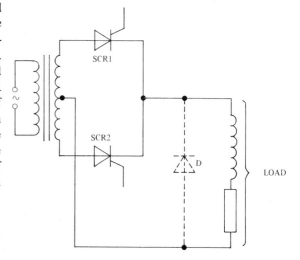

Fig. 9.34 Use of a 'free-wheeling' diode to by-pass the holding current

incorporated to by-pass this current and allow the SCR to turn off (fig. 9.34).

9.18 A bistable SCR circuit

A bistable SCR circuit requiring few components and capable of switching large direct currents is shown in fig. 9.35. Assume that when the dc voltage is applied neither SCR is conducting even though each is forward-biased. When a firing pulse is applied to the gate of SCR2 it is switched into a conducting state, current flows through the load R_L and the capacitor C is charged with the polarity shown in fig. 9.35. If a firing pulse is now applied to SCR1 to render it conducting, the charged capacitor is connected directly across SCR2 and reverse-biases it for an instant. Hence, as SCR1 turns on, SCR2 turns off. The capacitor C is now charged with a polarity which is the reverse of that shown in fig. 9.35 so is available to switch SCR1 off, when SCR2 is rendered conducting.

A minimum value of the commutating capacitance C is needed for reliable switching. For resistive loads,

$$C \geqslant \frac{1.5 \, t_{off} I}{E} \, \mu F$$

where t_{off} is the turn-off time of the SCR in microseconds, I is the maximum load current in amperes and E is the dc applied voltage. For inductive loads, C can be slightly less than this value but the diode D (fig. 9.35) should be included.

This circuit is particularly useful as an 'electronic crowbar' to remove dc power from a circuit merely by providing a firing pulse to the non-conducting SCR. On occasions, the resistor R_K can be chosen so that the current through SCR1 is below its holding current. Although SCR1 will switch on and serve to switch off SCR2, it cannot remain on and reverts almost immediately to a non-conducting state. Hence, the circuit can remain in an ambient state, consuming no power, until the firing pulse to SCR2 again provides current to the load resistor R_L.

9.19 Some further types of thyristor

(i) The DIAC, or *bidirectional trigger diode,* is a two-terminal three-layer device (fig. 9.36).

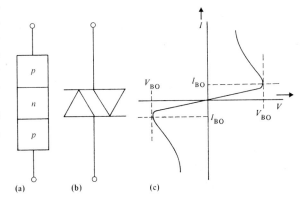

Fig. 9.36 The DIAC: (a) structure; (b) circuit symbol; (c) $V - I$ characteristic

The diac switches to a conducting state when the applied voltage exceeds the breakover voltage V_{BO}. The characteristic exhibits a negative resistance region above the breakover current I_{BO}. This negative resistance region extends over the full operating range of currents above I_{BO}, hence the concept of a holding current does not apply. Reversal of the applied voltage results in an almost identical reverse characteristic.

The most common use for the diac is to trigger SCRs or TRIACs in simple mains-operated phase control circuits.

(ii) The TRIAC: this term denotes a triode (three-electrode) ac switch. Sometimes called a *bidirectional triode thyristor,* it is a multi-junction, three-terminal device equivalent to two SCRs connected in inverse parallel but with a common gate. The circuit symbol and the V against I characteristic of a triac are shown in figs. 9.37a and b respectively.

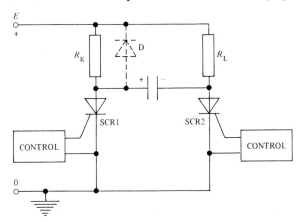

Fig. 9.35 A bistable SCR circuit

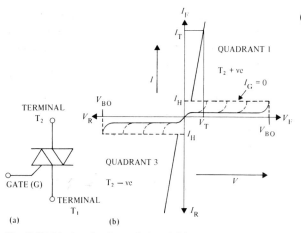

Fig. 9.37 **(a)** the circuit symbol and **(b)** the $V - I$ characteristic of a triac

Conduction between terminal T_1 and terminal T_2 is initiated by a positive or a negative current pulse to the gate electrode or by raising the applied voltage above the specified breakover value. The triac may be triggered with positive *or* negative gate current in both the first and the third quadrants (fig. 9.37). The triggering modes are therefore shown below where all voltages are relative to terminal T_1.

(a) T_2 + ve, gate + ve, first quadrant
(b) T_2 + ve, gate − ve, first quadrant
(c) T_2 − ve, gate + ve, third quadrant
(d) T_2 − ve, gate − ve, third quadrant.

Modes (a) and (d) are normally employed because they provide the most sensitive triggering.

Once the device is conducting, the gate electrode exerts no further control and conduction will continue while the current between the terminals exceeds the holding current I_H.

When two SCRs connected in inverse parallel are used in a phase control circuit, each SCR has an entire half-cycle in which to turn off. When a triac replaces this combination, it must turn off while the load current is passing through zero. For a resistive load and a 50Hz supply this creates no problems because the time available for turn off extends from the time that the terminal current falls below I_H until the re-applied voltage allows triggering in the reverse direction.

(iii) The QUADRAC is an integrated package containing a triac and a diac. Introduced to reduce component size and assembly costs, it may be triggered from a blocking to a conducting state for either polarity of the applied voltage. The circuit symbol is shown in fig. 9.38. The structure is passivated, protected against high voltage transients and the metal case is electrically isolated.

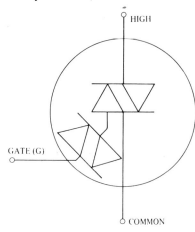

Fig. 9.38 The quadrac

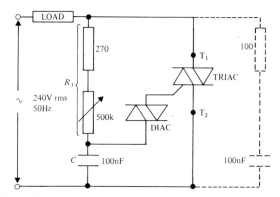

Fig. 9.39 Basic diac-triac phase control

The common terminal replaces terminal T_1 of the triac and again voltages are quoted with reference to this electrode.

A basic phase-control circuit utilising only four components (fig. 9.39) could be further simplified by having the diac and triac within the same quadrac encapsulation. This circuit is typical of the dimming controls marketed for domestic lighting. The variable resistor incorporates an on-off switch and the whole unit is mounted on a plastic panel for wall mounting in place of the conventional toggle switch. A filter circuit is usually incorporated to reduce radio-frequency interference.

The dotted portion of the circuit (fig. 9.39) is included if the load is inductive. When the voltage across the capacitor C_1 reaches the breakover value (V_{BO}) of the diac, C_1 is partially discharged into the triac gate. The triac is switched into a conducting mode for the remainder of that half-cycle. Triggering is accomplished in the (a) and (d) modes, but because these modes are not equally sensitive, non-symmetrical firing occurs, particularly at the low output end of the control. The simplicity of the control makes it attractive for non-critical applications.

Exercises 9

1. Referring to fig. 9.19, the circuit will not oscillate if the value of the resistor R is less than 3kΩ or greater than 3MΩ. Why should this be the case?
   If $\eta = 0.6$ and $C = 0.1\mu F$, calculate the range of frequencies over which the oscillator will function.

2. In the circuit designed to demonstrate phase control (fig. 9.20) one firing unit serves two SCRs. Examine the situation when two SCRs are connected in inverse parallel across the mains supply. Could one firing unit be used in this case and, if so, what form must the coupling take?

3. Examine the circuit diagram (fig. 9.40) and answer the following questions:
   (a) For what purpose would the circuit be used? Outline its mode of operation and sketch the output waveform.
   (b) Comment on the configuration of the bipolar transistor. What type of transistor would you choose for this role?
   (c) What would be the effect of decreasing the capacitance of C to $0.0047\mu F$?
   (d) What would be the effect of decreasing the voltage between the gate and the cathode of the PUT to 10V?

4. Examine the phase control circuit of fig. 9.20. Re-draw this circuit diagram with a PUT to replace the UJT to provide firing pulses for the SCR. Outline the mode of operation of your circuit.

5. A diac used to fire a triac in a phase control circuit has a break-over voltage V_{BO} of 30V. Calculate the smallest value of α assuming a 240V rms, 50Hz mains supply.

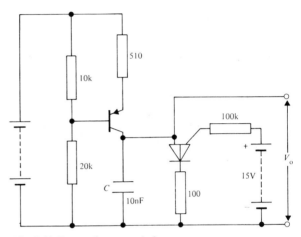

Fig. 9.40 Concerning example 3

6. Silicon-controlled rectifiers can be used to control the power to a load by
   (a) ac phase control
   (b) zero voltage switching.
   Describe, with the aid of diagrams, the principal features and advantages of each of these control systems.

7. Describe the operation of a bistable SCR circuit.

8. What advantages does the PUT offer over the conventional UJT?

9. Why is radio-frequency interference always associated with an ac phase control system?

10. Discuss the methods of initiating and terminating conduction in an SCR.

11. Define and explain the significance of the following terms:
    (a) a 'free-wheeling' diode;
    (b) the 'holding current' I_H of an SCR;

(c) form factor;
(d) peak point voltage of a UJT;
(e) intrinsic stand-off ratio η of a UJT.

12. How would you measure the current flowing in a circuit in which ac phase control was being used?

13. Current pulses from an oscillator circuit are often used to fire an SCR in a phase control system. Why is it necessary to synchronise the oscillator with the mains voltage and how is this achieved with a UJT relaxation oscillator?

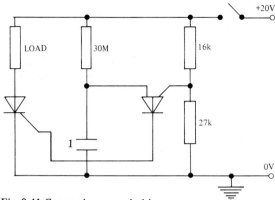

Fig. 9.41 Concerning example 14

14. Examine the circuit (fig. 9.41) and answer the following questions:
    (a) What value of η, the intrinsic stand-off ratio, has been selected?
    (b) What is approximately the time interval between closing the switch S and power being supplied to the load?

15. Discuss, with circuit diagrams, the principle of phase control using thyristors.
    Describe the action of a suitable trigger circuit providing manual control of the phase angle and outline the requirements of control circuits using low-level signals produced either by a thermistor or a cadmium sulphide photo-resistor.

Appendices

A THE CONSTRUCTION OF SEMICONDUCTOR DEVICES AND CIRCUITS

1. Introduction. The *active devices* in electronics are chiefly the various types of semiconductor diode and transistor. The *passive devices* in all kinds of circuitry are resistors, capacitors and inductors.

Particularly in the cases of low-power and medium-power electronic circuits, the connecting together of passive and active components has departed from the methods whereby discrete components having appropriate terminals or leads were linked together by conducting wires with soldered connections. The first departure from old-established practice came with the advent of the printed circuit board (PCB) of which an example is shown in Plate 1.

Plate 1 Printed circuit board – topside and underside

Three further advances in the construction of electronic circuits and sub-systems are classified as:

(a) *The monolithic integrated circuit* (monolithic IC). This is based on a single chip (slice) of silicon which is typically 0.2mm thick and with surface dimensions as small as 1.25mm x 1.25mm. Within the surface of this tiny chip are formed a number of active and passive components together with their interconnections to give a complete circuit (e.g. of an operational amplifier) or a system of circuits (e.g. a set of logic gates for performing computational operations). Three classes of monolithic ICs are somewhat artificially classified as those involving *small-scale integration* (SSI) in which there are a dozen or less passive and active components within the circuit, *medium-scale integration* (MSI) in which there are twelve to a hundred or so components on each chip, and large-scale integration (LSI) where several hundred or even a few thousand components, inter-connected to provide a complete electronic sub-system, are formed on a single chip (e.g. forming the 'heart' of an electronic calculator).

(b) *Thin-film circuits.* These are deposited by the evaporation of selected materials in a chamber at low pressure (a 'vacuum' of 10^{-3} Pa or less) or prepared by cathodic sputtering in an inert atmosphere at a pressure of a few Pa to form thin films on glass or ceramic substrates. Thin films may be roughly classified as of thickness less than $10\mu m$. Only the passive components, resistors and capacitors, may be formed in this way. The coated areas of the substrate concerned are defined by apertures in suitable masks or by selective etching using photoresists. The required conducting leads and lands are produced by the deposition of gold or aluminium. Resistors are formed by the deposition of appropriate patterns of tantalum or nickel-chromium alloys. Capacitors are made by the deposition of dielectrics such as silicon oxide, tantalum oxide, glass and other insulating materials on to appropriate underlying conducting areas of metal forming one plate and a counter-electrode on top forming the other plate.

(c) *Thick-film circuits.* The substrate most commonly used is of alumina (aluminium oxide, Al_2O_3) of 96% purity. The thick film (thickness greater than $10\mu m$, usually) is produced by screen-printing on to the substrate an appropriate paste which is subsequently dried and heated in a furnace carefully regulated to provide a specific temperature-time schedule. The pastes used for the conducting leads (strips) contain precious metals such as gold, platinum, silver or palladium. Subsequent to the printing through a nylon or metal screen of the conducting leads there is printed the resistors (using appropriate resistor pastes) and the capacitors, which are usually based on a dielectric formed from a glaze containing barium titanate. As in the case of thin film circuitry, connectors and passive components (including inductors in the case of thick film circuits) can be formed but not active components.

Hybrid circuits are those based on monolithic IC's and on thin and thick film circuits where separate discrete components are attached between specific contact lands on the substrate. They are usually encountered where the magnitude of the passive or active component needed exceeds that practically or economically possible by monolithic IC or film deposition methods.

Active semiconductor devices depend, in the first place, on the production of extremely pure silicon and, to a lesser extent, germanium. A feature of bipolar semiconductor devices is the *p-n junction* of which there are two chief types: the *step junction* and the *graded junction*. Junctions within active semiconductor devices are primarily made by either an alloying process or a diffusion process. Of these, the silicon planar diffusion process is the most important. However, and especially in the monolithic IC, the use of metal oxide semiconductor transistors (MOST's) is of increasing importance.

Accompanying the development of monolithic IC methods, both those based on bipolar junction transistors and MOST's, there has been a revolution in the methods of producing microminiature passive components.

Certain specialised types of semiconductor device (and the number of types promises to increase) are based on the method of *ion implantation*. The silicon slice is doped over areas selected by suitable masking by bombarding it *in vacuo* with positive ions of the appropriate impurity (e.g. pentavalent phosphorus or trivalent boron) which are accelerated to energies of up to 300 keV. This energy is an important factor in determining the depth of penetration into the silicon slice of the particular kind of ion.

2. The production of pure silicon or germanium. Silicon, which is much more important than germanium in semiconductor device and circuit manufacture, needs to have in its intrinsic (pure) form a number of intrinsic carriers n_i of about 1.5×10^{10} per cm^3 at room temperature as compared with about 2.4×10^{13} per cm^3 for germanium. Hence, intrinsic silicon is much more difficult to prepare than intrinsic germanium because its purity has to be some 1000 times greater.

Silicon obtainable from silicon dioxide (SiO$_2$) or silicon tetrachloride (SiCl$_4$) by normal metallurgical processes has to be further purified to reduce the number of foreign atoms present in it to considerably less than 1 in 10^{10}, demanded for the manufacture of semiconductor devices.

Purification beyond that possible by normal metallurgy can be undertaken by the process of *zone-refining* (fig. A1) and by *single-crystal growing methods*.

In the zone-refining process a bar of the solid element (which must be a conductor of electricity) within a graphite or silica boat (not suitable for producing very pure silicon, but satisfactory for germanium) is placed in a silica tube which is pumped to a low pressure and then filled with an inert gas such as argon. A short length of this bar (the zone) is heated to above the melting point of the element (958°C for Ge and 1420°C for Si) by eddy-currents produced in it by an induction coil (a water-cooled copper tube of one or two turns) carrying a large radio-frequency current and placed around the silica tube. The induction coil is traversed from one end of the bar to the other (a process which can be repeated). Alternatively, several induction coils are spaced at intervals along the outside of the silica tube and the bar of the element is drawn slowly through this tube in such a manner that only a few zones are molten at any one time. As the bar is progressively melted from one end to the other, the impurities in it are concentrated within the molten region whereas behind this region, the element will re-crystallize in purer form. This occurs because the impurity atoms as they cool (lose energy) do not find locations within the specific energy sites peculiar to the solid crystalline element which forms; rather they occupy the much less well-defined energy sites within the molten region.

The container boat can be omitted in an arrangement in which the rod of the solid element is supported vertically instead of horizontally within the silica tube. This gives rise to the useful *floating zone process* (which is necessary for silicon because this element reacts with all known crucible materials) in which the molten region (of which the traverse is now from top to bottom) is held in place by surface tension forces. This method demands a more precise control of temperature (to ensure satisfactory surface tension forces) than that needed with a horizontal furnace. The advantage is in producing *n*- or *p*-type silicon because a doping material addition at one end can be uniformly distributed along the silicon bar in a single cycle of zone-refining.

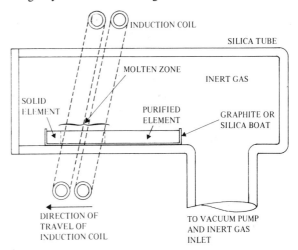

Fig. A1 The zone-refining process

3. Crystal growing. The pure silicon resulting from zone-refining is substantially polycrystalline. The action in most semiconductor devices depends essentially on single crystal formation. A monocrystalline ingot of silicon (or germanium) is made from polycrystalline pure material by the Keck method (a vertical floating zone process, as described in outline in section 2) or by the Czochralski method and developments therefrom. The latter makes use of a very small piece of a single crystal of silicon with the correct lattice orientation which is lowered on to molten pure silicon within an electrically-heated crucible maintained within an inert gas atmosphere. As the pull rod supporting the mount for this 'seed' crystal is withdrawn (and rotated) very slowly (over a period of several hours), a single crystal is prepared which may, for example, be 300mm long and 30 to 60mm in diameter. To manufacture *p*- or *n*-type silicon single crystals in this way, suitable trivalent or pentavalent doping elements are added in the appropriate concentration to the melt in the heated crucible before the crystal is grown.

4. Forming *p-n* junctions. Processes used are *rate-growing, alloying, diffusion* and *ion implantation*. The rate-growing process does not have the importance as a method of manufacturing *p-n* junctions that it had in the early 1950s when it was introduced. It is based on the fact that the equilibrium between the dopant and the silicon (or germanium) depends on the rate at which a crystal is formed from the melt, so that it is possible to incorporate an excess of *n*-type semiconductor in the crystal at one location and, adjacent to it, form an excess of *p*-type semiconductor.

The alloying method of forming *p-n* junctions is illustrated by describing the preparation of a *p-n-p* germanium transistor (fig. A2).

A large crystal of germanium is doped with pentavalent antimony to give *n*-type semiconductor material of resistivity about $5 \times 10^{-2} \Omega$m. It is cut along a specific crystal plane into slices of thickness about 0.3mm. From these slices are cut by means of a diamond-impregnated cutter hundreds of wafers each, say, 3.0 x 7.0 x 0.3mm. During this cutting the slices are mounted in wax (subsequently removed) to prevent the wafers from disintegration. A pellet or disc of indium (trivalent) is placed on each side of each wafer of *n*-type germanium. The wafers are then heated in a hydrogen atmosphere in a furnace to a carefully regulated temperature over a specific time period. The indium melts at 156°C and dissolves some of the germanium (m.p. 957°C) in contact with it. On subsequent cooling (also controlled) the germanium-indium alloy region is *p*-type. The indium pellet which forms the collector electrode on one side of the wafer is about three times the diameter of that which forms the emitter. A nickel tag is soldered to the germanium wafer to form the base connection B whereas the emitter E and collector C connections are by wires soldered to the respective indium pellets on either side of the wafer. The *p-n-p* assembly is etched to remove surface contamination, washed in de-ionized water, dried, and finally mounted in water-repellent grease in a light-tight capsule.

The preparation of *p-n* junctions by a *diffusion process* is of great importance because, first, it enables the concentration and concentration gradient of the dopant over small regions to be controlled precisely and, second, it leads to the manufacture of the discrete silicon planar transistor (and diode) and to monolithic integrated circuits involving an array of transistors (and/or diodes) and passive components.

The ion implantation process is described briefly in section 1.

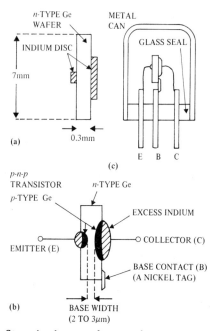

Fig. A2 Steps in the manufacture of a *p-n-p* germanium transistor **(a)** initial structure; **(b)** structure has been heated to produce alloying; **(c)** encapsulation

5. Outline of the preliminary procedure involved in the preparation of silicon planar semiconductor devices. Without at first considering the manufacture of a specific type of active or passive component by the use of a diffusion process, it is of value to indicate in outline the preliminary stages involved.

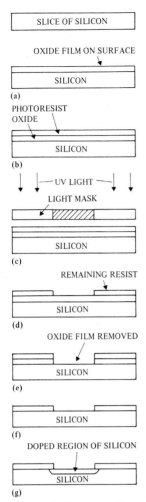

Fig. A3 Preliminary stages in the preparation of silicon planar devices.
(a) Oxide on surface: formed by heating: oxide film about 1μm thick. (b) Photoresist spread over silicon oxide; (c) Light mask in place: incidence of u.v. light polymerises resist in exposed regions. (d) Resist developed by washing-out unexposed parts and baking to harden thoroughly remaining resist. (e) Slice etched to remove silicon oxide in unprotected areas. (f) Resist removed (g) Exposure to gas stream containing appropriate impurity with heating causes impurity to diffuse into silicon in exposed areas

Referring to fig. A3, the initial requirement is a slice of single crystal silicon. This may be intrinsic silicon, p-type or n-type or have an epitaxial layer (section 7) of p- or n-type silicon on its surface. The stages are:

(a) The silicon slice is heated to about 1100°C in an oxidising atmosphere so that its surface becomes coated with a film of silicon oxide (mostly silica, SiO_2, but also containing some silicon monoxide, SiO) of thickness 0.5 to 1.0μm. Suitable oxidising atmospheres are steam, wet oxygen or dry oxygen. The times required to form the oxide film are 1 to 2 hours for steam, somewhat longer for wet oxygen but considerably longer for dry oxygen. The silicon oxide film is almost impervious to the diffusion of boron (trivalent) and phosphorus (pentavalent) atoms. However, these elements diffuse (at approximately equal rates) into silicon.

(b) It is required to diffuse acceptor (e.g. boron) or donor (e.g. phosphorus) atoms into certain specific regions within the silicon. These regions are defined by a mask. It is the silicon oxide film which is to act as the mask. Within this silicon oxide film it is therefore necessary to cut openings (windows) of specified geometrical shapes and locations, i.e. the oxide layer has to be removed over the desired areas. To do this a photolithographic etching process is used. This entails coating the oxide layer with a uniform film of a photoresist (e.g. Kodak photoresist KPR or others such as KMER, KTFR or AZ 1350). To ensure uniformity of thickness of this spread photoresist film, the slice is spun at about 500 rpm. To ensure satisfactory adhesion of this photoresist film, the silicon oxide surface to which it is applied must be extremely clean. After coating, the photoresist film is baked at about 50°C for about 15 minutes.

(c) A black and white photographic plate stencil is prepared by a separate photographic procedure. This involves preparing first a master drawing in black and white of the particular mask geometry needed, where the white areas represent the windows (to be made eventually in the silicon oxide film) and the black regions represent the opaque areas. This mask is photographically reduced (by 250 to 500 times linear) to form a negative. This stencil negative is placed over the photoresist film and exposed to ultra-violet light. The regions of the photoresist beneath the windows in this stencil hence become polymerised. Specially designed high precision apparatus is demanded to ensure accurate registration between the stencil and the photoresist film.

(d) The stencil is removed and the unpolymerised (unexposed) areas of the photoresist are washed away, e.g. by trichloroethylene. The remaining photoresist (corresponding to the regions where the silicon oxide is to remain) is subsequently hardened thoroughly by baking at a temperature above 100°C.

(e) The slice is etched by immersion in a solution of hydrofluoric acid which removes the areas of silicon oxide unprotected by the remaining hardened photoresist. Silicon itself is not attacked by hydrofluoric acid.

(f) The polymerised photoresist is removed by a special stripping process, and the slice is then thoroughly washed.

(g) The diffusion is undertaken.

6. The diffusion process. The diffusion at a given temperature of impurity atoms into silicon is represented mathematically by the equation

$$n_x = n_o \left[1 - \mathrm{erf}(x/2 \sqrt{Dt})\right] \qquad (A1)$$

where n_x is the number of impurity atoms per unit volume which diffuse to a depth x below the surface of the slice, n_o is the number of impurity atoms per unit volume at the surface (i.e. $n_x = n_o$ at $x = 0$), t is time, D is the diffusion constant for the impurity in silicon at the given temperature and erf is the error function, recalling that

$$\mathrm{erf}\, y = (2/\sqrt{\pi}) \int_o^y \exp(-\lambda^2)\, d\lambda.$$

The diffusion coefficient D (it is expressed in the unit $m^2 s^{-1}$) increases rapidly with increase of temperature, as would be expected because the average speeds of the atoms of the diffusant increase greatly with temperature. To obtain reasonable diffusion times of impurity atoms in silicon, the temperature of the silicon needs therefore to be about 1100°C and must also be accurately controlled because of the rapid variation of D with temperature. Widely used impurities which are diffused into silicon are phosphorus (n-type donor) and boron (p-type acceptor). These are compatible with silicon and have the same diffusion coefficients over a wide temperature range. Arsenic as an n-type impurity has a diffusion coefficient about ten times smaller than phosphorus at the same temperature but this is an advantage in obtaining a more precisely controlled diffusion depth. Gallium is unsuitable because it diffuses readily through silicon oxide.

A typical diffusion furnace (also used for the preliminary production of the silicon oxide film) is of the 'open tube' type. The silicon slices are mounted within a silica tube of about 75mm diameter of which the central region over a length of some 500mm is heated in an oven to temperatures up to 1300°C, and where any particular temperature is maintained constant (usually by thyristor control of the power supply) to within ±0.5°C. For oxidation of the silicon, steam is introduced. For the introduction of phosphorus as a diffused donor impurity, liquid phosphorus oxychloride is vaporized, mixed with nitrogen and passed through the furnace over the silicon slices which are heated to 1000°C. To introduce boron as an acceptor impurity, liquid boron tribromide is vaporized and introduced in nitrogen as the carrier gas into the furnace.

The number of atoms of silicon per unit volume is 5×10^{28} atoms/m³. The maximum number of phosphorus atoms that can be introduced into silicon is about 10^{27} atoms/m³, corresponding to a maximum concentration of 2%. For boron this maximum concentration is lower at about 1%, though for arsenic it is somewhat larger than for phosphorus.

7. Epitaxial growth. Widely used in the preparation of monolithic integrated circuits, in epitaxial growth a film is formed on the surface of a substrate of single crystal material in such a manner that this film has the same crystal orientations as those of the substrate. In the case of silicon as the substrate, this means that a film of extra silicon (which can be intrinsic, n-type or p-type) can be formed on a silicon slice and the crystal lattice orientations are preserved throughout. The composite material (silicon substrate plus silicon film) then acts as a single crystal — a feature demanded in the manufacture of semiconductor devices such as junction diodes and transistors.

An advantage of epitaxial growth is that it enables a layer of silicon of required conductivity (decided by the concentration of donor or acceptor atoms in it) to be formed on the surface of a silicon slice.

To obtain satisfactory epitaxial growth on silicon (intrinsic, n-type or p-type) the substrate silicon slice must have a surface which has been prepared by lapping and polishing to be extremely flat and has also been cleaned scrupulously. This is essential to avoid undesirable electrical interface phenomena between the epitaxial film and the substrate. A satisfactory epitaxial film growth rate requires that the silicon slice be heated to about 1200°C and satisfactory formation of this film is best with the silicon wafers locally heated by placing them on a graphite platform of which the temperature is raised by eddy-currents produced by an induction coil surrounding the open-tube silicon furnace. To provide the silicon which forms the epitaxial film, hydrogen is bubbled through silicon tetrachloride ($SiCl_4$) and this gas is passed over the heated silicon wafer in the furnace. At the hot silicon substrate surface, the silicon tetrachloride dissociates and silicon is deposited epitaxially, i.e. so that its crystal form and orientation is decided by that of the substrate. To obtain the required electrical conductivity of the epitaxial layer, the necessary n-type or p-type impurity is introduced at the required concentration into the stream of hydrogen which carries the silicon tetrachloride vapour. The rate of deposition of such epitaxial layers on to silicon is about $1\mu m$/minute up to a thickness of about $12\mu m$.

8. The manufacture of a discrete planar n-p-n transistor.

A typical procedure is illustrated by the schematic diagram of fig. A4.

An outline of the stages in the preparation is as follows:

A slice of n^+-type silicon about 0.4mm thick and of resistivity about $10^{-4}\Omega m$ is mechanically and electrically polished to provide a flat smooth surface. On this surface is grown an n-type epitaxial layer of thickness about $7.5\mu m$ and resistivity about $10^{-2}\Omega m$.

A silicon oxide film about $1\mu m$ thick is formed on the surface of this epitaxial layer. This requires heating the slice of silicon in a furnace to about 1000°C for two hours in steam (section 6). The photoresist method is used to provide only a window within this silicon oxide layer.

By the diffusion of trivalent boron through this window a p-type base region of depth about $2\mu m$ is formed within the n-type silicon epitaxial layer (fig. A4a). This is usually done in two stages: first, with the slice at 850°C in a furnace when boron is deposited at the surface; second, the slice is transferred to another furnace in which it is heated to 1150°C in a stream of nitrogen to allow the boron to

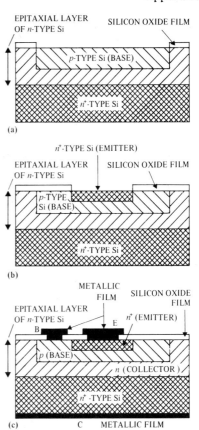

Fig. A4 Indicating the manufacture of a silicon planar transistor

diffuse in to the depth of $2\mu m$ approximately. In the latter stages of this diffusion, steam is also admitted to the furnace to provide a new layer of silicon oxide on the surface.

Making use of a new photoresist procedure, a second window is etched into the p-type base region produced. Phosphorus is diffused through this second window to form an n^+-type emitter region of depth about $1.5\mu m$ within the p-type base region (fig. A4b). Again, steam is introduced during the latter part of the time so that a fresh silicon oxide layer is formed.

A third photoresist sequence enables windows to be etched in the silicon oxide over the base and emitter regions. The slice is then transferred to a vacuum chamber in which aluminium is volatilised. Aluminium is thereby deposited over the whole surface of the slice. By the use of a fourth photoresist sequence and appropriate masking, this aluminium film is removed except where it is required for emitter and base contacts (fig. A4c). To ensure that the aluminium makes satisfactory low resistance contacts with the emitter and base silicon, heating to about 550°C is carried out in an inert atmosphere (e.g. of argon) so that the aluminium alloys with the silicon.

Beginning with a silicon slice of 50mm diameter, 7000 silicon planar transistors can be produced in the same sequence of processes just outlined, where each transistor occupies an area of about 0.5 x 0.5mm within the slice surface.

Having thus produced several thousand individual planar transistors on the single silicon slice, appropriate dividing lines are scribed by a diamond-tipped tool on this slice so that the discrete transistor chips can be readily separated. The discrete transistor is mounted on a base collector contact and fitted into a small can with the appropriate base and emitter leads. This base collector is via a deposited metal film and the n^+-silicon layer to the n-type silicon collector (fig. A4c).

9. Silicon planar transistors within a monolithic integrated circuit.

As described in section 8, the collector of the discrete planar transistor is the underside of the substrate. This cannot be permitted when incorporating a number of planar transistors within a single integrated-circuit chip because their collectors would all be connected electrically whereas usually they need to be electrically isolated from one another. To ensure this, the silicon planar transistor incorporated within monolithic ICs has its collector contact brought out to the top surface instead of to the bottom of the substrate. The collector region is now surrounded by a 'moat' (isolated region) of p-type silicon (fig. A5). Neighbouring transistors within the chip are now electrically isolated because they are effectively separated by reverse-biased diodes.

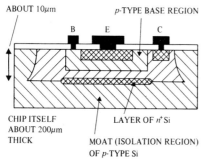

Fig. A5 Silicon planar transistor construction (n-p-n type) as arranged within a monolithic integrated circuit

This top surface contact introduces an undesirably large resistance between the metallic contact itself and the collector region. This is because the path afforded to the passage of collector current is inevitably through the thin n-type collector region. To reduce this resistance greatly (and so also reduce V_{CE} for the saturation collector current) a layer of n^+ silicon of low resistivity is diffused into the collector (fig. A5).

10. Metal-oxide-semiconductor transistors within monolithic integrated circuits.

The metal-oxide-semiconductor transistor (insulated gate field-effect transistor: mosfet) is an alternative to the bipolar junction transistor. In many applications it is a strong rival and may dominate the field, for example in the integrated circuit systems used in logic circuitry for computer technology.

A p-channel mosfet fabricated within a monolithic IC (fig. A6) is made by silicon oxide growing and diffusion processes similar to those used in the manufacture of bipolar silicon planar transistors. Beginning with a substrate of n-type silicon, two simplifications arise immediately: (a) an epitaxial layer is not needed; (b) only one diffusion is necessary to form both the source and the drain regions (both of p-type silicon) whereas four diffusions are required for an IC bipolar transistor.

Great care is needed, however, in the production of the silicon oxide layer. That over the gate electrode has to be precisely controlled in thickness because it is an important factor in determining the mutual conductance $\dfrac{\partial I_D}{\partial V_G}$ of the mosfet. This mutual conductance increases as the thickness of the oxide layer over the gate is reduced. A thickness of $0.1\mu m$ is typical.

The advantages of mosfets over bipolar transistors in integrated circuits rest chiefly on the fact that about three times as many devices (including capacitors of a given value) can be incorporated within the same area of a silicon chip. Although excessive micro-miniaturisation is not necessarily an end in itself (though it clearly reduces time constants to nanosecond values), the pure silicon starting material is very expensive, so that more components per unit area results in economy. The ability to include more active and passive components in a given area arises because the diffusion and masking processes are less numerous, the mosfets are electrically isolated from one another (an isolation 'moat' is not necessary) and resistor and capacitor formations are often easier.

The stages in the manufacture of a mosfet within a silicon slice are:

(a) a substrate slice of *n*-type silicon is surface oxidized;
(b) windows are etched (photoresist method) in the oxide in those regions where the *p*-type source and drain are to be formed by the diffusion of boron;
(c) the boron diffusion is undertaken and the slice is re-oxidized;
(d) windows are etched (second photoresist process) in the silicon oxide for formation of the gate and contacts;
(e) the precisely controlled thin layer of pure silicon oxide is formed over the gate region of the silicon;
(f) a third photoresist sequence enables the contact windows to be formed;
(g) aluminium is deposited on the slice by evaporation *in vacuo*;
(h) a fourth final photoresist process is undertaken to enable subsequent removal of the aluminium except in those locations where it is needed for contacts and electrical connections.

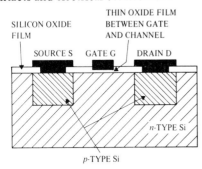

Fig. A6 A mosfet (MOS transistor or MOST)

11. The fabrication of passive components in integrated circuit technology. The passive components that can be formed in integrated circuits are resistors and capacitors, though coupling capacitors are avoided if possible by arranging direct-coupling between stages. Inductors cannot readily be formed, though small values can be fabricated utilizing thick-film methods (section 15) and simulation of the current lag on voltage typical of inductive circuits is possible by circuit methods not involving inductors themselves.

Resistors can be provided in monolithic IC technology, but having values restricted by the technique used. Methods available are:

(a) The *p*-type base region of an *n-p-n* planar silicon transistor for IC (fig. A5) can have a resistance of as much as 50kΩ, provided that it is sufficiently thin and made of *p*-type silicon of sufficiently high resistivity. It can therefore be used as a resistor provided that the limitations on its location in the circuit connections are acceptable and the performance of the transistor of which it forms a part is within limits decided by the resistance value required and its power dissipation.

(b) A 'diffused resistor' may be prepared (usually at the same time in the manufacture as the diffusion of a separately masked base region of an *n-p-n* planar transistor) in the silicon substrate chip of a monolithic IC, by taking advantage of the fact that the resistivity of the silicon can be pre-determined over a wide range by variation of the donor or acceptor impurities diffused into it. For a *p*-type diffusion of 100Ω/square (a typical value for a base region) a resistor stripe (fig. A7a) of width typically 25μm will have a resistance of 100Ω per 25μm of length. Resistors up to 1kΩ can therefore be reasonably accommodated as a

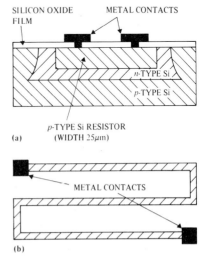

Fig. A7 Resistors in stripe form **(a)** straight, **(b)** meandered, prepared by the diffusion of a trivalent element (usually boron) into silicon to form *p*-type silicon

straight stripe (2.5mm long) but larger values (up to 20kΩ, requiring a stripe length of 50mm) make use of a meandered geometry (fig. A7b). The minimum value practicable for a short wide stripe is about 20Ω. Alternatively, for small resistor values down to 2Ω, the n^+ emitter diffusion region may be used.

Diffused resistors cannot be reproduced to an accuracy of better than about ±10% and have a temperature coefficient of resistance of about 0.2%/°C, which is large compared with that of a discrete resistor. However, it is possible to ensure that two diffused resistors formed adjacent to one another in an IC have a ratio of values kept to within ±2%, and where this is maintained over a fairly wide change of ambient conditions. Indeed, in the fabrication of ICs, it is common practice to ensure a circuit design is dependent upon resistance value ratios rather than absolute values.

(c) A bipolar transistor in an emitter follower circuit has a resistance R between its emitter and earth. The input resistance of this transistor is approximately $h_{FE}R$. Hence a fairly large effective resistance is obtainable.

(d) Simple modification (e.g. connection of the drain to the gate) of a mosfet (fig. A8) gives a resistance dependent upon the gate voltage, but this is not a drawback in many circuit arrangements and can even be used with advantage. Such mosfet or MOS resistors have better electrical isolation between components than is achievable by the use of bipolar transistors; moreover, the area of the silicon chip used is only about 1% of that required for diffused resistors of the same value.

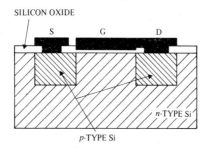

Fig. A8 A resistor based on the connection of the drain (D) to the gate (G) of a mosfet: an MOS resistor

Capacitors restricted in maximum value to about 100pF (otherwise they occupy too much chip area) are provided by two main methods:

(i) By making use of the capacitance of a *p-n* junction. This junction must have a reverse bias across it, otherwise the resistive loss is considerable. The value of the capacitance so obtained diminishes with increased width of the depletion layer associated with the *p-n* junction and so with increase of the reverse bias voltage. At low signal levels, this capacitance variation with voltage is often of little importance; moreover, with larger signals it can often be taken into account in the circuit design and, in some instances, used with advantage.

The capacitance of the collector-base *p-n* junction in a bipolar planar transistor is about 240pF/mm² at zero bias whereas that of the emitter-base junction at zero bias is about 1600 pF/mm². Short-circuiting the emitter and collector in a planar bipolar transistor in an IC (fig. A9) gives two junctions in parallel with a capacitance of about 1000pF/mm². A disadvantage is the high series resistance involved, but this can sometimes be utilized in the IC design.

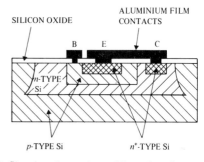

Fig. A9 Shorting the emitter (E) and collector (C) in a bipolar transistor in an integrated circuit to obtain a capacitance

(ii) A capacitor may be formed by making use of silicon oxide as a dielectric. One electrode 'plate' of the capacitor is in the form of an n^+-region of the silicon substrate surface whereas the second electrode is formed by depositing aluminium *in vacuo* on to the top surface of the silicon oxide film on this n^+-region of the substrate. Such a capacitor is capable of withstanding a higher voltage than a *p-n* junction. However, their capacitance per unit area is smaller than that of the *p-n* junction type, but there is an advantage in that the capacitance is not dependent on the voltage applied.

The mosfet (MOS transistor) also lends itself readily to silicon oxide dielectric capacitor construction.

12. Interconnections in integrated circuits. Electrical connections between one component and another in an IC are by aluminium film or by p^+ or n^+ regions. Cross-overs may well be involved. At such cross-overs it is often essential to avoid electrical contact. The necessary insulation can be arranged in the masking and processing sequences so that a layer of silicon oxide is formed between the connections at the cross-over. Alternatively, an 'underpass' connection may be made in which a high conductivity p^+ or n^+ region is diffused into the silicon slice (fig. A10).

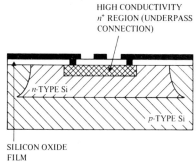

Fig. A10 An underpass connection

13. Encapsulated monolithic integrated circuit elements. A wide variety of IC modules (or packages) is commercially available. Included are the following types of amplifier: operational, differential, sense (comparators), audio-frequency (af) including stereo, intermediate frequency (if) for radio and television receivers, and radio-frequency (rf). Also voltage regulators, modulators and demodulators (for both amplitude and frequency), oscillators, multivibrators, flip-flops and logic elements such as diode-transistor (DTL), transistor-transistor (TTL), resistor-transistor (RTL) and a number of further circuit elements and sub-systems for digital computers, control circuitry, interfacing and so on.

The typical sequence of processes in the manufacture of a number of such monolithic IC elements or sub-systems based on bipolar transistors and the formation of the necessary resistors and capacitors on a *p*-type silicon slice of diameter about 50mm, thickness about 0.4mm and of known resistivity, is as follows:

(a) Slice surface is lapped and polished.
(b) Silicon oxide film of thickness about 2μm is formed.
(c) Photoresist process 1 is undertaken to define the initial diffusion regions.
(d) The *n*-type material is diffused in. This may be phosphorus or the slowly diffusing arsenic, which enables the diffusion depth to be more accurately regulated.
(e) This oxide is removed and an *n*-type epitaxial layer of thickness about 7μm and of known resistivity is grown on the slice surface in an epitaxial reactor furnace.
(f) The surface is re-oxidised to a thickness of about 2μm.
(g) Photoresist process 2 is undertaken to define the isolation region in the epitaxial layer.
(h) *p*-type material, usually boron, is diffused into the epitaxial layer.
(i) Photoresist process 3 is undertaken to define the base regions.
(j) *p*-type material, boron, is diffused in to form the base regions.
(k) Photoresist process 4 is undertaken to define the emitter regions.
(l) *n*-type material, phosphorus, is diffused in to form the emitter regions.

[the masks used during stages (i), (j), (k) and (l) are cut and arranged, as appropriate, to enable the resistors and capacitors required to be produced].

(m) Photoresist process 5 is undertaken to define the electrical contact areas.
(n) The slice is transferred to a vacuum coating plant in which aluminium is evaporated to produce a film of thickness about 1μm.
(o) Photoresist process 6 is undertaken to define the required electrical interconnection geometry.
(p) The aluminium film is removed from those locations other than where needed to form contact pads and electrical interconnections.
(q) The slice is subjected to a fully automatic probe testing schedule, usually computer-controlled.

(r) The slice is scribed with a diamond-tipped tool to form 'break' rulings between rows in two perpendicular directions of the individual circuit elements (which are usually each from 1.25 to 2.5mm square). The slice is then broken into the many chips, on each of which is formed the transistors (diode-connected transistors, if required), resistors, capacitors, aluminium contacts and interconnections making up the complete circuit or sub-system.

All these successive stages involving, for example, six photoresist sequences (fewer for mosfet circuitry) may occupy a period of a month or more. In this apparently long manufacturing time there has, however, been made a large number of individual circuit modules or sub-systems on single chips. For example, if each chip is 1.25mm x 1.25mm, simple calculation shows that several hundred chips can be formed from one slice. Moreover, as several slices can be processed simultaneously in the same furnaces and vacuum coating plant, many thousands of chips are produced in a month, and each chip may contain in extreme cases within its surface several thousand passive and active components.

Plate 2 shows a much magnified version of the components, connections and contact pads in a single chip of relatively simple form. This chip is mounted in a suitable capsule of which the standard forms are shown in fig. A11 where (c) is based on the TO-5 transistor capsule, (b) shows an 8 pin dual-in-line package of dimensions typically 6.5 x 3.5 x 1.25mm, (a) is a 14 pin dual-in-line package which may be plastic or ceramic.

Wire bond connections have to be made between the contact pads (aluminium) on the chip itself and the leads to the capsule where these leads appear outside the capsule and are connected up (usually on printed circuit boards) by standard soldering methods.

The making of wire bond connections has demanded the development of special techniques because of the extremely small areas of the contact pads on the chip. Typically, thermal compression bonding (also called 'nail-head' bonding) is used to do this (fig. A12, in which all the dimensions are greatly magnified). Gold wire of diameter about 25μm is employed. This gold wire is fed through a vertical heated capillary tube, its lower end, emerging from the capillary, is melted in a tiny hydrogen gas flame

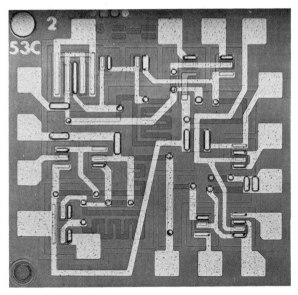

Plate 2 Components, connections and contact pads in a monolithic IC silicon chip (courtesy of Mullard Ltd)

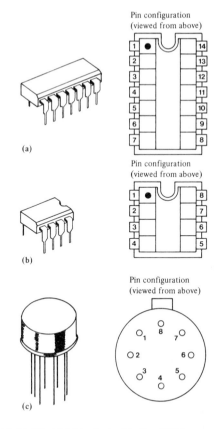

Fig. A11 (a) TO 116 14 pin dual-in-line (DIL) encapsulation – an A package (b) 8 pin version of dual-in-line package – V package (c) TO 99 8 pin reduced height TO 5 package

applied momentarily to form a small ball, then the capillary is lowered so that this gold ball is brought to bear upon the chip contact pad whilst the chip is in the capsule and heated to about 320°C. The tiny gold ball becomes welded to the chip aluminium contact pad on applying pressure. Then, to join the gold wire to the appropriate lead header within the capsule, the capillary — with the gold wire still inside it — is raised, moved laterally to be above the capsule lead header, lowered and the gold wire is welded by thermo-compression to this header. The capillary is raised again and the gold wire is hydrogen-flame cut which not only finishes off the connection but also forms a new tiny gold ball for the next contact to be made. Ultrasonic aluminium bonding is also used with the advantages that aluminium instead of gold connecting wire is used and also no heating is needed.

14. Thin film hybrid circuits. In microelectronics, passive components and interconnections may be made by the deposition of thin films and where the active components are miniature encapsulated transistors, usually of the planar silicon type. The thin films (thicknesses usually between 25nm and 2.5μm) are prepared either by evaporation *in vacuo* or sputtering of an appropriate metal on to a substrate of insulating material, frequently glass or sintered alumina (Al_2O_3) in the form of a square of side 6mm to 25mm. Resistors are made in stripe or meandered form by the evaporation *in vacuo* of either nickel-chromium alloy or tantalum and values between 10Ω and 1MΩ are possible. Capacitors are formed by sandwiching tantalum oxide or silicon oxide dielectric films between appropriate plate electrodes. For connecting leads, gold or aluminium is used.

Tantalum and its compounds are perhaps the most widely used materials. Capacitors of good uniformity and controllable dielectric thickness are often made from anodically grown oxide films on tantalum, where the thickness of the dielectric is usually 0.2μm, giving a capacitance of 1000pF/mm^2.

15. Thick film circuitry. Thick films (defined approximately as of thickness exceeding 10μm) are formed from precious or semi-precious metals or metal alloys by direct painting or screen printing of a suitable suspension of the metal onto a substrate. This substrate needs to have a good quality surface finish flat to within about 0.6μm and is most frequently of 96% pure alumina (Al_2O_3), which has a maximum working temperature of 1400°C, a bulk resistivity exceeding $10^{12}\Omega$m, a surface resistivity of 3 x $10^{11}\Omega$/square and a coefficient of thermal conductivity at 25°C of 25.2 J m^{-1} s^{-1} K^{-1}. Other ceramics are also used amongst which beryllia (BeO) has the advantage over alumina from the point of view of heat dissipation to a cooling agent, that it has a thermal conductivity of 227 J m^{-1} s^{-1} K^{-1} but it is mechanically weaker, more costly and toxic. Standard sizes of alumina substrates are 25 x 25mm, 27.5 x 20mm, 100 x 100mm with thicknesses usually 0.5 to 1mm.

The conducting paste is based on gold, platinum-gold, palladium-gold, palladium-silver, or silver. In a suitable organic medium (which can be diluted, as required) a suspension is made consisting of the finely divided metal with glass frit. This conducting paste is printed through a metal or nylon screen in which

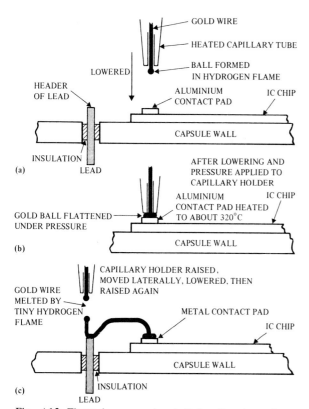

Fig. A12 Thermal compression ball bonding to make a connection by gold wire between an IC chip contact pad and a capsule lead header: **(a)**, **(b)** and **(c)** represent successive stages

open mesh areas are located and shaped to define the pattern of conducting pads and leads required on the ceramic substrate. The conductor pattern is then dried, fired in a carefully regulated furnace following a specific temperature-time schedule to develop the appropriate conductivity of the conductor film and to bond it satisfactorily to the substrate surface. Subsequently, the required pattern of resistive areas is printed on to the substrate so as to be registered correctly in relation to the conducting contact pads and leads. Drying and firing is then repeated to ensure the appropriate resistor characteristics and intimate bonding to the substrate surface.

To form capacitors, a film of dielectric is printed on top of the appropriate pre-printed conductor plate and then a counter-electrode with appropriate connections is printed on top of the dielectric. The dielectric is usually a glass in a special glaze containing barium titanate (relative permittivity is 1200 to 3000). Discrete components (active and passive) are soldered to appropriate contacts to form a hybrid circuit. The circuit is coated with resin and finally protected by encapsulation which is in the form of a transfer moulded housing or a hermetically sealed metal or ceramic enclosure.

The disadvantage of thick film circuitry as against monolithic IC is that the size is considerably greater and a hybrid construction is needed to incorporate passive components. The advantages are that in thick film form resistors can be from a few ohms up to several megohms, capacitors of several 1000pF can be produced by screen printing and even small value inductors can be printed. Furthermore, in hybrid connection, capacitor chips of up to several μF can be incorporated, also variable resistors, magnetic, optical and ferroelectric devices whilst the semiconductor devices need not be confined to planar silicon transistors as in monolithic IC but can include those based on such materials as gallium arsenide, indium antimonide and cadmium sulphide. The fairly high thermal conductivity of the alumina substrate also enables the ready attachment of metallic conducting fins or even thermoelectric or fluid-based cooling devices.

16. The manufacture of thyristors. Two methods of preparation are by the alloy-diffusion technique and the all-diffusion technique.

In the *alloy-diffusion method*, the starting point is a slice of an *n*-type single crystal of silicon, prepared as described in sections 2 and 3, where the *n*-type dopant is usually phosphorus. Into the surface of this slice is diffused gallium (trivalent). This diffusion is undertaken by placing within a quartz tube the *n*-type silicon slices and a piece of gallium in a silicon block (fig. A13). This quartz tube is first exhausted of air by a vacuum pump, an inert gas is admitted at a pressure of a few pascal and the tube is sealed by a quartz plug. The sealed tube is then placed in a furnace which is electrically heated and regulated so that the silicon slices are at 1250°C whereas the gallium is at 1050°C. The gallium volatilizes and its vapour reaches the hot silicon slice surfaces into which it diffuses. The silicon slices are then cut into circular wafers each of which will be of *p-n-p* form (fig. A14). The anode contact of the thyristor is then made to one of the *p*-regions (the bottom one in fig. A14). This contact has to be of low resistance and ohmic. To ensure this, the contact itself is made of molybdenum or tungsten of which a disc is soldered to the appropriate *p*-region using aluminium. On heating in a furnace the *p-n-p* silicon wafer with aluminium solder sandwiched between its bottom face and the molybdenum disc, the *p*-region in the vicinity of the aluminium (trivalent) becomes high conductivity p^+ silicon. The exposed surface of the molybdenum disc is previously gold-plated so that it can be readily soldered to a copper heat-sink.

The *p-n-p* wafer, now conveniently on the molybdenum anode contact, has its edges bevelled (fig. A15) to prevent breakdown at the edges on the application of a high-voltage supply.

To prepare the second *n*-type region of the *p-n-p-n* structure needed for a thyristor and where this *n*-type region is associated with the top cathode contact (a molybdenum disc), a foil (about 50μm thick) of gold-antimony alloy is sandwiched between the top *p*-region (as in fig. A15) and the molybdenum cathode contact. On heating in a furnace, the molybdenum cathode contact is soldered on and the pentavalent antimony in the alloy converts the nearby *p*-type silicon into *n*-type (fig. A16). The necessary gate wire is of aluminium (trivalent) which is ultrasonically welded to the top face of the *p*-type region adjacent to the *n*-type region formed near the cathode contact. Aluminium is needed to ensure an ohmic non-rectifying contact to this *p*-type silicon.

After etching and coating with a protective

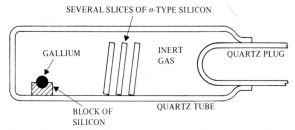

Fig. A13 A closed-tube furnace for the diffusion of gallium into silicon

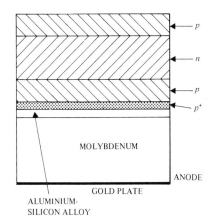

Fig. A14 A *p-n-p* wafer from a gallium-doped *n*-type silicon slice with molybdenum anode contact disc soldered to bottom by means of an aluminium disc

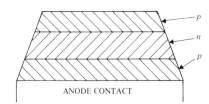

Fig. A15 *p-n-p* wafer with bevelled edges

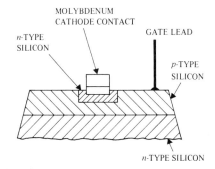

Fig. A16 Formation of the *n*-type cathode region, the molybdenum cathode contact and the gate lead of an *n-p-n-p* thyristor

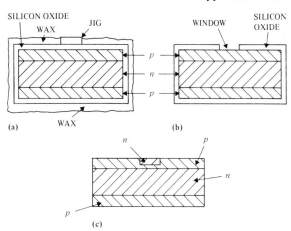

Fig. A17 Forming a window in the *p-n-p* wafer of a large thyristor: **(a)** a jig-mask is placed on the oxidised wafer to cover the window opening required. The wafer is sprayed with wax and the mask removed. Immersion in hydrofluoric acid removes the silicon oxide but not the wax. **(b)** After removing the wax, the required window in the silicon oxide remains through which **(c)** phosphorus (pentavalent) is diffused in a suitable furnace

lacquer, the device is soldered to a copper heat-sink and mounted.

In the *all-diffusion method*, the final *n*-type cathode region is produced by diffusion instead of by alloying. This requires a masking method based on the use of silicon oxide. A window is 'cut' by the photoresist method described previously (section 5) or, for larger thyristors, by a jig-mask and wax procedure (fig. A17).

The *n*-type cathode region is formed in a furnace within the top *p*-type silicon region on diffusion of phosphorus (pentavalent) through the window within the silicon oxide. After such diffusion it is necessary to remove the layer of phosphosilicate glass which forms on the wafer. Finally, the required electrical contact areas on the wafer are coated with nickel and appropriate molybdenum or tungsten plates are soldered to these nickel contacts. Encapsulation and mounting is then as for the previously described alloy-diffused thyristor.

The dimensions of the various layers in a thyristor are of critical importance in relation to satisfactory functioning. A few important points in design are:

The thickness of the *n*-type region between the *p*-type regions is a compromise choice: too thick a layer will reduce the gate sensitivity and increase the forward voltage drop; too thin a layer will decrease the reverse voltage applicable. For a 1000V thyristor the thickness of this basic *n*-type region is typically 125μm, and *n*-type silicon of resistivity about 0.5Ωm is used.

The current carrying capacity of the thyristor will clearly depend on the diameter of the circular silicon wafer from which it is constructed. A diameter of 7.5mm is typical for a 15A device.

B FEEDBACK IN AMPLIFIERS

1. The principle of feedback. Following the notation introduced in section 5.4, an amplifier not provided with any feedback is said to be in 'open-loop'. The gain of the amplifier on open-loop is A_{OL}: A_{VOL} is the voltage gain in open-loop, but the *V* is omitted here because either voltage gain or current gain or both may be involved, though voltage gain is the commonest requirement.

Feedback is involved when a fraction of the output is fed back to the input. An amplifier operating with such feedback is said to be in 'closed-loop': the closed-loop gain is A_{CL}. This feedback is arranged by a specific feedback network whereby a fraction β of the output is arranged to be conveyed to the input. β is known as the 'feedback factor' and βA_{OL} is the feedback loop gain, or simply the 'loop gain'.

Consider fig. B1 involving an amplifier of open-loop gain A_{OL} to which the input signal (which may be voltage or current, though the former is more usually concerned) is s_i and from which the output signal on open-loop is s_o. The feedback network provides at the 'summing junction' an additional signal βs_o. The effective signal s_e to the amplifier itself is then given by

$$s_e = s_i + s_o \quad (i)$$

where

$$s_o = A_{OL} s_e \quad (ii)$$

Hence,

$$s_e = s_i + A_{OL} s_e \quad (iii)$$

On closed-loop, with feedback present, the gain A_{CL} is given by

$$A_{CL} = s_o/s_i \quad (iv)$$

From equations (ii) and (iv),

$$A_{CL}/A_{OL} = s_e/s_i \quad (v)$$

From equation (iii)

$$s_e = s_i/(1 - \beta A_{OL}) \quad (vi)$$

From equations (v) and (vi), therefore,

$$A_{CL} = A_{OL}/(1 - \beta A_{OL}) \quad (vii)$$

If the amplifier is such that the feedback loop provides at the summing junction a signal which is in antiphase with the input signal s_i, βA_{OL} is negative and $|A_{CL}|$ is less than $|A_{OL}|$. This is the condition for *negative feedback*.

On the other hand, if the signal so fed-back is in-phase with the input signal s_i, βA_{OL} is positive and $|A_{CL}|$ is greater than $|A_{OL}|$. This is the condition for *positive feedback*.

In the case of negative feedback, if βA_{OL} is considerably larger than 1, the closed-loop gain A_{CL} is given from equation (vii) by

$$A_{CL} \simeq 1/\beta.$$

The closed-loop gain, i.e. the gain with feedback, is therefore primarily dependent on the feedback factor β, i.e. on the values of the passive components in the feedback network. The closed-loop gain is consequently independent of the active device(s) in the amplifier circuit. Changes of the parameters of

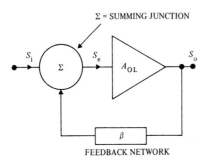

Fig. B1 The principle of feedback

these active devices consequently have little influence on the gain. This is the outstanding advantage of negative feedback, indeed, it is almost universally employed in amplifier design as it leads to great stability of operation involving independence of active device parameter changes due to change of this device itself, or supply voltage alterations, or temperature changes.

It is assumed so far that A_{OL}, A_{CL} and β are real numbers. This is tantamount to the assumption that the output from the amplifier is 180° out-of-phase (for negative feedback) or in-phase (for positive feedback) with the actual input signal, and that β depends only on the ratio of resistances. These are the usual circumstances but conditions can be otherwise in that the gain may be of complex form $[A \exp (j\theta)]$ if the phase difference introduced between output and input is θ; moreover β may be complex if the feedback network includes reactive components, so β has to be represented by $\beta \exp (j\theta')$.

The advantages obtained from the use of a resistive negative feedback network in amplifier design may be listed as follows:

(a) The gain is largely independent of the performance of the active components used (e.g. transistors). Such active components, especially the semiconductor devices, have non-linear characteristics, suffer noise, are temperature sensitive, have characteristics dependent on the supply voltage and, above all, are not the same for two specimens of the apparently identical device.

(b) The gain is also largely independent of changes in the values of the passive components in the amplifier, other than those within the actual feedback network.

(c) The gain depends almost entirely on the values of the feedback network components. In the usual case where these are solely resistive, components can be chosen with close tolerances and having values unaffected by normal variations in the ambient conditions.

In some cases, the feedback network can comprise specific reactive as well as resistive components so that a negative feedback amplifier having a specified frequency response, i.e. gain against frequency characteristic, can be designed.

2. The application of voltage feedback to a voltage amplifier. The input impedance Z_i of a voltage amplifier needs to be very large compared with the resistance of the source of input voltage, v_i. The amplifier provides an open-loop voltage gain of A_{VOL} to an open-circuit output impedance Z_o.[1] The output load resistance is R_L, across which there is developed an output voltage v_o.

To provide a feedback voltage, one way is to connect across R_L two resistances R_1 and R_2 in series and feedback in series with the input voltage that voltage which is generated across R_2. The feedback fraction β is then clearly $R_2/(R_1 + R_2)$. The equivalent circuit of this arrangement is readily deduced from Thévenin's theorem to be as in fig. B2.

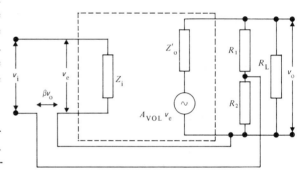

Fig. B2 The equivalent circuit for a voltage amplifier with voltage feedback

To ensure that the feedback is negative, βA_{VOL} must be negative. This use of the potential divider based on resistors R_1 and R_2 inevitably means that β is positive and real. A_{VOL} must therefore be negative so the voltage variation across R_L must be in antiphase with the input voltage.

3. The application of current negative feedback to an amplifier. By far the most common ways of achieving negative current feedback are

(a) In the case of the jgfet, include a source resistor R_S without a by-pass capacitor across it (section 3.11).

(b) In the case of the junction transistor in common-emitter connection, include a resistor R_E in the emitter lead (section 4.6d), again without a by-pass capacitor across it.

C MILLER THEOREM

In the statement of this theorem, the term 'node' in any electrical circuit refers to a junction point at which the algebraic sum of the currents in the various circuit branches meeting at this junction is zero (i.e. Kirchhoff's current law, KCL, is obeyed).

The theorem may be conveniently stated as follows:—

In any arbitrary electrical circuit (fig. C1a), suppose the voltages with respect to earth E at nodes 1 and 2 are v_1 and v_2 respectively where node 1 is connected to node 2 by an impedance Z. If $v_1/v_2 = K$, this circuit can be replaced by one (fig. C1b) in which the impedance Z is removed and replaced by an impedance of $Z_1 = Z/(1-K)$ joining node 1 to E together with an impedance of $Z_2 = ZK/(K-1)$ joining node 2 to E.

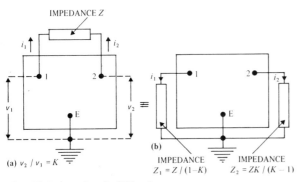

Fig. C1 Concerning the Miller theorem

A justification of this theorem may be considered on the basis that the current i_1 from node 1 through Z is given by $(v_1 - v_2)/Z$ and, because $v_2 = Kv_1$,

$$i_1 = v_1(1-K)/Z \text{ in fig. C1a} \qquad (i)$$

In the equivalent circuit (fig. C1b)

$$i_1 = v_1/Z_1 = v_1/Z(1-K) = v_1(1-K)/Z \qquad (ii)$$

As the current i_1 is the same in equations (i) and (ii) it follows, in accordance with the Miller theorem, that the current drawn from the node 1 in the arbitrary circuit (fig. C1a) is the same as that in the equivalent circuit (fig. C1b).

A similar treatment is valid on considering the current i_2 from node 2.

In electronics, the situation which occurs frequently is where node 1 is at the input of an amplifier stage of voltage gain A_v while node 2 is at the output of this amplifier stage. v_1 is now the input voltage v_i to this stage and v_o is the output voltage, where $K = A_v = v_o/v_i$. If the input and output are linked by an impedance Z the input impedance to the amplifier due to this feedback link is $Z/(1-A_v)$.

An example is the behaviour of a jgfet common-source (CS) amplifier stage (section 3.13). Suppose the input voltage v_i is alternating sinusoidally at a frequency $f = \omega/2\pi$. The actual physical input capacitance is that between the gate and the source, i.e. C_{gs}. But this capacitance is linked to the output (the drain circuit) by the capacitance C_{gd} between the gate and the drain, and this link capacitance has an impedance (reactance) of $1/\omega C_{gd}$. There is consequently a feedback link between output and input and where A_v, the voltage gain, is negative because of the inverting nature of the CS amplifier.

The Miller theorem shows that the effective input impedance Z_i is not simply the reactance $1/\omega C_{gs}$ because this is increased due to the feedback link by an amount

$$Z/(1-K) = (1/\omega C_{gd})/(1+A_v) = \frac{1}{\omega C_{gd}(1+A_v)}$$

The effective input capacitance C_e is consequently given by

$$C_e = C_{gs} + C_{gd}(1+A_v)$$

where $A_v = A_L$ in the mid-frequency region.

Making use of the values given in example 3.13 where $C_{gs} = 5\text{pF}$, and $C_{gd} = 5\text{pF}$, it is seen that, since $A_L = |40|$,

$$C_e = 5 + 5(41) = 210\text{pF},$$

i.e. the effective input capacitance is 42 times the actual physical input capacitance.

The Miller theorem is named after J. M. Miller who, in 1919, first considered analytically the dependence of the input impedance of a triode valve on the load in the anode circuit.

Index

Acceptor energy level, 13
    technique, 12
Active devices, 215
Alphanumeric characters, 157, 191
Amorphous semiconductor, 22
Amplifiers, 102
    Bipolar transistor, 77, 87, 88, 94, 96
    boot-strapping technique, 57
    buffer, 54
    cascode, 57
    differential (*see* Differential amplifiers)
    feedback, 230
    jgfet, 47, 49, 51, 54, 60
    long-tailed pair, 62
    operational (*see* Operational amplifiers)
    paraphase, 61
Analogue-to-digital (A to D) converter, 132
Avalanche effect, 30, 31
Avogadro constant, 7

Bardeen, J, 42
Binary arithmetic, 156
    numbers, 156
    operation, 155
Bipolar transistors, 69
    amplifiers, 77, 87, 88, 94, 96
    bias circuits, 75
    characteristics, 71
    connections, 69, 70
    constant current source, 96
    Darlington pair, 95
    emitter follower, 94
    forward current transfer ratio *or* gain, 71
    hybrid or *h* parameters, 84
    leakage currents, 73
    n-p-n, 69
    parametric equations, 86
    p-n-p, 69
    stabilised voltage supply, 97
    static characteristics, 70, 72
    tester, 74
    thermal runaway, 73
Bismuth telluride, 21
Bistable circuits, 144, 166
Boltzmann constant, 7

Boolean algebra, 160
    de Morgan's theorem, 161
    dualisation laws, 161
Boot-strapping, 57
    applied to jgfet circuit, 57
Brattain, W H, 42
Buffer amplifier, 54

Cadmium sulphide, 21
Carbon, crystalline form, 9
Cascode amplifier, 57
Channel electron multipliers (channeltrons), 187
Conduction band, 9
Controlled silicon rectifier (*see* Silicon controlled rectifier)
Covalent bonds, 7
Crystal growing, 217
    Czochralski method, 218
    Keck method, 217
Current carriers, 11
    concentrations in intrinsic silicon, 11
    life-times, 15
Cut-in voltage, 25

Darlington pair, 95
Decade counter, 146
Destriau effect, 190
Detectivity, 175
Difference (*or* differential) amplifiers, 103
    based on jgfets, 60
    common mode rejection ratio (CMRR), 61
    differential voltage gain, 61
    general case, 62
Diffusion current in a semiconductor, 17
    length, 18
Digital instruments, 151
    current-to-frequency converter, 152
    integrating logic, 152
    ramp logic, 152
Digital-to-analogue (D to A) converter, 132
Diodes, 24
    basic circuit, 26
    characteristics, 25
    double-ended (*see* Unijunction transistor)

free-wheeling, 210
gold-bonded, 24
Gunn, 40
impatt, 40
incremental resistance, 25
light-emitting, 40, 190
parametric, 36
p-n junction, 24
point-contact, 24
rectifiers, 26
reverse saturation currents, 24
threshold voltages, 25
tunnel, 36
varactor, 36
variable capacitance, 36
voltage reference, 31, 33
voltage regulator (see Zener)
Zener, 30
Donor energy level, 13
technique, 12

Effective masses of electrons and holes, 10
Einstein, A,
equation relating mobility and diffusivity, 17
photoemissive effect equation, 182
Electroluminescence, 190
Destriau effect, 190
injection type, 190
Electrostriction, 136
Energy-band structure, 13
n-type silicon, 13
p-type silicon, 13
p-n junction, 20
Epitaxial growth, 220
Equality detectors, 165
Extrinsic semiconductor, 12
n-type, 12, 13
p-type, 12, 13

Feedback, 102
attenuated, 109
closed-loop, 104
current negative, 231
negative, 204, 230
open-loop, 104
positive, 133, 230
principle, 230
voltage, 231
Fermi-Dirac distribution, 11
Fermi energy level, 11
Field effect transistor, 42
insulated-gate (igfet), 42, 43, 63
action, 64
characteristics, 65, 66
construction, 63
n-channel depletion type, 43

n-channel enhancement type, 43
p-channel depletion type, 43
p-channel enhancement type, 43
protecting the gate, 66
within monolithic integrated circuits, 222
junction-gate (jgfet), 42, 43
behaviour at high frequencies, effects of capacitances, 55
characteristics, 45
common-drain (CD) amplifier, 53
common-source (CS) amplifier, 47, 49, 51
differential amplifier, 60
drain conductance, 46
gate-source bias provision, 47
load-line, 50
low frequency response, 58
parameters, 46, 48
reduction of input capacitance, 57
source follower, 54
switch, 45
voltage-controlled resistor, 44
Forbidden gap, 9
in some semiconductor compounds, 22
Form factor, 208 (see Silicon controlled rectifier)
Forming p-n junctions, 218
alloying, 218
diffusion, 218, 220
ion implantation, 218
rate-growing, 218

Gallium antimonide, 22
arsenide, 21, 191
arsenide phosphide, 191
phosphide, 21, 191
Germanium, 7, 9, 12
production of pure, 217
properties of pure, 12
Group IVA elements, 7, 8
Gunn diodes, 40

Hole, 9
Hybrid circuits, 216
thin film, 227

Image converter tube, 188
Image intensifier tube, 188
Impatt diode, 40
Indium antimonide, 21
phosphide, 22
Inequality detectors, 165
Input capacitance, reduction in a jgfet, 57
Insulated gate field effect transistors (igfet), 42, 22 (see Field effect transistors)

Integrated circuits, 60, 61, 66, 69, 167
    encapsulation of, 225
    epitaxial growth, 220
    fabrication of passive components, 223
    interconnections within, 215
    large-scale integration (LSI), 216
    medium-scale integration (MSI), 216
    monolithic, 216
    MOS, 167
    small-scale integration (SSI), 216
Ion implantation, 216

Junction-gate field effect transistors (jgfet), 42 (*see* Field effect transistors)

Lamp-dimming circuit, 204
Large-scale integration (LSI), 66
Law of mass action, 14, 24
Lead selenide, 21
    sulphide, 21
    telluride, 21
Light-actuated silicon controlled rectifier (LASCR), 182
Light-emitting diodes (LEDs), 145
    in a display of alphanumeric characters, 191
    in a display matrix, 191
Load-line, 50
    bipolar transistor in CE connection, 76, 141
    jgfet, 50
Logic circuits, 155
    AND, 157
    AND-OR-INVERT (AOI), 166
    Boolean algebra, 157, 158, 160
    complementary metal oxide semiconductor (CMOS), 169
    diode-transistor logic (DTL), 163
    enable, 164
    exclusive OR, 165
    fan-in, 166
    fan-out, 166
Logic gates, 157
    integrated, 167, 168
    NAND, 163
    negative logic, 158
    noise immunity, 166
    NOR, 163
    NOT (or INVERTER), 157
    OR, 157
    positive logic, 158
    totem-pole output driver, 170
    transistor-transistor logic (TTL), 167
    use of diodes, 158
Long-tailed pair, 62

Majority carriers, 15
Mass action, law of, 14
Memory circuit, 166
    basic digital, 166
    flip-flop, 166
Miller effect, 106, 232
    in high frequency performance of a jgfet, 56
Miller theorem, 106, 232
Minority carrier injection, 21
Minority carriers, 14
Mosfet, 42, 64, 222 (*see* insulated gate field effect transistors)
    protecting the gate, 66
Multivibrators
    astable, 147
    as voltage-to-frequency converters, 148
    bistable, 144
        binary (divide-by-two function), 145
        binary decade counter based on, 146
        Schmitt trigger circuit, 149
        with symmetrical base triggering, 145
    very low frequency,
        based on Darlington pairs, 148
        n- channel jgfets, 148

Noise equivalent power (NEP), 175

Operational amplifiers, 60, 102
    ac, 111
    attenuated feedback in, 109
    band-width, 204
    booster, 111
    common-mode gain, 118
    common-mode rejection ratio (CMRR), 118
    constant current source, 110
    crystal-controlled oscillator, 117
    current booster, 111
    difference (or differential), 110
    differential gain, 118
    differentiator, 114
    finite gain error, loop gain and feedback factor, 119
    frequency response, 121
    ideal differential, 103
    inverting configuration, 105
    linear dc applications, 106
    linearity, 106
    logarithmic, 117
    Miller integrator, 113
    modular, 102
    multivibrator (astable), 115
    multiplication, 108
    negative feedback in, 104
    non-inverting configuration, 105, 109
    non-linear applications, 113
    output current and output resistance, 109

overload protection, 107
slew rate, 125
squaring circuit, 117
summation, 108
summing junction, 104
switching applications, 107
transfer function, 106
triangular waveform generator, 114
unity gain sign inverter, 107
    voltage follower, 110
wave-shaping applications, 113
Opto electronics, 173
Oscillation, Barkhauzen criterion for, 133
Oscillators,
    crystal-controlled, 136, 138
    phase-shift, 133
    quadrature, 140
    relaxation, 203, 206
    sinusoidal, bipolar transistor, 134
        jgfet, 134
        operational amplifier, 139
        two-stage bipolar transistor amplifier, 139
    Wien bridge type, 138

Parametric diode, 36
Paraphase amplifier, 61
Passive devices, 215
Periodic table of the elements, 7, 8
Photoconductive cells, 173
    cadmium selenide, 173
    cadmium sulphide, 173
    charge multiplication, 174
    indium antimonide, 176
    infra-red, 175
Photoconductivity, 173
Photodiodes, 180
    avalanche, 180
    p-i-n structure, 180
    silicon, 180
Photoemissive cells, 184
    gas-filled, 185
    vacuum, 184
Photoemissive effect, 182
    classification of materials, 183
    Einstein equation, 182
Photoexcitation, 15
Photofet, 181
Photomultiplier tubes, 186
    box-and-grid type, 186
    venetian blind type, 186
Photons, incidence on a semiconductor, 15
Photoresists, 219
Phototransistors, 180
    Darlington-connected, 182
Photovoltaic barrier layer cells, 178
    effects, 177
Piezoelectric effects, 136

Pinch-off voltage of an n-channel jgfet, 45
p-n junction, 19
    application of a bias voltage, 20
    basic equation, 24
    capacitance, 35
    diode, 24
    graded, 35, 216
    open-circuited, 19
    short-circuited, 21
    step, 35, 216
p-n junction diode, 24
    as a rectifier, 26, 27
    characteristics, 25, 26
Positive hole, 9
Power control utilising thyristors, 193
Printed circuit board (PCB), 215
Programmable unijunction transistor (PUT), 205
    phase-control, 207
Pulse, 132
    decay time, 133
    duration, 133
    height, 132
    leading edge, 132
    repetition frequency, 132
    rise time, 133
    trailing edge, 132

Quartz crystal, 137

Rectifier circuits,
    full-wave, 27
    half-wave, 27
    reservoir capacitor, 28
Relaxation oscillators, 203, 206
Reverse saturation current through a p-n junction, 21, 25
Ripple voltage, 29

Schmitt trigger circuit, 149
Semiconductors based on compounds, 21
Semiconductor devices and circuits construction, 215
Semiconductors of III-V and II-VI types, 21
Shockley, W,
    equation for current through a p-n junction, 24
    equation for reverse saturation current due to holes, 24
    invertion of fet, 42
Silicon, 7
    crystalline form, 9
    intrinsic, 9, 11
    production of pure, 217
    properties, 12

Silicon controlled rectifiers (SCRs), 193
    ac phase control, 196, 198, 201
    bistable circuit, 211
    characteristics, 194
    commercial firing unit, 207
    commutation in ac circuits, 210
    experiments, 197
    structure, 194
    theory of operation, 194
    use with high-power incandescent lamps, 210
    zero voltage switching, 200
Smoothing filters for rectifier circuits, 28
Solar cells, 179
Source follower with jgfet, 54
Surface photoemission (*see* photoemissive effect)
Switch,
    binary operation, 155
    bipolar transistor, 140
    speed-up capacitor, 143
Switching circuit, combinational, 162

Thermocouple ammeter, 199
Thick film circuits, 216, 227
Thyristors, 193 (*see* Silicon controlled rectifiers)
    bidirectional trigger diode (DIAC), 193, 211
        triode thyristor (TRIAC), 193, 211
    manufacture, 228
    QUADRAC, 193, 212
    silicon controlled switch (SCS), 193
Tunnel diodes, 36
    chief applications, 38
    energy band structure, 37
    voltage-current characteristic, 38

Unijunction transistor (UJT), 201
    characteristics, 202
    intrinsic stand-off ratio, 202
    programmable (PUT), 205
    relaxation oscillator, 203
Unipolar device, 43

Valence band, 9
    electrons, 8
Varactor diodes, 36
Variable capacitance diode, 36
Voltage reference diodes, 33
Voltage regulator circuits, 31
Voltage-to-frequency converters, 148

Waveform generators, 132
Waveform information, 132
    exponential, 133
    linear voltage ramp, 133
    rectangular, 132
    sawtooth, 133
    square, 133
    triangular, 133

Zener diodes, 30
    applications, 34
    breakdown voltages, 31
    calculating best value for the limiting resistance, 32
    in parallel, 33
    in series, 33
    voltage regulator, 32
Zone refining, 217